W0254379

NICOTINIC ACID

CLINICAL PHARMACOLOGY

Series Editor

MURRAY WEINER
Department of Pharmacology
University of Cincinnati
College of Medicine
Cincinnati, Ohio

Volume 1 Nicotinic Acid: Nutrient-Cofactor-Drug, *by Murray Weiner* and *Jan van Eys*

Other Volumes in Preparation

NICOTINIC ACID
Nutrient-Cofactor-Drug

MURRAY WEINER
University of Cincinnati
College of Medicine
Cincinnati, Ohio

JAN VAN EYS
University of Texas System Cancer Center
M.D. Anderson Hospital and Tumor Institute
Houston, Texas

MARCEL DEKKER, INC. New York and Basel

Library of Congress Cataloging in Publication Data

Weiner, Murray, [date]
Nicotinic acid.

(Clinical pharmacology ; v. 1)
Includes bibliographical references and index.
1. Niacin--Metabolism. 2. Niacin--Physiological effect. 3. Niacin--Therapeutic use. I. Van Eys, Jan. II. Title. III. Series. [DNLM: 1. Nicotinic acids. W1 CL764H v.1 / QU 193 W423n]
QP772.N48W44 1983 574.19'242 83-12772
ISBN 0-8247-7015-3

MARCEL DEKKER, INC.
270 Madison Avenue, New York, New York 10016

Current printing (last digit):
10 9 8 7 6 5 4 3 2 1

PRINTED IN THE UNITED STATES OF AMERICA

PREFACE

This volume organizes and correlates the available information about nicotinic acid and metabolically related compounds in a manner designed to be useful to students, research scientists, and medical practitioners. It is intended to facilitate the finding of clues to the solution of one specialist's problems in the work of other types of specialists.

The book is organized so that the reader can identify and selectively study areas of particular interest. It is intended as a practical aid to those with an interest in nicotinic acid and related compounds, rather than as an exhaustive review of the subject. Emphasis is placed on correlating—rather than cataloging—clinical, animal, and biochemical data which reflect our current understanding of this uniquely active chemical species.

Murray Weiner
Jan van Eys

CONTENTS

INTRODUCTION

When nineteenth-century chemists first synthesized nicotinic acid by the oxidation of nicotine and related compounds, they had no idea that they were involved with a molecule of fundamental importance to essentially all forms of life. To the biochemist, this simple aromatic compound with both an acidic and basic substituent is an indispensable component of the electron-transferring function of pyridine nucleotides essential to the action of so many of the important enzymes of life. To the nutritionist, nicotinic acid and its amide are interchangeable, essential nutrients commonly referred to as niacin and truly "vital amines" (vitamins) for man, beast, and microorganisms. To the pharmacologist, nicotinic acid is indeed a most versatile and at times exasperating drug. It has a spectrum of actions that seem unrelated to each other, and that have a bewildering variety of relationships to its pharmacokinetics. Nicotinamide, which is as effective as nicotinic acid as a nutrient, has essentially none of the drug effects of nicotinic acid in humans.

The history of nicotinic acid is as fascinating, complex, and unique as the substance itself. Its introduction to the scientific world was in the context of a synthetic drug derived from the oxidation of "nicotines" by C. Huber in 1867 (1). In 1873, H. Weidel (2) described the elemental analysis and crystalline structure of salts and other derivatives of nicotinic acid in some detail. It was nearly another 50 years before nicotinic acid was isolated for the first time from natural sources, still with little understanding of its significance.

Hundreds of years earlier, the disease pellagra had been recognized as an entity which was associated with populations ingesting large amounts of corn (maize), that popular yellow grain imported from the New World into Europe by Columbus. In twentieth-century America, the clinical studies of Goldberger, which were first reported in 1915 (3), were widely accepted as demonstrating that pellagra was a dietary deficiency disease, curable by nutrients later referred to as the pellagra preventive (P. P.) factor and identified as niacin. In Europe, however, the idea persisted that the disease was primarily the result of a "toxin" (4). Goldberger's clinical observations were well ahead of the demonstration of the consequences of niacin deficiency in animals and the first experimental evidence suggesting a critical role for niacin in an essential biochemical mechanism. The pharmacological aspects of nicotinic acid and their potential therapeutic utility evolved more recently.

In considering the current views of nicotinic acid, several concepts applicable to the study of nutrients and drugs in general deserve attention. The fact that nicotinic acid is both a nutrient and an active pharmacological agent with toxic potential exemplifies the pitfalls of the convenient but counterproductive tendency to substitute categorization for understanding. "Nutrients" are good, essential things which we all need, while "toxins" are, of course, bad by definition. Yet, nicotinic acid is both. It can kill and cure, depending on how much is taken in what manner by whom, and under what circumstances.

The nutrient, therapeutic, and toxic potential of nicotinic acid are all influenced by a wide variety of concomitant factors. This reality brings into focus another general concept deserving attention. Standard teaching dictates that studies be designed with only one variable, so that the differences observed will be clearly attributable to that one variable. Thus, the results of a clinical or laboratory study of two groups whose treatment differs only in regard to the administration of niacin can confidently be interpreted as reflecting the action of niacin. But the action reflected under the specific conditions of the experiment may be quite different from events to be observed under other conditions. In each experiment, innumerable factors are fixed, or assumed to be fixed, at a level or in a manner which is usually arbitrary and often set primarily for the convenience of the experimenter. Extrapolation from such fixed conditions to the universe must be made with great caution.

In the real world, nutritional deficiencies are rarely "pure." Personnel involved in treating liberated prisoners of war after the Japanese surrender in 1946 remarked how similar the clinical manifestations of nutritional deficiency were within a camp, and how different they were at times from one camp to another, even though all were severely malnourished. There were emaciated, paralyzed victims of starvation who, nevertheless, breathed easily (typical of dry beriberi), while in a neighboring camp, essentially all the prisoners were dyspneic and swollen with edema (typical of wet beriberi), with no significant neurological problems. And both of these remarkably different clinical conditions were presumably the result of thiamine deficiency, and therefore curable by thiamine. Very likely, other substances contained in (or not contained in) the materials ingested by the people in the neighboring camps must have been responsible for the marked clinical differences. In one particular camp, most of the victims could not read because of functional central scotoma, with no ophthalmoscopically detectable abnormality. Such scotoma had not been reported to be a consequence of a particular nutrient deficiency. Was it due to a "toxin" which is toxic only under certain types of compromised nutrition?

The consequences of niacin deficiency are often described with great specificity. Yet, they vary considerably with species. Thus, the classic clinical triad of "dermatitis, dementia, and diarrhea" characteristic of human pellagra is a far cry from black tongue in dogs and slow growth in chicks. In other words, the significance of a nutrient depends on the nature and status of the subject, just as the consequences of an infection depend on the resistance and reactivity of the host.

In evaluating the role of a nutrient such as niacin, it is critically important to be cognizant of the possibility of the other coexisting deficiencies or suboptimal quantities of other nutrients, the co-administration of other substances with a biochemical conversion potential, such as tryptophan, the presence of otherwise tolerable amounts of potentially "toxic" substances, and a wide variety of internal physiological or environmental stresses, each of which may influence the clinical picture. How, then does one define the clinical consequences of a specific nutrient like niacin, or quantify the "daily requirement," or determine the "cause" of a complex pathological state? The problem occurs frequently, not only in nutrition, but in drug therapy as well.

In the last analysis, there is no substitute for an understanding of the underlying biochemistry and interacting mechanisms which cause and control disease. But as long as that understanding remains incomplete, clinical action in relation to both drugs and nutrients must be consistent with clinical facts, whether or not they are fully understood. The more clearly we understand the biochemical mechanism, the better we will be able to correctly catalog clinical observations so as to best apply our science to the maintenance and recovery of good health.

Murray Weiner
Jan van Eys

1. C. Huber, *Ann. der Chemic und Pharmacie. 65*:271(1867).
2. H. Weidel, *Ann. der Chemic und Pharmacie. 165*:328(1873).
3. J. Goldberger, C. H. Waring, and D. G. Willits, *Public Health Report 30*:3117(1915).
4. C. A. Elvehjem, *Physiol. Rev. 20*:249(1940).

PART I

NICOTINIC ACID AS A NUTRIENT

The need for a dietary source of precursors for the nicotinamide found in the coenzymes nicotinamide adenine dinucleotide and nicotinamide adenine dinucleotide phosphate is now well accepted. The relationship between dietary factors and human deficiency disease is usually the first level of understanding of the nutritional-biochemical role of a nutrient. The way in which the need for nicotinamide can be satisfied is very complex. The interpretation of data obtained through nutritional experimentation provided many insights into the biochemical pathways of synthesis and degradation of nicotinamide and nicotinic acid. Even the seemingly simple conversion of nicotinic acid to nicotinamide is very complex indeed.

What makes the story of nicotinic acid as nutrient so fascinating is the widespread incidence of niacin deficiency in the past. The epidemiology of pellagra was and still is a major concern of the World Health Organization. The animal models for the disease are imperfect. Fascinating observations could be made on the comparative enzymology between species and the consequent alterations in dietary requirements.

The ways in which the nutritional requirements for niacin can be satisfied are so multiple and complex that the process is very susceptible to outside influences, be it drugs, dietary balance, or protein quality.

It is clear therefore that the following review is selective. History is stressed, and nutrition is reviewed, but so is dietetics. These two aspects of food and dietary components are only too often confused (1). The utilization of structural analogues as rational chemotherapeutic agents, explored early in nicotonic acid research, is also reviewed.

The complexity of nicotinic acid as a nutrient remains a source for exciting opportunities for modulation of normal metabolism. Yet this area is not as well developed as we might hope. This review is directed to the development of new ideas in nicotinic acid research.

REFERENCES

1. J. van Eys, in *Molecular Interrelations of Nutrition and Cancer* (M.S. Arnott, J. van Eys, and Y.M. Wang, Eds). Raven Press, New York, 1982, p. 449.

1

THE DISCOVERY OF NICOTINIC ACID AS A NUTRIENT

I. NOMENCLATURE

Nicotinic acid has been associated with vitamin-like effects ever since it was found in an antiberiberi concentrate of rice polishings (1). It was repeatedly associated with vitamin-like effects but it was not identified as a true nutrient until 1937 when it was shown to benefit the black tongue syndrome in dogs(2). Before that time, in 1935, nicotinamide was found to be an integral part of the coenzymes for metabolism, coenzyme I (3) and II (4,5), later named diphosphopyridine and triphosphopyridine nucleotide (DPN and TPN) and now called nicotinamide adenine dinucleotide and nicotinamide adenine dinucleotide phosphate (NAD and NADP).

That nicotinic acid or its amide was actually a nutrient was proven when it was shown that the nature and quality of the protein ingested influenced the degree of nicotinic acid dependence.

Once nicotinic acid was recognized as a nutrient, a standard nomenclature was adopted. *Niacin* is:

> the generic description for pyridine 3-carboxylic acid and derivatives exhibiting qualitatively the biological (nutritional) activity of nicotinic acid. Thus phrases such as "niacin activity" and "niacin deficiency" represent preferred usage. The compound pyridine 3-carboxylic acid..., also known as niacin or vitamin PP, should be designated nicotinic acid (6,7).

II. NUTRITIONAL HISTORY

A. Nicotinic Acid as Nutrient

Pellagra: Pellagra is a nutritional deficiency disease which was found worldwide until recently. The disease was first described as "Mal de la Rosa" by Casal in 1735 (8). The term "pellagra" was introduced by Frapolli in 1771 from *pelle agra,* "rough skin" (8). While all textbooks of medicine contain good descriptions of the disease, a classic textbook provides an especially detailed description (9). It is usually described as a disease of the classic "four D's": dermatitis, diarrhea, dementia, and death. Since the advent of a cure, pellagra is usually called the disease of the "three D's" since death ought not to be part of the picture anymore. However, the three D's represent the last stage of the disease. The early stages of the disease are nonspecific, and include lassitude, anorexia, weakness, and maldigestion. Mental symptoms such as anxiety and irritability occur early. If left untreated, inflammation of mucous membranes ensues, beginning with soreness of the tongue, progressing through glossitis to stomatitis, esophagitis, urethritis, proctitis, and vaginitis. The stomatitis may be so painful that food intake becomes impossible. From the intestinal inflammation diarrhea follows.

The dermatitis is primarily manifested by sun sensitivity. At first, sunburn appears to be present. If the deficiency is slow in onset, thickening, scaling, hyperkeratinization, and pigmentation will occur. In either case, sun-exposed areas are primarily affected. An open neckline or loose shirt gives rise to lesions known as "Casal's necklace." In more advanced states, skin lesions develop in moist, contact areas, such as axillae, groin, and the area under the breast.

The depression will, if untreated, progress to disorientation, hallucinations, and delirium. Even acute encephalopathy has been described.

In clinical settings pure niacin deficiency is rare, and anemia of associated deficiencies, especially folic acid, is frequently seen.

Shortly after 1900, pellagra became a major problem in the southern United States. The U.S. Public Health Service estimated a total of 5000 cases in the year 1911 (10). Joseph Goldberger, who was directed to investigate the etiology managed to produce pellagra in healthy subjects, convicts at the Mississippi State Penitentiary, by feeding them a deficient diet (11). Twelve inmates received the classic southern diet; eighty served as controls. Of the eleven who completed the eight-month period, five developed pellagra, thereby proving the dietary etiology. In 1937, the effectiveness of nicotinic acid in the treatment of pellagra was clearly demonstrated (12,13). At face value that might have indicated that the problem was solved. However, there were many questions and observations about pellagra that were not answered by the explanation of a simple nicotinic acid deficiency. These centered around two major facts: (1) dietary protein quality and quantity seemed more important than nicotinic acid content, and (2) corn played a peculiarly specific etiological role. Some of these questions were accentuated by the study of the animal model of pellagra, black tongue in dogs (14). Goldberger also was impressed by the similarity of that disease to human pellagra (15,16).

Black tongue in dogs: This disease is characterized by a constellation of vomiting, loss of appetite, and diarrhea, with ulceration of the gastrointestinal tract. A detailed review is given by Follis (10). It was considered an infectious disease by veterinarians until Chittenden and Underhill induced the disease by a vegetarian diet (14). Goldberger also produced black tongue in dogs (15,16), but the different symptoms produced by Underhill and Goldberger responded to different dietary regimens—the Underhill type with liver and cod liver oil, and the Goldberger type with meat and yeast (17). Again, the confusing data seemed to be reconciled by the studies of Elvehjem, who found that the active ingredient in liver fractions that cured the Goldberger black tongue was nicotinamide (18,19). Dogs could be kept healthy with the addition of nicotinic acid or nicotinamide to the diet (19,20).

However, as in human pellagra, the discovery of this association of nicotinic acid and black tongue did not solve the mystery entirely. First of all, animals could be saved temporarily from impending death by salt solutions (21). While some animals never got diarrhea, there was still the likelihood that diarrhea represented a response to fluid therapy. However, a serious dilemma was that the dogs with black tongue did not have lower levels of nicotinamide-containing coenzymes (22). More importantly, a purified diet deficient in nicotinic acid failed to induce black tongue in dogs (23).

For a time these discrepancies were thought to be the result of a double deficiency of folic acid and nicotinic acid. Weanling puppies fed purified diets developed a black tongue-like syndrome that did not reproducibly respond to nicotinic acid until folic acid supplementation was given (24,25). Interestingly, it was not until 1979 that tryptophan was demonstrated to be an essential nutrient for puppies (26).

The etiological role of corn diets and nicotinic acid deficiency in other diseases of animals, such as black tongue in dogs, became known shortly thereafter (25,27,28). A partial explanation of the relationship between corn and pellagra and black tongue involves the role of tryptophan, although, as will be described later, that is also too simplistic an answer.

Hartnup's disease: The relationship between a pellagra-like syndrome and tryptophan can be clinically demonstrated in the genetic disorder called Hartnup's disease, named after the first family described (29). These patients have a typical pellagra, but the defect is the lack of absorption or increased renal excretion of tryptophan. The symptoms come intermittently, but grow milder with increasing age. There is a marked skin rash, usually made worse by sunlight. Neurological symptoms are prominent with unsteadiness, clumsiness, nystagmus, tremor, and diplopia, as in cerebellar ataxia. Psychiatric disturbances are seen up to frank deliria.

The disease is associated with poor nutrition, and nicotinamide deficiency is a major etiological component (30). Though the unabsorbed tryptophan is metabolized in the gut and does thereby generate neuroactive amines, the disease is diagnosed through the fact that in addition to malabsorption of tryptophan, general transport of many other amino acids is defective, leading to a characteristic aminoaciduria with losses of alanine, serine, threonine, valine, leucine, isoleucine, phenylalanine, tryosine, tryptophan, histidine, glutamine, and asparagine (31).

B. Epidemiology of Pellagra

Introduction: The role of nicotinic acid in the human and animal deficiency disease and the biochemical basis of the need for nicotinic acid in the diet were firmly established by the finding of nicotinamide as a component of the fermentation coenzymes. However, the relationships of protein to pellagra and the special role of corn in the etiology needed to be further elucidated. These relationships were known early as a result of the careful study of the epidemiology of the disease.

Protein deficiency: The early data from the serious studies undertaken by Goldberger in the southern United States clearly indicated that the poor diet was responsible. One feature of the diet was the absence of meat and milk; they were in fact preventative. Therefore, Goldberger initially proposed that pellagra was an amino acid deficiency (32). In fact, one patient treated with tryptophan had a dramatic response. Goldberger's associate Tanner was quoted as saying that the response (33):

> ...surpassed anything I have ever seen in a case of pellagra in an equal period of time.

The similarity between Hartnup's disease and pellagra seemed to establish the direct relationship between protein and pellagra and strengthened the identification of tryptophan as the nutrient which could explain protein deficiency as a factor in pellagra. However, the tryptophan and nicotinic acid content of various grains does not explain how pellagra is associated with corn and millet but not with rice or wheat (Table 1).

It has been proposed that the dietary requirement for tryptophan/niacin varies depending on the amino acid balance, especially the quantity of leucine and isoleucine, but this is not universally accepted. The special role of corn in the etiology stimulated many speculations and investigations.

Corn and pellagra: The literature on corn and pellagra is so voluminous that a complete review could not be presented within the boundary of a single monograph. Excellent reviews do exist, however. For the older literature, Harris's book is a good objective summary (35), while the major literature is discussed and interpreted by Chick (36,37). A less complete but interesting review is presented by Mainzer (38), and yet another has been contributed by Roe (39).

Table 1 The Content[a] of Nicotinic Acid and Certain Amino Acids in Some Foodstuffs

Foodstuff	Nicotinic acid	Tryptophan	Leucine	Isoleucine
Rice	1.2	1.2	8.0	6.0
Wheat	5.5	1.1	6.5	3.5
Maize	1.4	0.8	14.9	6.4
Jowar (Sorghum vulgare)	1.8	1.2	12.9	6.1

[a]Grams per 100 gram of protein.
Source: Ref. 34.

The role of corn as an etiological agent in pellagra has been recognized practically since the discovery of the disease by Casal (40). Corn was unknown in Europe until Columbus's time, as is evidenced by his elaborate descriptions of the plant. This explorer took the grain home from America and introduced it in Europe. To quote Columbus himself: "—y hay ya mucho en Castilla.—" "...it grows abundantly in Spain" (41).

Corn, once introduced, spread rapidly over Europe, and so did pellagra. Marzari (42) was the first author to claim that corn and pellagra were definitely associated, even though he was not the first to suggest a possible relationship. Pellagra has on numerous occasions been considered to be an infectious disease, but this aspect will not be reviewed. Goldberger actually tried to transfer the disease from an infected prisoner to himself, and of course, failed (48). Once corn and pellagra were linked, two lines of thought evolved: (1) Corn cannot suffice as the major food item. In other words, corn was deficient in some factor, but the concept of deficiency disease was then nonexistent; (2) Corn is toxic.

There was good evidence for the first concept, as was strongly expressed by Morelli (44), who maintained that a diet consisting of solely one food item, not just a diet of corn, will cause pellagra. This may be untrue as far as classical pellagra is concerned, but it represents a sound view as to the necessity of a varied diet. It was generally accepted in those days that pellagra could be cured with a varied diet, which again suggested a deficiency. This concept was outlined in Funk's postulate which states that corn is deficient in some "vitamine" (45). More pertinent (but less recognized) is the Wilson postulate which maintained that corn is deficient in protein of good biological value (46). Indeed, Funk revised his statement to read: "It may either be a vitamin deficiency, a protein deficiency, or both" (47). About this time Goldberger produced pellagra in humans experimentally with suitable diets (48). Knowledge of this association existed as far back as the early nineteenth century in Italy. A notable cause is cited by Cerri (35), who described the clinical history of a farmer's boy whose usual rural diet was cornbread and polenta. This boy took a position in the city, where his pellagra was promptly cured with the city diet. Once cured, the boy returned to the farm and again lived on cornbread and polenta. When the return to the city and farm was repeated several times, along with the expected cure and recurrence of the pellagra, the boy was convinced of the necessity of city life for his health, while Dr. Cerri was convinced of the causation of pellagra by cornbread and polenta.

Examples such as the previous one make Goldberger's work appear less revolutionary, but it served to demonstrate the dietary origin of pellagra, and culminated in the famous dog experiment of Elvehjem (19). This experiment resulted in the recognition of the effectiveness of nicotinic acid in curing the experimental counterpart of pellagra, canine black tongue. Further elucidation of the role of corn in dietary deficiency was provided by the work of Krehl and his associates (49), with their recognition of the curative effect of tryptophan. However, knowing the daily requirements for niacin and tryptophan, it was hard to conceive how corn, which is not any lower in these nutrients than most cereals, would provoke a deficiency. Once this issue was raised, the ideas took several directions. Of course the possibility of a toxic factor was still considered. This will be discussed later.

First of all the idea existed that one was not dealing with a niacin-tryptophan deficiency but rather with a lysine deficiency. This was expressed in very strong terms as late as 1950 by a group of French investigators: "Le role stimulant du tryptophane ou de la niacin, signalé par des auteurs americains, ne pourrait se manifester qu'après avoir comblé le déficit du facteur limitant, en l'espéce la lysine" (50). In other words, the effect of tryptophan is but the result of an artificial stimulation. This idea is no longer maintained.

There was also the possibility that tryptophan or niacin exists in corn in a bound form, unavailable to the mammalian organism (51). There have been numerous claims that alkali treatment of corn abolishes its pellagragenic effect. The most striking indication is the absence of pellagra in Central America, where corn is routinely steeped in lime water before consumption. This has been experimentally tested with varying results. Laguna and Carpenter (52) claim that rats fed lime-treated corn do not develop a deficiency, while their controls do. This could not be repeated unequivocally by a Mexican group of workers (53), but was repeated by Pearson et. al. (54) and Harper et. al. (55).

A further difficulty is that corn in Central America is not as much a main food item as it is in endemic pellagra regions, since Central Americans generally consume a liberal quantity of beans which might well abolish any effect of corn because of their large content of tryptophan (56). Calculations by Bressani et. al. (57), based on actual niacin and tryptophan analyses of Guatemalan beans, show that this combination of corn and beans provides 44% more tryptophan than the minimum requirement suggested by Rose (58) and 11% more niacin than is pellagra preventive (59). There are, however, claims for a specific effect of corn based on experiments. The extraction of corn with alkali is supposed to liberate more nicotinic acid (60,61) than is detectable by chemical or microbiological methods in whole corn. All these observations can be explained on the basis of Krehl's claim for the existence of an "alkali-labile" niacin precursor in corn, "alkali-labile" meaning that alkali makes the compound more available for rat growth. These reports do not exclude the possibility that a toxic factor is destroyed or removed during treatment with alkali. It is noteworthy that the alkali-labile niacin precursor is also present in wheat (61,62), a grain which does not provoke a niacin deficiency and which is actually pellagra preventive.

Repeated attempts at isolation of this bound nicotinic acid of cereals have been made, culminating in a report by Das and Guha that the material was crystalized from wheat, rice, and corn (63). They used the name niacinogen for their material, but the name niacytin is more generally used. They claimed that alkali-labile nicotinic acid had a molecular weight of 1150-1700 daltons, 60% polysaccharide, and 40% peptide again with other aromatic substances (65). The chemical nature of niacin, and indeed its relevance to human pellagra, must await further structural determinations to establish the actual form in which niacin is bound.

Furthermore, millet is also associated with pellagra. Although alkaline hydrolysis was said to increase the nicotinic acid content of millet, this could not be confirmed, and the growth of puppies on lime-treated or untreated millet was similar (67). Millet is variable in tryptophan and lysine content. Certain varieties of millet, when supplemented with lysine, can overcome the niacin-tryptophan deficiency (68).

The lime treatment of corn does abolish the niacin deficiency inducing the effect on rats of the standard casein/corn test diet (54,55), but boiling without lime is enough (54). Heating and alkali treatment liberate a bound nicotinic acid (54,55), but the bound niacin seems water-unstable (54); thus, it cannot explain the lime or heat effect, even if the structure of bound niacin in corn is different from that found in wheat. Finally, cats, who cannot rely on tryptophan for their niacin requirement, do not grow better when corn is lime-treated (69).

Another explanation of the action of lime water on corn is that it alters the amino acid ratio of the cereal. This point had to be considered because of the reports that an imbalance in the dietary amino acid ratio will provoke niacin deficiency. This cannot explain the lime effect, however (55). The earlier observations leading to this hypothesis of an altered amino acid ratio were the discovery that a diet containing 9% casein resulted in a niacin-tryptophan deficiency when supplemented with tryptophan-deficient proteins like zein or gelatin (70). This has been attributed to the inclusion of certain specific amino acids such as threonine (72,73,75), phenylalanine (73), and cystine (74). Later high levels of valine and lysine were also reported to have an effect (75), although an actual niacin deficiency due to suppression of intestinal synthesis was also entertained (73). This latter possibility had some support from the observation that dietary carbohydrate has a strong influence on the development of the deficiency (76). Moreover, it is known that the type of carbohydrate in the diet determines to a great extent the intestinal flora. The concept of an amino acid imbalance has gained additional evidence and is now a major hypothesis for the effect of corn. The effect of corn has been duplicated independently of the source of protein by simulating the composition of the classical corn diet (77). This has already been alluded to and will be discussed in further detail later. However, this result does not explain the effect of the lime treatment of corn.

It should be unnecessary to stress that such experiments utilizing natural grains are often difficult to duplicate because of the wide variety of protein content and quality of corn (78) or millet (79), and the undetermined varieties that grow in the areas which have nutritional disease toxins in the corn.

While insight into the relationship between protein composition and pellagra was being explored, the old idea of a toxic factor in corn still persisted in three main lines of thought.

The idea which was first entertained was that corn is particularly susceptible to molds, and thus, indirectly supplies toxins from fungi. This possibility was first suggested by Pari (80) who blamed a fungus he called *Uredo maydis*, now known as *Ustilago maydis* (81). The same idea was repeated in the report of Gherardi (82) who claimed that apparently normal corn can cause pellagra but that moldy corn will cause severe pellagra. It is not surprising that a pellagrin might get sicker from eating spoiled food. The idea persisted until 1910, when Cammuri made the same claim with a more modern approach (83). Barrow et al., finding that corn bran slowed growth of rats and that niacin could overcome the inhibition, advanced a toxin as explanation (84). A claim that Borrellidin, a fungus product, causes growth retardation in rats which can be counteracted with niacin or tryptophan, falls along the same lines (85).

It has also been considered that corn is difficult to digest, and that the undigested fragments act as toxins (86,87). In addition, it has been postulated that these undigested fragments make the intestinal bacteria, especially *Escherichia coli,* more virulent (88-90). It has been suggested that lime treatment primarily softens the outer layer of the corn kernel and makes the protein more digestible (91).

Many authors, however, believed in the existence of a direct toxin in corn. A group of investigators, especially in Austria, observing the marked photosensitivity of the pellagrous patient, considered the possibility of a photosensitizing substance in the grain. Aschoff (92) was a strong advocate of this concept, although he had no convincing evidence. Experimental verification was sought by other investigators, notably Horbeczewski (93) and Raubitschek (94). Both investigators reportedly extracted a toxic substance (toxit for mice) from corn which worked photodynamically. Horbaczewski claimed, however, that the action could not be wholly photodynamic since gray mice were as susceptible as white mice. He claimed, furthermore, that the factor was not associated with zein or with colored matter (probably carotenoids) in the grain (95). Raubitschek noted the same effect under similar conditions using an alcoholic extract from corn (94). He insisted that these findings were independent of Horbaczewski, and he called Horbaczewski's results, though published more than one year before his: "die erste Bestatigung meiner Befunde." [The first confirmation of my findings.] The result of Horbaczewski and Raubitschek has been confirmed to some extent by Unmusz (96). No further references to a photosensitizing compound in corn can be found in the literature.

There are reports of a more or less "purely" toxic factor. Many older investigators believed in a toxic factor because the feeding of corn produced sick experimental animals. The concept of dietary deficiency had not arisen. An interesting example is the work of Ballner, who described the effect in guinea pigs of corn feeding, but the syndrome which he ascribes to corn toxicity looked strikingly like scurvy (97,98). More definitive reports have appeared, however. The oldest and most affirmative claim comes from Selmi (99), who reported to have isolated from corn a toxic substance which he called "acroleina ammoniacale." He reported for his material the empirical formula $C_{12}H_{20}O_3$. Unfortunately his directions are not clear. Harris also concluded that corn is toxic to humans, and further postulates that this toxicity is "inherited" (35). How this is to come about is not clear, unless he means that the susceptibility of the patient is hereditary in nature. Later reports of other investigators are also vague. Stockman and Johnston isolated toxic materials from several grains which they associated with pellagra, but this toxicity seems to be unspecific (100). Another Englishman, Leutsky, claimed that the toxicity of corn is due to the mineral composition, and that there toxic effects could be duplicated by feeding a mineral mixture patterned after the mineral composition of the grain (101). However, his salt mixture, which contained Na, K, Mg, Ca, Fe and Cl, SO_4, and PO_4, was strikingly normal. The only abnormal feature was a high magnesium level.

More specific claims of the existence of toxic substances include the report that some symptoms of pellagra in rats could be aggravated with mandelonitril (102). However, there is a difficulty here: a niacin deficiency in rats is not at all comparable to human pellagra, just as black tongue in dogs does not resemble the human disease. Therefore, even if one can duplicate the skin changes of pellagra in experimental animals, this is not likely to be the experimental counterpart of pellagra. Another specific compound which was claimed to be the toxic factor was indole-3-acetic acid (103), but this claim was later withdrawn (104). The corn is reported to be toxic for the meal worm, *Tenebrio molitor*, but this toxicity is unconnected with niacin or tryptophan metabolism (105).

Two unretracted claims remain in the recent literature, and unfortunately, both have remained preliminary reports. The first, by Borrow et al., (106), states that maize bran fed to mice at a level of 10% is toxic. The second report is the well-known observation of Woolley (107), which claims that an alkaline chloroform extract of corn yields a niacin deficiency in mice. Neither of these reports has been followed up by the authors. Kodicek (108) mentions a personal communication from Woolley to the effect that the corn factor was purified 100,000-fold. This statement was already incorporated in the original note by Woolley.

Attempts at duplicating these experiments provided a partial explanation. In mice given a protein-restricted diet, niacin as well as tryptophan is a dietary essential. The amount of niacin required is a function of the amount of fat in the diet. Adding 9% cottonseed oil to the diet resulted in growth retardation that could be overcome by additional niacin. The chloroform extracts, prepared as described by Woolley, have the same effect as cottonseed oil. Therefore, the extract behaved as a "toxin," but not in a specific way related to the origin in corn, but rather just because of the consequent fat content of the diet (109). The effect of high fat on the niacin requirement has been noted also for rats (110).

The original claims for a specific antinicotinic acid principle were made during the height of discoveries of specific structural analogues of vitamins or antivitamins. The data were interpreted in the tenor of the times (111).

The high leucine content of corn (Table 1) may to a degree be responsible for the specific effect of corn. In that sense one could say a "toxin" is present.

Conclusions: The epidemiology and clinical behavior of the dietary syndromes related to nicotinic acid illustrate the complex biochemical relationships of nicotinic acid. No special toxins nor bound form of niacin needs to be postulated in corn or other grains. While it has been claimed that the traditional maize-treating techniques allowed the rise of Mesoamerican civilizations (112), newer varieties of corn have allowed growth of children for whom corn is the only staple (113,114).

REFERENCES

1. Y. Suzuki, T. Simanura, and S. Odale, *Biochem. Zeitschr. 43:*89 (1912).
2. C. A. Elvehjem, R. J. Madden, F. M. Strong, and D. W. Woolley, *J. Am. Chem. Soc. 59:*1767(1937).

3. H. von Euler, H. Ahlers, and F. Schlenk, *Z. Physiol. Chem. 124:* 113(1936).
4. O. Warburg and W. Christian, *Biochem. Zeitschr. 274:* 112(1934).
5. O. Warburg and W. Christian, *Biochem. Zeitschr. 275:* 464(1935).
6. AIN Committee on Nomenclature, *J. Nutr. 105:* 134(1975).
7. E. H. Todhunter, in *A Guide to Nutrition Terminology for Indexing and Retrieval,* U. S. Government Printing Office, Washington, D.C., 1970, p. 270.
8. R. H. Major, in *Classic Descriptions of Disease,* Thomas, Springfield, Ill., 1932.
9. T. D. Spies, in *Clinical Nutrition* (N. Joliffe, F. Tisdall, and P. R. Cannon, eds.), Hoeber, 1950, p. 531.
10. R. M. Follis, Jr., in *Deficiency Disease,* Thomas, Springfield, Ill., 1958, p. 316.
11. J. Goldberger and J. Wheeler, *Public Health Rep. 30:* 3336(1915).
12. T. D. Spies and M. A. Blankenhorn, *JAMA 110:* 622(1938).
13. J. M. Ruffin and D. T. Smith, *Southern Med. J. 32:* 40(1934).
14. R. H. Chittenden and F. P. Underhill, *Am. J. Physiol. 44:* 13 (1917).
15. G. A. Wheeler, J. Goldberger, and V. Blackstock, *Public Health Rep. 37:* 1063(1922).
16. J. Goldberger and G. A. Wheeler, *Public Health Rep. 43:* 172(1928).
17. D. T. Smith, E. L. Persons, and H. I. Harvey, *J. Nutr. 14:* 373 (1937).
18. C. T. Koehn and C. A. Elvehjem, *J. Biol. Chem. 118:* 693(1937).
19. C. A. Elvehjem, R. J. Madden, F. M. Strong, and D. W. Woolley, *J. Biol. Chem. 123:* 137(1938).
20. C. A. Elvehjem, R. J. Madden, F. M. Strong, and D. W. Woolley, *J. Am. Chem. Soc. 59:* 1767(1937).
21. P. Handler and W. Y. Dann, *J. Biol. Chem. 145:* 145(1942).
22. W. J. Dann and P. Handler, *J. Nutr. 22:* 409(1941).
23. P. Handler, *Proc. Soc. Exper. Biol. and Med. 52:* 263(1943).
24. W. A. Krehl and C. A. Elvehjem, *J. Biol. Chem. 158:* 173(1945).
25. W. A. Krehl, I. J. Tepley, and C. A. Elvehjem, *Proc. Soc. Exper. Biol. and Med. 58:* 334(1945).
26. J. A. Milner, *J. Nutr. 109:* 1351(1979).
27. W. A. Krehl, L. T. Tepley, and C. A. Elvehjem, *Science 101:* 283 (1945).
28. M. Chick, T. F. Macrae, A. J. P. Martin, and C. J. Martin, *Biochem. J. 32:* 10(1938).
29. D. N. Baron, C. E. Dent, M. Harris, E. W. Hart, and J. B. Lepson, *Lancet 1:* 421(1956).
30. K. Halvorsen and S. Halvorsen, *Pediatrics 31:* 29(1963).
31. J. B. Jepson, in *The Metabolic Basis of Inherited Disease* (J. B. Stanbury, J. B. Wyngaarden, and D. S. Frederickson (eds.), 3rd Edition, McGraw-Hill, New York, 1972, p. 1486.
32. J. Goldberger and W. F. Tanner, *Public Health Rep. 37:* 462(1922).

33. R. C. Williams, *The United States Public Health Service*, Washington, D. C., 1951.
34. C. Gopalan and K. S. J. Rao, *Vitamins and Hormones 33:*505(1975).
35. H. F. Harris, *Pellagra*, Wiley, New York, 1919.
36. H. Chick, *Nutr. Abstr. Rev. 20:*523(1951).
37. H. Chick, *F. A. O. Nutritional Studies No. 9*, Food and Agricultural Organization of the United Nations, Rome, Italy, (1953).
38. F. Mainzer, *Klin. Wochenschr. 28:*729(1950).
39. D. A. Roe, *A Plague of Corn. The Social History of Pellagra*, Cornell University Press, Ithaca, New York, 1973.
40. G. Casal, in *Obra Posthuma* (M. Martin, ed.), Madrid, 1762.
41. R. H. Major, in *Letters of Christopher Columbus with other Documents Relating to his Four Voyages to the New World*, 2nd Edition, London, 1870.
42. G. B. Marzari, *Saggio medico-politico sulla pellagra, o scorbuto*, Italico, Venezia, 1810.
43. V. P. Sydenstricker, *Am. J. Clin. Nutr. 6:*409(1958).
44. C. Morelli, *Le pellagra nei suoi rapporti medicine sociali*, Firenze, 1856.
45. C. Funk, *Münch. Med. Wochenschr. 60:*2614(1913).
46. W. H. Wilson, *J. Hyg. 20:*1(1921).
47. C. Funk, in *The Vitamins*, Williams and Wilkins, Baltimore, 1922.
48. J. Goldberger and G. A. Wheeler, *U. S. Publ. Health Serv. Hyg. Lab. Bull. 120*, part 1 (1920).
49. W. A. Krehl, L. J. Teply, P. S. Sarma, and C. A. Elvehjem, *Science 101:*489(1945a).
50. R. Jacquot, T. Terroine, J. Creveaux, F. Charconnet, and J. Adrian, *Arch. des Sciences Physiol. 4:*115(1960).
51. W. A. Krehl, C. A. Elvehjem, and F. M. Strong, *J. Biol. Chem. 156:*13(1944).
52. J. Laguna and K. J. Carpenter, *J. Nutr. 45:*21(1951).
53. R. O. Cravioto, G. H. Massieu, O. Y. Cravioto, and F. de M. Figueroda, *J. Nutr. 48:*453(1952).
54. W. N. Pearson, S. J. Stempfel, J. Salvador-Valenzuela, M. H. Utley, and W. J. Darby, *J. Nutr. 62:*445(1957).
55. A. E. Harper, B. D. Punelear, and C. A. Elvehjem, *J. Nutr. 66:*163(1958).
56. H. Chick, *Nutr. Abstr. Rev. 20:*523(1951).
57. R. Bressani, E. Marcucci, C. E. Robles, and N. S. Scrimshaw, *Food Res. 19:*263(1954).
58. W. C. Rose, *Fed. Proc. 8:*546(1949).
59. G. A. Goldsmith, H. P. Sarret, U. D. Register, and J. Gibbens, *J. Clin. Invest. 31:*533(1952).
60. D. K. Chaudhuri and E. Kodicek, *Biochem. J. 47:*xxxiv(1950).
61. D. K. Chaudhuri and E. Kodicek, *Nature 165:*1022(1950a).
62. W. A. Krehl and F. M. Strong, *J. Biol. Chem. 156:*1(1944).
63. M. L Das and B. C. Guha, *J. Biol. Chem. 235:*2971(1960).

64. P. K. Sarkar, H. P. Ghosh, and B. C. Guha, in *International Congress of Biochemistry VI*, Vol. 5, New York, 1964, p. 449.
65. J. B. Mason, N. Gibson, and E. Kodicek, *Br. J. Nutr. 30:*297 (1973).
66. H. P. Ghosh, P. K. Sarkar, and B. C. Guha, *J. Nutr. 79:*451 (1963).
67. B. Belavady and C. Gopalan, *Ind. J. Biochem., 3:*44(1966).
68. A. S. Mangay, W. N. Pearson, and W. J. Daily, *J. Nutr. 62:*377 (1957).
69. J. E. Braham, A. Villareal, and R. Bressani, *J. Nutr. 76:*183 (1962).
70. W. A. Krehl, P. S. Sarma, and C. A. Elvehjem, *J. Biol. Chem. 162:*403(1946).
71. S. A. Singal, V. P. Sydenstricker, and J. M. Littlejohn, *J. Biol. Chem. 176:*1051(1948).
72. S. A. Singal, V. P. Sydenstricker, and J. M. Littlejohn, *J. Biol. Chem. 176:*1063(1948a).
73. L. V. Hankes, L. M. Henderson, W. L. Brickson, and C. A. Elvehjem, *J. Biol. Chem. 174:*873(1948).
74. L. V. Hankes, L. M. Henderson, and C. A. Elvehjem, *J. Biol. Chem. 180:*1027(1949).
75. L. M. Henderson, O. J. Koeppe, and H. H. Zimmerman, *J. Biol. Chem. 201:*697(1953).
76. W. A. Krehl, P. S. Sarma, L. J. Teply, and C. A. Elvehjem, *J. Nutr. 31:*85(1946a).
77. O. J. Koeppe and L. M. Henderson, *J. Nutr. 55:*23(1955).
78. H. H. Mitchell, T. S. Hamilton, and T. R. Beadlles, *J. Nutr. 48:* 461(1952).
79. J. Adrian and C. Sayerre, Les Plantes alimentaires de l'ouest Africain II. Composition de Mils et Sorghas du Senegal. *Government Géneral de l'Afrique Occidentale Francaise (Dakar)*
80. A. G. Pari, *Sperimentale 25:*345(1870).
81. H. A. Wallace and E. N. Bressman, in *Corn and Corngrowing*, Wiley, New York, 1949, 165 o.v.
82. G. Gherardi *La pellagra nolle Marche e in modo specialla nelle provincia di Pesaro*, Urbino, Pesaro, 1905.
83. L. V. Camurri, *Centralbl. Bakteriol. 53:*538(1910).
84. A. W. Borrow, L. Fowdon, M. M. Stedman, J. C. Waterlow, and R. A. Webb, *Lancet 1:*752(1948).
85. J. M. Cooperman, S. H. Rubin, and B. Tabenkin, *Proc. Soc. Exper. Biol. Med. 76:*18(1951).
86. E. Neusser, *Wien. Med. Wochenschr. 37:*132(1887).
87. E. Neusser, in *Die Pellagra in Oesterreich und Rumanien Nien*, 1887.
88. V. de Giaxa, *Ann. de Inst. d'Ig. Sper. de Univ. di Roma 2:*1 (1892).
89. V. de Giaxa, *Ann. de Inst. d'Ig. Sper. de Univ. di Roma 3:*367 (1893).

90. V. de Giaxa, *Ann. de Inst. d'Ig. Sper. de Univ. di Roma 13:*367 (1903).
91. A. P. de Groot, P. Slump, L. van Beek, and W. J. Feron, in *Evaluation of Proteins for Humans* (C. E. Bodwell, ed.), AVI, Westport, Conn., 1975, p. 270.
92. L. Aschoff, in *Ueber die Wirkungen des Sonnelichtes auf den Menschen,* Speyer und Karner, Leipzig, 1908.
93. J. Horbaczewski, *Das Oesterreichiesche Sanitätswesen 22* (Beilage zur #31):1(1910).
94. H. Raubitschek, *Centralbl. Bakteriol. 57:*193(1911).
95. J. Horbaczewski, *Das Oesterreichische Sanitätswesen 24:*417(1912).
96. O. Unmusz, *Z. Immunitätsforsch. 13:*461(1912).
97. F. Ballner, *Das Oesterreichische Sanitätswesen 22:*165(1910).
98. F. Ballner, *Das Oesterreichische Sanitätswesen 22:*201(1910a).
99. A. Selmi, *Atti della R. Accad. di Linoei; Scionze Fioiche, Hatenatiche e Naturali, So Sorie, Sc Serie 1*(2):1099(1876).
100. R. Stockman and J. M. Johnston, *J. Hyg. 33:*204(1933).
101. C. M. Leutsky, *Lancet 233:*1421(1937).
102. A. Clark, *J. Trop. Med. Hyg. 41:*315(1938).
103. E. Kodicek, K. J. Carpenter, and L. J. Harris, *Lancet 251*(2): 491(1946).
104. E. Kodicek, K. J. Carpenter, and L. J. Harris, *Lancet 253*(2): 616(1947).
105. H. Lipke, and G. Fraenkel, *J. Nutr. 55:*165(1955).
106. A. Borrow, L. Fowden, M. M. Stedman, J. C. Waterlow, and R. A. Webb, *Lancet 254*(1):752(1948).
107. D. W. Woolley, *J. Biol. Chem. 163:*773(1946).
108. E. Kodicek, *B. J. Nutr. 2:*373(1949).
109. W. N. Pearson, J. S. Valenzuela, and J. van Eys, *J. Nutr. 66:* 277(1958).
110. N. V. Shasti and M. C. Nath, *J. Vitaminol. 18:*200(1972).
111. D. W. Woolley, *Physiol. Rev. 27:*308(1947).
112. S. H. Katz, M. L. Hediger, and L. A. Valleroy, *Science 184:*765 (1974).
113. G. G. Graham, D. V. Glover, G. Lopez de Romaña E. Morales, and W. C. MacLean, Jr., *J. Nutr. 110:*1061(1980).
114. G. G. Graham, R. P. Placko, and W. C. MacLean, Jr., *J. Nutr. 110:*1070(1980).

2

NUTRITIONAL BIOCHEMISTRY

I. INTRODUCTION

The emphasis in this chapter will be on human nutritional biochemistry. Much relevant literature on biochemistry in animal model types will be reviewed, but the biochemistry of nicotinic acid in plants and animals will only be touched upon briefly.

The fact that tryptophan could replace the requirement for nicotinic acid in many species implied the possibility of a sparing action or interconversion. Interconversion proved to be the correct answer. There are three sets of essential amino acids-derived vitamin pairs: tryptophan-nicotinic acid, methionine-choline (1), and lysine-carnitine (2). The vagaries of sequence and timing of discovery have decreed that niacin is considered a vitamin, and choline is not—at least not anymore—and carnitine never has been considered as such, although it was called vitamin B_t to designate its essential nature for the confused flour beetle, *Tenebrio molitor* (3).

To discuss the nutritional role of nicotinic acid, its metabolism must be summarized. This includes both the conversion of tryptophan to niacin and the incorporation of nicotinic acid and its amide into coenzymes, and the conversion of nicotinic acid and nicotinamide to excretory metabolites.

I. THE INTERRELATIONSHIPS BETWEEN TRYPTOPHAN AND NICOTINIC ACID

A. The Tryptophan-Niacin Pathway

The average daily intake of tryptophan in the American diet is 800-1000 mg (4,5). The minimal daily requirement is estimated to be 300 mg for adults (6). Much of the tryptophan is excreted as indole derivatives in stool and urine. However, the primary degradation of tryptophan which is of concern here is its conversion to niacin. Figure 1 summarizes the pathway. The unraveling of this pathway required putting together a large number of independently discovered reactions and metabolic transformations. There is a great deal of waste in the pathway when viewed as a biosynthetic process for niacin (Fig. 2). As will be discussed later, the yield of niacin is only 1 mg for 60 mg of tryptophan metabolized, or a molar yield of only 2.75%. The major pathway for tryptophan after its metabolism to 3-hydroxykynurenine is degradation to carbon dioxide and water (Fig. 3). Only because the step beyond α-amino-β-carboxy-muconic-ε-semialdehyde to α-amino-ε-semialdehyde is rate limiting, is niacin produced in quantity. Cats have fifty times higher activity of this enzyme (7). Therefore, they cannot utilize tryptophan to satisfy their niacin requirement (8). The intermediates are excreted, frequently in a further metabolized form, not on the pathway of interest here. The various pathways for tryptophan metabolism are illustrated in Fig. 2. They will not be discussed in detail.

The first step in the metabolism is the conversion of tryptophan to *N'*-formylkynurenine, but kynurenine was the first conversion product discovered in the urine of rabbits fed high doses of tryptophan (9). This was later reconfirmed (10). The intermediate role of a compound like *N'*-formylkynurenine was suggested by Ellinger and Matsuoka (11), but it was not shown to be an intermediate until tryptophan 2,3-dioxygenase (tryptophan pyrrolase) was described (12). In mammalian liver this enzyme is hormonally induced by glucocorticoids and by tryptophan itself (13). It is a rate-limiting enzyme in the sequence when not induced. It is also found in rabbit intestine, but only in the distal one-sixth of the small intestine (14). The enzyme requires hematin which must be added for maximal activity (15).

The next step in tryptophan to niacin conversion involves the action of a formamidase or formylase to convert *N'*-formylkynurenine to kynurenine (16). This enzyme is usually in excess, and thus *N'*-formylkynurenine is not normally an excretory product after tryptophan loading. Kynurenine is converted to anthranilic acid (17), kynurenic acid (18), or 3-hydroxykynurenin (19). In addition it is excreted as such and in an acetylated form, at least in vitamin B_6-deficient rats (20).

These conversions are all interrelated, and several vitamin B_6-dependent enzymes are involved. The conversion of kynurenine to anthranilic acid occurs through the enzyme kynureninase which yields alanine together with anthranilic acid. This enzyme acts also on 3-hydroxykynurenine to form alanine and 3-hydroxyanthranilate. The anthranilic acid is a metabolic deadend in mammals, while 3-hydroxyanthranilic acid is a precursor of nicotinic acid.

Anthranilic acid is not excreted in a free form in significant quantities in humans, but primarily as the acylglucuronide or the glycine conjugate—anthranilic acid hippurate (21). Brown and Price noted that these conjugates, together with kynurenine and actylkynurenine, account for most of the diazotizable aromatic amines in urine (22).

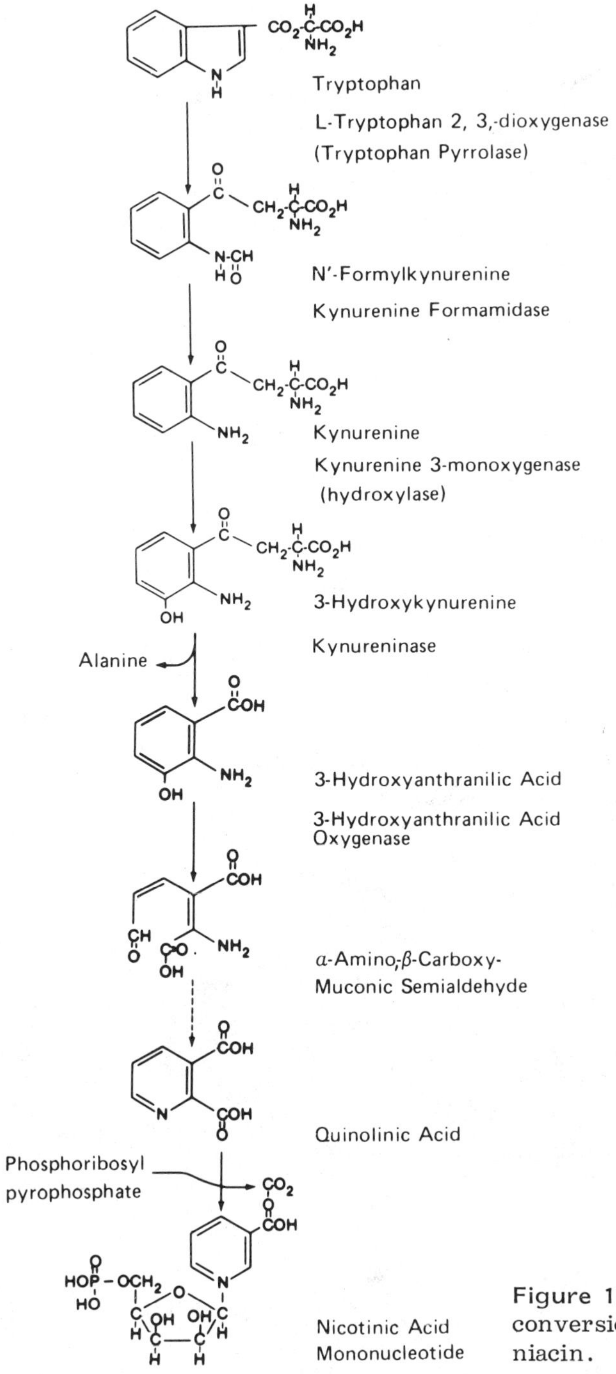

Figure 1 The pathway of the conversion of tryptophan to niacin.

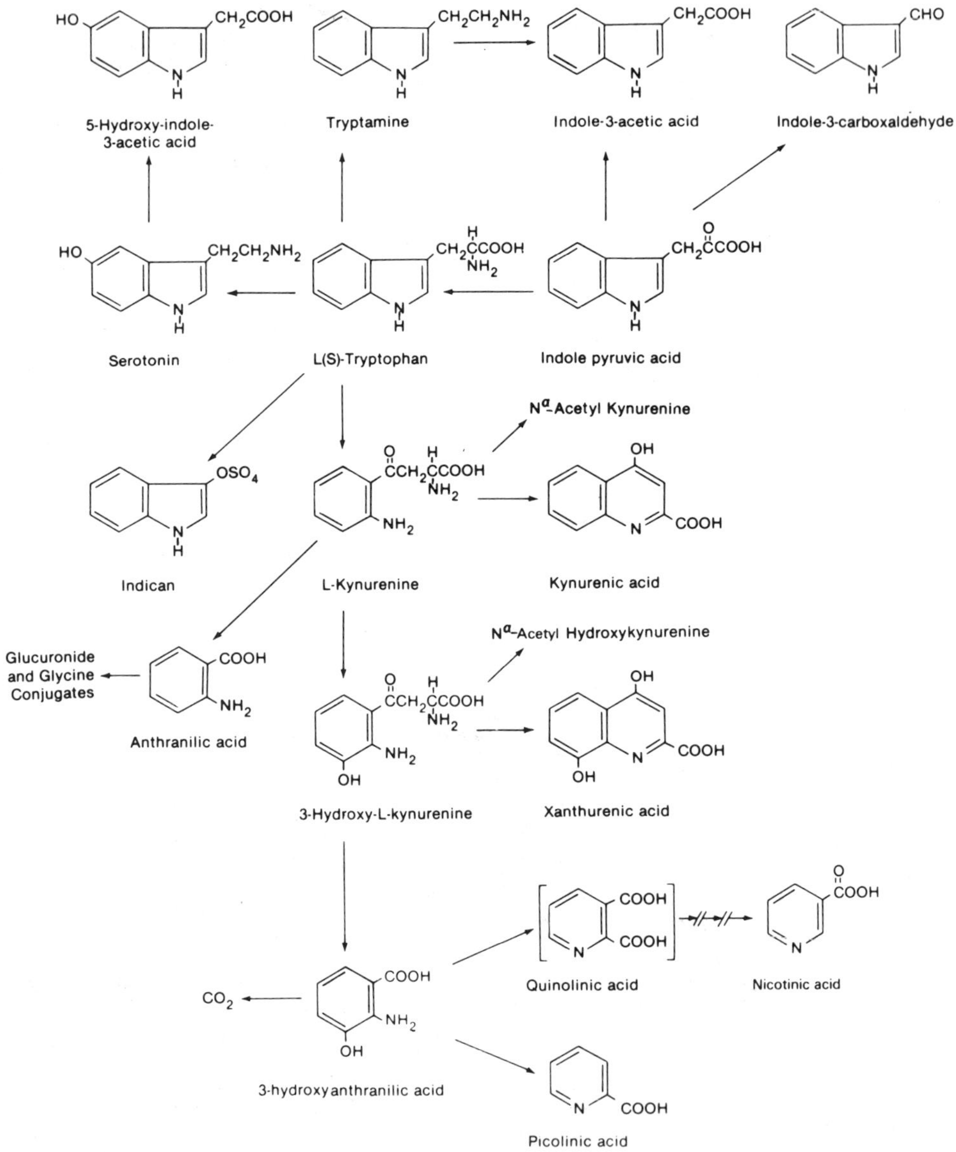

5-Hydroxy-indole-3-acetic acid

Tryptamine

Indole-3-acetic acid

Indole-3-carboxaldehyde

Serotonin

L(S)-Tryptophan

Indole pyruvic acid

Nα-Acetyl Kynurenine

Indican

L-Kynurenine

Kynurenic acid

Nα-Acetyl Hydroxykynurenine

Glucuronide and Glycine Conjugates

Anthranilic acid

3-Hydroxy-L-kynurenine

Xanthurenic acid

CO_2

3-hydroxyanthranilic acid

Quinolinic acid

Nicotinic acid

Picolinic acid

Figure 2 Competing pathways of tryptophan metabolism in animals and man.

Excess kynurenine can be further metabolized to kynurenic acid, which is a spontaneously cyclized product from the α-keto analogue of kynurenine, formed from a transamination with α-ketoglutarate through kynurenine aminotransferase (20,23,24).

The next step in metabolism is the hydroxylation of kynurenine by kynurenine 3-monooxygenase (hydroxylase). This is a classical oxidase, membrane bound, dependent on reduced nicotinamide adenine dinucleotide phosphate in membrane preparations. In addition, the enzyme has a flavin adenine dinucleotide requirement (25). That the conversion of kynurenine to hydroxykynurenine exists in humans was made likely by the discovery of hydroxykynurenine in human urine (26) and the increased excretion after kynurenine administration (27). It is found in significant quantities in normal human urine (28) without tryptophan loading.

Hydroxykynurenine has a metabolic fate similar to that of kynurenine. It can be deaminated through transamination with α-ketoglutarate to yield the α-keto analogue which spontaneously cyclizes to xanthurenic acid (24). This transamination requires vitamin B_6 as pyridoxal phosphate (PLP). Similar to kynurenine, some hydroxykynurenine is excreted as the N-acetylated derivative (20).

Most important, 3-hydroxykynurenine yields 3-hydroxyanthranilic acid through the action of kynureninase, again in a PLP-dependent step. This enzyme is found in the liver of all mammalian species tested. The same enzyme has both kynurenine and hydroxykynurenine as substrates, but hydroxykynurenine is the preferred substrate, at least in hog liver (29). 3-Hydroxyanthranilic acid is an intermediate in the conversion of tryptophan to niacin as well as in the total degradation of tryptophan to carbon dioxide (30,31). This sequence proceeds via a first step catalyzed by 3-hydroxyanthranilic acid oxygenase, which generates α-amino-β-carboxymuconic-ε-semialdehyde. This substance is either metabolized further by decarboxylation to α-aminomuconic-ε-semialdehyde, or it spontaneously cyclizes to quinolinic acid. The second semialdehyde can spontaneously cyclize to picolinic acid or be further degraded via α-aminomuconic acid, α-ketoadipic acid, glutaryl-coenzyme A, and on to carbon dioxide (Fig. 3) (32).

The role of quinolinic acid as an intermediate in nicotinic acid biosynthesis was for a while unclear, since the question of quinolinic acid was a complicated one. On the one hand, there was strong evidence that the compound was a product of tryptophan metabolism. Tryptophan labeled with ^{15}N in the indole ring yields quinolinic acid labeled with ^{15}N in the pyridine ring, as indicated by studies in whole rats (33). Furthermore, 3-hydroxyanthranilic acid had been shown to be converted by liver homogenates to quinolinic acid (34,35). There is little doubt that 3-hydroxyanthranilic acid is a niacin precursor, as indicated from isotropic experiments (36). Furthermore, all compounds known to be on the conversion scheme from tryptophan to niacin in animals on a deficient diet (37). The activity of these compounds in-

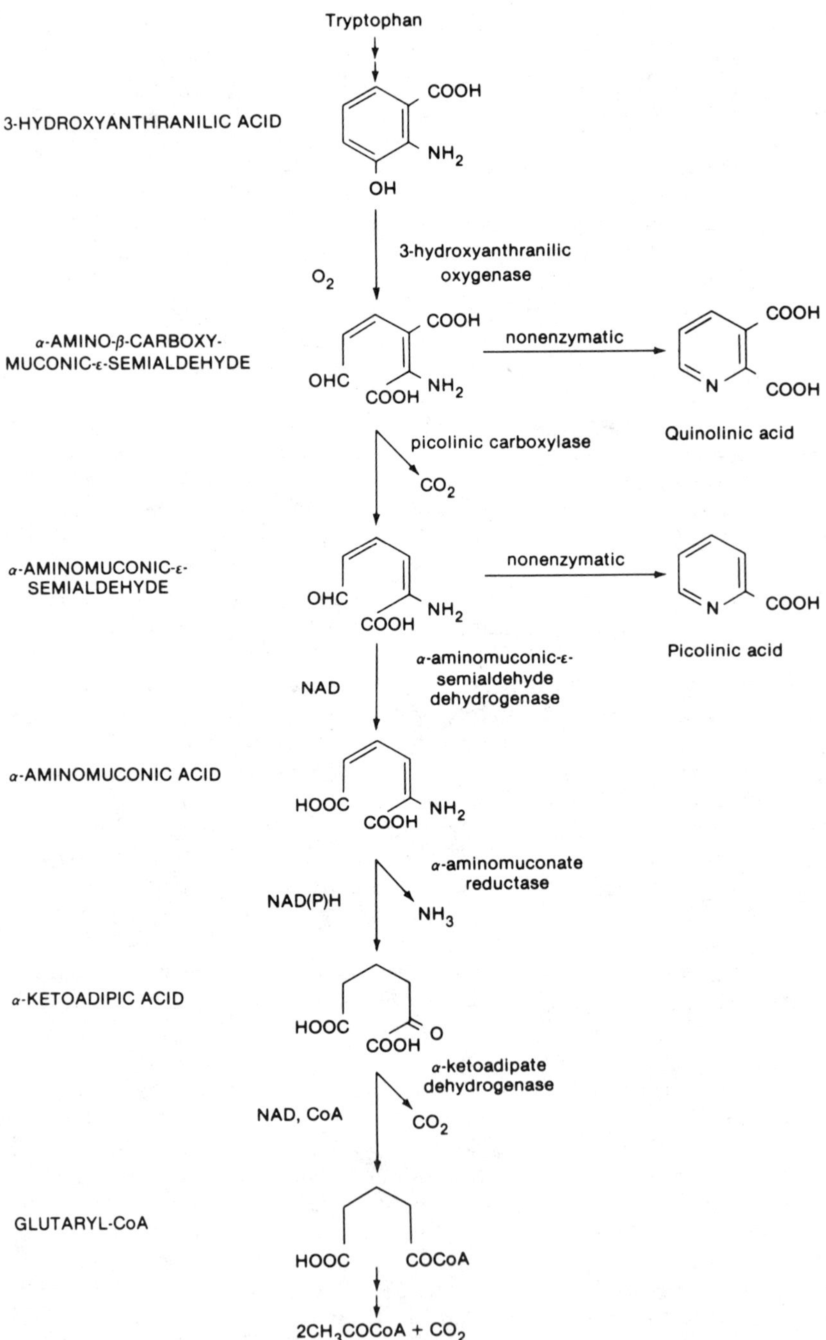

Figure 3 Metabolism of tryptophan to CO_2.

creases from tryptophan to niacin. However, the response to quinolinic acid is but a fraction of the response to tryptophan (37,38). Yet quinolinic acid seemed an attractive precursor for nicotinic acid since it is easily converted to the latter by acid decarboxylation (39). The solution was the finding of a specific enzyme that converts quinolinic acid to nicotinic acid— the enzyme quinolinate phosphoribosyltransferase. There is simultaneous decarboxylation in this step (40). The enzyme seems to be specific for quinolinate (41,42). It is very widespread in nature (42,43), and has been crystallized from hog liver (45).

The pathway from tryptophan to nicotinic acid leads directly to the functional nicotinic acid nucleotides. Free nicotinic acid is formed by the reversible pyrophosphorolysis of the mononucleotide. Conversely nicotinic acid enters the pathway via a specific nicotinate phosphoribosyltransferase (44–47) to form nicotinic acid mononucleotide. Therefore, the last steps in the conversion of tryptophan to nicotinic acid would be:

Quinolinic acid + phosphoribosyl pyrophosphate →
nicotinic acid mononucleotide + pyrophosphate + CO_2
Nicotinic acid mononucleotide + pyrophosphate → nicotinic
acid + phosphoribosyl pyrophosphate

Quinolinic acid → nicotinic acid + CO_2

However, practically little conversion occurs. Nicotinamide is converted to the mononucleotide by a specific nucleotide pyrophosphorylase or is deamidated before final incorporation in newly synthesized coenzymes. This will be discussed in detail later.

The quantitation of the flow through the pathway is frequently measured. The capacity of the pathway can be tested through a tryptophan load, using DL-tryptophan. D-Tryptophan yields D-kynurenine (48) which cannot be converted to niacin. Different species differ in their capacity to utilize D-tryptophan. For instance, the rat can do so readily, but the chick does so poorly (49). This difference is explained in part by the ability of rats to convert the D to the L isomer (50). The rabbit (51), mouse (52), and human (53,54) likewise utilize D-tryptophan poorly while the dog has an intermediate ability to utilize the D enantiomer (55).

In view of this great variation among species in their capacity for utilizing D-tryptophan, data are only reliable if L-tryptophan is used. Price et al. advocated a tryptophan loading dose of 2.0 g for an adult which is roughly two to two and one-half times the average intake.

With such a loading experiment, data as illustrated in Table 1 are obtained (56). Males excrete more anthranilic acid and kynurenine (as such or as metabolites) as do females, but the yield of new metabolites after tryptophan loading is the same.

It must be mentioned here that the administration of tryptophan entails real and theoretical hazards. If the amino acid is administered in the fasting state, some people develop lightheadedness and dizziness, and headache (58). This effect can be duplicated in monkeys (59).

Table 1 Average Urinary Excretion of Metabolites by Female (F) and Male (M) Control Subjects[a]

Parameter		KA	XA	ISA	AAG	O-AH	ACK	KYn	HKYN	MPCA	4-PA	CR
Basal	F	18.8	8.1	371	4.2	21.3	9.8	9.8	21.4	89.3	3.87	0.17
		± 4.0	± 2.0	±145	±1.2	± 4.5	± 2.1	± 3.2	±14.4	± 31.4	±1.4	±0.20
	M	13.2	10.1	383	5.4[c]	26.5[d]	12.7[d]	13.5[d]	19.3	103.2	4.10	1.99[c]
		± 1.8	± 2.4	±121	±1.6	± 5.4	± 3.2	± 4.2	± 9.1	± 28.0	±0.56	±0.21
Post-tryptophan	F	59.7	29.8	500	6.9	40.3	14.5	27.5	43.1	130.8	3.34	1.15
		±19.1	± 8.2	±270	±2.1	±11.2	± 3.2	±13.3	±14.0	± 36.1	±1.16	±0.22
	M	56.6	29.4	544	9.1[d]	49.6[b]	17.1	35.5	37.6	135.6	3.89	1.98
		±21.5	±25.4	±312	±2.4	±14.2	± 4.4	±21.0	±20.4	± 37.3	±1.33	±0.41
Yield[e]	F	46.4	21.7	129	2.7	19.0	4.7	17.7	21.7	41.5	−0.53	−0.03
		±19.7	± 8.5	±196	±1.8	±10.2	± 3.5	±12.8	±13.3	± 31.7	±0.82	±0.11
	M	43.4	19.3	161	3.7	23.1	4.4	22.0	18.3	32.4	−0.20	−0.01
		±20.6	±24.8	±303	±1.8	±22.7	± 3.5	±18.2	±21.1	± 29.9	±1.24	±0.36

[a]Values are expressed as average micromoles of metabolite excreted (± standard deviations) per 24 hr. Abbreviations used are KA, kynurenic acid; XA, xanthurenic acid; ISA, indoxylsulfate (indican); AAG, anthranilic acid glucuronide; O-AH, *O*-aminohippuric acid; 4-PA, 4-pyridoxic acid; CR, creatinine. There were 19 male and 18 female subjects studied for each metabolite except for XA, ACK, HKYN, 4-PA, and CR, where the number of subjects ranged from 6 to 17.

[b,c,d]Indicate male values statistically different from corresponding values for females with $P < 0.05$, 0.02, and 0.01, respectively.

[e]The yield is the net change after tryptophan loading with 2.0 g of L-tryptophan, i.e., post-tryptophan value minus basal value for each subject.

Source: Ref. 56.

L-Tryptophan can induce hypoglycemia in rats (60) and can be lethal through this hypoglycemic effect (61). Likewise, deficiency in tryptophan causes a decreased glucose tolerance (62).

The actual conversion to niacin from a tryptophan loading dose is overshadowed by the conversion of tryptophan to deadend metabolites, anthranilic acid, kynurenine, and xanthurenic acid. However, in the normal nutritional state, the adequacy of niacin nutriture depends on the conversion of tryptophan to niacin. This pathway is very sensitive to a variety of nutritional, hormonal, and pathological alterations as well as to iatrogenic effects of drug therapy. Therefore, such effects need to be discussed.

B. Factors Influencing the Tryptophan–Niacin Conversion

Vitamin B_6: The pathway from tryptophan to niacin has many steps dependent on vitamin B6 (Fig. 4). Therefore, any pathological condition affecting the B_6 nutriture is potentially pellagragenic, especially during marginal protein and niacin intake. Since vitamin depletion does not deplete coenzymatic function uniformly, there is no a priori prediction what the in vivo effect of vitamin B_6 depletion is on this pathway. The data indicate that there is a significant effect on nicotinic acid nutrient status.

When animals are depleted of vitamin B_6, a disturbance in the tryptophan pathway after tryptophan loading can readily be demonstrated. This has been done in rats (63–67), mice (66,68), hamsters (67,69), dogs (70), pigs (71), monkeys (72,75), and guinea pigs (67). Humans have frequently been studied also (67,74–82), usually with dietary restrictions or sometimes with antimetabolites (79,82). There are marked species differences in a quantitative sense, but an effect is clear. Table 2 shows the clear increase in the metabolites of tryptophan in women on a B_6-deficiency diet. Other data could readily be quoted from the literature. Data on women were chosen because of the effect of birth control pills. This will be discussed later.

However, the data in Table 2 show only those metabolites which are not directly on the pathway to niacin. In view of the low conversion ratio of tryptophan to niacin usually given as 60:1 (6)—although it may be lower yet (81)—it could be conceivable that the conversion to nicotinic acid might be unaffected, even if the excretion of metabolites increases. Data have been controversial. Quinolinic acid could be used as an index substance. Henderson et al., demonstrated that rats have a decreased quinolinic acid excretion after tryptophan loading (64) which could be confirmed by Yeh and Brown (67). However, in humans, quinolinic acid excretion actually increases as a percent of tryptophan load in vitamin B_6-deficient states in humans (67,75,82). Hamsters and guinea pigs have intermediate effects but the changes were said to be insignificant (67). 3-Hydroxyanthranilic acid was also said to be elevated in humans on a vitamin B_6-deficient diet (79,83).

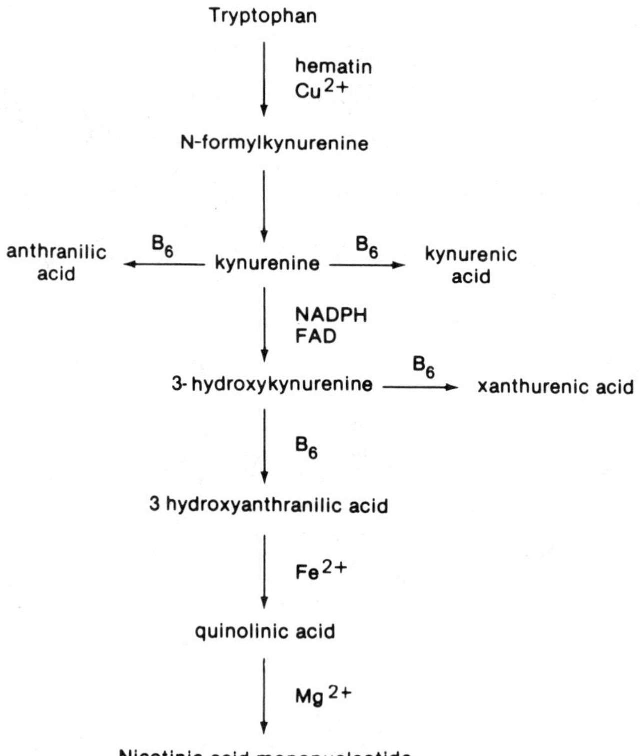

Figure 4 The interaction between vitamin B_6 and metals and the conversion of tryptophan to niacin.

When no tryptophan load was given but only a graded amount of tryptophan, there seemed to be no effect of vitamin B_6 deficiency on niacin metabolites (69), and supplementation by vitamin B_6 had no effect on excretion using different amounts of leucine (84).

The explanation is probably to be found primarily in the difference of degree of deficiency of vitamin B_6, since all subjects in the quoted studies were marginally deficient or only biochemically deficient, such as shown by a decreased urinary vitamin B_6 excretion (80). Frank pellagra was not induced except in the studies using deoxypyridoxine (82). However, clinically such situations do arise and therefore need to be summarized. There are many concerns in the literature about the marginal B_6 deficiency in American human populations, and this, too, needs to be mentioned. However, it is in the poorly nourished, sick

Table 2 Mean Basal and Post-Tryptophan Excretion of Four Tryptophan Metabolites

Tryptophan metabolite		Self-chosen diet	Experimental period[a]				
			Predepletion	Depletion		Repletion	
			1	1	2	1	2
			(μmoles per 24 hours)[b]				
Hydroxy-	pre[c]	38.2± 3.51	29.0± 4.87	46.7± 5.12	63.8± 4.34	35.8± 8.15	21.9± 6.56
kynurenine	post[c]	47.1± 5.57	46.21 6.56	460.4±143.4	1444.1±385.1	56.3±10.8	40.8±10.6
Kynurenine	pre	19.8± 1.94	15.6± 0.65	25.5± 3.96	38.6± 5.29	28.7± 4.87	15.0± 1.55
	post	49.5±13.21	49.4±11.76	673.9±257.5	1170.4±195.5	73.6±19.6	48.8± 9.27
Xanthurenic	pre	2.6± 0.10	2.6± 0.28	2.4± 0.14	3.2± 0.59	1.8± 0.2	1.8± 0.17
acid	post	4.4± 0.40	4.8± 0.57	266.5± 1.4	953.9±262.1	6.9± 2.17	4.6± 0.44
Kynurenic	pre	8.3± 1.00	5.5± 1.14	6.6± 1.67	6.9± 1.67	10.8± 1.34	10.9± 1.72
acid	post	37.0± 4.69	41.1± 4.69	125.0± 18.2	193.0± 29.3	39.81 3.89	43.3± 9.41

[a]The lengths of the experimental periods were 1 week for predepletion, 2 weeks for depletion and 2 weeks for repletion. Numbers under each experiment period refer to the number of the week.

[b]Each result is the mean ± SEM for the five subjects.

[c]The terms *pre* and *post* mean before and after the oral administration of 2 g L-tryptophan (9,800 μmoles) on days 6, 13, 20, 27, and 34.

Source: Ref. 80.

patient that the deficiency becomes significant. In such a patient the degree of deficiency approaches that found in a rat.

There is a further problem in interpreting the experimental data, in that tryptophan induces the initial step of the kynurenine pathway (85), so that tryptophan loading may in effect generate an artificial picture of the actual situation under ordinary tryptophan intake (86). In fact, tryptophan administration increases protein synthesis (87,88). Finally, there are pitfalls in the assay of the metabolites because of the interference by drugs, such as sulfa derivatives, diphenylhydantoin, and reserpine (89).

The effect of oral contraceptives: The initial claim of an effect of oral contraceptives on the tryptophan–niacin interconversion came from Rose, who reported that estrogen caused increased excretion of 3-hydroxykynurenine and xanthurenic acid after a tryptophan load test and that the effect was reversible with vitamin B_6 administration (90). In patients with breast carcinoma, the altered hormonal balance may also result in abnormal tryptophan metabolism (91). These studies were later amplified (78,83,92), and other workers found the same effects (93,94). Utilizing other parameters, primarily vitamin B_6-dependent erythrocyte transaminases, researchers could show that there was a vitamin B_6 defect in oral contraceptive users, which could be overcome by vitamin B_6 administration (95–97). Increased quantities of tryptophan metabolites were also found in these studies (95–97), but all the parameters except xanthurenic acid excretion rapidly returned to normal (95,96), suggesting either a higher sensitivity of the pathway to marginal B_6 deficiency or a special problem in the pathway. The amount of vitamin B_6 required was not excessively high in the women on oral contraceptives who were made B_6 deficient (98). These contradictory results are most likely explained by the hormonal effects on the tryptophan pyrrolase level, and therefore the increased metabolite excretion does not represent a B_6 defect. There is no real reason for concern over the adequacy of niacin nutrition in women taking oral contraceptives, though some increase in daily requirements for vitamin B_6 might be considered prudent (98).

Tuberculosis and antituberculosis drugs: In contrast to the effect of oral contraceptives on niacin, the effect of tuberculosis on the niacin nutriture is real. First of all, in the past, tuberculosis was especially prevalent among the same population that had the monotonous pellagragenic corn diet as major sustenance. The decreased intake from malaise frequently precipitated overt symptoms.

However, the problem is markedly aggravated by the antituberculous drug isoniazid (isonicotinic acid hydrazide, INH), (Fig. 5). This drug, like all hydrazides, can react with pyridoxal and pyridoxal phosphate to induce a real or functional vitamin B_6 deficiency. Increased urinary excretion of vitamin B_6 can be demonstrated in patients on isoniazid

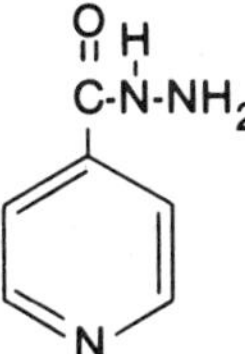

Figure 5 Isonicotinic acid hydrazide.

therapy (99). Isoniazid is normally excreted as an N-acetylated derivative and the ability to execute that acetylation has genetic polymorphism (100). This polymorphism may be the cause of differential susceptibility to isoniazid, though it is not a major factor in B_6 excretion (101), and therefore, it is thought that the antipyridoxine effect is primarily due to coenzyme inactivation.

These interactions are relevant to niacin nutriture. Cases of pellagra have been reported on isoniazid therapy which responded to combined vitamin B_6 and niacin treatment (102). In populations where pellagra is endemic, the problem is a real one. In India INH has been called a potential "final straw" in the development of pellagra (103). A similar situation pertains among the Bantu (104). That this effect is not seen among all population groups with endemic tuberculosis is possibly explained by the already mentioned remarkable polymorphism of isoniazid inactivation among populations (Table 3) (105).

The interaction between niacin and isoniazid is further complicated by the exchange reaction that can occur between nicotinamide and INH on the cofactor nicotinamide adenine dinucleotide catalyzed by the nicotinamide adenine dinucleotide glycohydrolase (NADase) to form the INH

Table 3 Racial Variation in Rate of Isoniazid Inactivation

Population	Slow Inactivators (%)
Hindu Indians	60
Jews	55-75
American Caucasians	45-60
American Negroes	45
Thais	28
Chinese	15
Japanese	10
Eskimos	5

Source: Ref. 105.

analogue (106,107). This effect will be described in detail later. Suffice it to say that, to a degree, isoniazid is a nicotinic acid antagonist also.

Riboflavin deficiency: In the enzymatic sequence converting tryptophan to niacin there is a riboflavin requirement: Kynurenine 3-monooxygenase, which converts kynurenine to hydroxykynurenine, has flavin adenine dinucleotide as a prosthetic group (108). Data suggest that riboflavin deficiency thus affects the tryptophan–niacin pathway at this step. Rats on a riboflavin-deficient diet excreted increased kynurenine and kynurenic acid and reduced pyridine derivatives after tryptophan or kynurenine administration (64), but not after hydroxykynurenine administration (109). Other workers noted the defect in niacin metabolites from tryptophan in riboflavin-deficient rats (110), and the elevated excretion of metabolites of tryptophan before the hydroxykynurenine step (111,112).

These observations are again linked to oral contraceptives. After cortisol administration the abnormal increased excretion of tryptophan metabolites was normalized not by niacin alone, but by riboflavin plus niacin (92). Again, the riboflavin status was diminished in women on oral contraceptives in a Thai population that were potentially undernourished, as demonstrated by low level red cell glutathione reductase activity (113). In Zambia, women on oral contraceptives excreted less riboflavin than women on Depo-Provera (medroxyprogesterone acetate) (114). It is unlikely that these effects significantly affect the niacin nutriture of the populations (115). The same can be said for riboflavin as was said for vitamin B_6, as it relates to pellagra and niacin nutriture. However, ariboflavinosis is still seen among alcoholics (116), and alcoholism predisposes significantly to pellagra (117). Alcoholism is, however, such a complicated factor in malnutrition that a simple, single-nutrient cause and effect is highly unlikely.

Copper deficiency: Rats made deficient in copper also have an apparent block in tryptophan conversion to niacin after a load test (119). Again, the block seemed to be between kynurenine and hydroxykynurenine. The enzyme involved is an oxygenase but not a copper-containing one. Penicillamine has been widely used clinically in the past as a copper chelator in Wilson's disease and for its sulfhydryl-reducing properties in a variety of clinical conditions. Both the L (119) and D isomers show vitamin B_6 antagonism in the rat (120). Penicillamine has been demonstrated to affect the tryptophan–niacin pathway in normal (121) and schizophrenic humans (122). It is now unclear whether these effects are copper or pyridoxine related. No penicillamine-induced pellagra has ever been reported. When copper-deficient rats were tested for tryptophan metabolites without a tryptophan load, the levels were found to be decreased (119). This is in accord with the supposition that copper is a component of rat liver tryptophan 2,3-dioxygenase (123).

Gopalan has speculated upon copper as a factor in the skin lesions of pellagra (124) because the pathological findings resemble copper deficiency, but this remains speculation.

Cancer: Cancer has a complex effect on the nutritional state. An overall protein–calorie malnutrition is frequent (125) with consequent cachexia if left untreated. However, specific relationships between cancer and the tryptophan–niacin pathway occasionally have been postulated.

Carcinoid tumors metabolize tryptophan in high quantities. Pellagra has been reported as a complication of this excessive conversion (126). Patients with bladder cancer have an incomplete metabolism of tryptophan (127–130). It was thought that this might explain the association of smoking and bladder cancer, but smoking does not affect the pathway (131). The abnormal metabolism correlates with the kynureninase activity (132). Tryptophan administration changes the excretion of transfer RNA metabolites (133–135), as does nicotinamide (134,135). There have been claims for a cause-and-effect relationship, in that tryptophan metabolites have been considered carcinogenic (129,130), but this has never been substantiated. No evidence refuting the hypothesis has been advanced (129).

Tryptophan depletion through injection of a degradative enzyme is now thought to be antineoplastic. Indolyl-3-alkane-α-hydroxylase shows antineoplastic effects on mouse sarcomas and carcinomas (136, 137).

Diabetes: Abnormalities of tryptophan metabolism in diabetes have been reported (138–141). The experimental model of alloxan diabetes in the rat shows the same effect: little N^1 methylnicotinamide is formed from tryptophan after diabetes is induced (142). This is apparently due to an increased conversion of ε-semialdehyde to α-aminomuconic-ε-semialdehyde and carbon dioxide (Fig. 3) (143,144). The activity of the responsible enzyme is inversely proportional to the amount of tryptophan available for niacin biosynthesis (145).

It is of interest that this intermediate, like its precursor, spontaneously cyclizes to a pyridine, α-picolinic acid, normally only a minor metabolite. Recently there has been a resurgence in interest in this isomeric pyridine carboyxlic acid because of its ability to bind zinc. That tryptophan metabolism may have a role in zinc absorption has been suggested (146–148), but the evidence is weak (149,150).

It also has been claimed that tryptophan metabolites are diabetogenic (151,152), but this is not the usually accepted etiology of this disease.

Other disorders: A very large number of disorders have been associated with pellagra or just disordered tryptophan–niacin interconversion, usually without discussion of the diet of the patients, the degree of stress, or even hormone manipulation. Porphyria, scleroderma, rheumatoid arthritis, celiac disease, sprue, and other malabsorption syndromes have been noted to have tryptophan metabolism abnormalities

(153). Special mention must be made of thyrotoxicosis, in which energy requirements are increased, most likely through the mechanism of an increased vitamin B_6 requirement (154). Schizophrenia has attracted a great deal of attention because of the possible involvement of tryptamine and serotonin in that process. However, no conclusive evidence exists for an etiological role of a disturbance in the tryptophan–niacin pathway (155).

C. Synopsis of Information on the Tryptophan–Niacin Pathway

The above review is far from complete since an exhaustive summary of the metabolic fate of tryptophan would go too far afield. There is a considerable flexibility in the system when the emphasis is on nicotinic acid. In general, the flux through the pathway adapts to the amount of tryptophan administered. However, tryptophan is easily a limiting dietary essential. Therefore, if tryptophan is used as a source of nicotinic acid, any factor that diminishes the conversion from tryptophan to niacin can precipitate pellagra. This is a clinically important factor in less nutritionally affluent countries than America and can occasionally be seen in individual patients in otherwise well-fed populations (156).

Many of the alleged disturbances found in various disease states are a direct consequence of the effects of the hormonal milieu on the induction of the first rate-limiting enzyme in the conversion. The literature on that phenomenon is voluminous and cannot be reviewed here. But stress changes the pathway, as do directly administered glucocorticoids and estrogens, and therefore great caution must be exercised in interpreting the data on intermediate excretion as an indication of the adequacy of the pathway.

REFERENCES

1. W. M. Griffith and H. M. Dyer, *Nutr. Rev. 26:*1(1968).
2. M. E. Mitchell, *Am. J. Clin. Nutr. 31:*293(1978).
3. S. Friedman and G. S. Fraenkel, in *The Vitamins* (W. H. Sebrell and R. S. Harris, eds.), Academic Press, New York, 1972, p. 329.
4. M. S. Reynolds, *J. Am. Diet. Assoc. 33:*1015(1957).
5. M. L. Orr and B. K. Watt, *USDA Home Econ. Res. Rept.* U. S. Government Printing Office, Washington, D. C., 1957.
6. Food and Nutrition Board. Committee on Dietary Allowances. *Recommended Dietary Allowances,* 9th revised ed., National Academy of Sciences, Washington, D. C., 1080, p. 43.
7. R. J. Sahadolnik, C. O. Stevens, R. H. Decker, L. M. Henderson, and D. V. Hankes, *J. Biol. Chem. 228:*973(1957).
8. A. C. da Silva, R. Fried, and A. C. de Angelis, *J. Nutr. 46:*399 (1952).

9. Z. Matsuoka and Z. Yoshimatsu, *Z. Physiol. Chem. 143:*206(1925).
10. Y. Katake and J. Iwao, *Z. Physiol. Chem. 195:*139(1931).
11. A. Ellinger and Z. Matsuoka, *J. Biol. Chem. 109:*259(1920).
12. W. E. Knox and A. H. Mehler, *J. Biol. Chem. 187:*419(1950).
13. W. E. Knox and M. M. Piras, *J. Biol. Chem. 242:*2959(1967).
14. S. Yamamoto and O. Hayaiski, in *Methods in Enzymology*, Vol. XVIII (H. Tabor and C. W. Tabor eds.), Academic Press, New York, 1971, p. 434.
15. P. Feigelson and O. Greengaard, *J. Biol. Chem. 214:*151(1964).
16. A. H. Mehler and W. E. Knox, *J. Biol. Chem. 187:*431(1950).
17. Y. Kotake and S. Otani, *J. Physiol. Chem. 214:*1(1938).
18. Y. Kotake, *J. Physiol. Chem. 195:*158(1931).
19. K. Mahino, K. Arai, and S. Oka, *Science 120:*1104(1954).
20. C. E. Dalgliesh, W. E. Knox, and A. Neuberger, *Nature 168:*20 (1951).
21. R. R. Brown and J. M. Price, *Fed. Proc. 15:*225(1965).
22. R. R. Brown and J. M. Price, *J. Biol. Chem. 219:*985(1956).
23. M. Mason and C. P. Berg, *J. Biol. Chem. 195:*515(1952).
24. O. Wiss, *J. Physiol. Chem. 293:*106(1953).
25. M. Okamoto, in *Methods in Enzymology*, Vol. XVII (H. Tabor and C. W. Tabor, eds.), Academic Press, New York, N. Y., 1970, p. 460.
26. K. Makino, K. Satoh, T. Fujiki, and K. Kawagushi, *Nature 170:* 977(1952).
27. K. Makino, K. Arai, and S. Oka, *Science 120:*1104(1954).
28. R. R. Brown, *J. Biol. Chem. 227:*649(1957).
29. K. Soda and K. Tanizawa, *Adv. Enzymol. 49:*1(1979).
30. R. K. Gholson, L. M. Henderson, G. A. Mourkides, R. J. Hill, and R. E. Koeppe, *J. Biol. Chem. 234:*96(1959).
31. R. K. Gholson, L. V. Hankes and L. M. Henderson, *J. Biol. Chem. 235:*132(1960).
32. Y. Nishizuka, A. Ichiyama, and O. Hayaishi, in *Methods in Enzymology*, Vol. XVII (H. Tabor and C. W. Tabor, eds.), Academic Press, New York, 1970, p. 463.
33. R. W. Schayer and L. M. Henderson, *J. Biol. Chem. 195:*657(1952).
34. A. H. Bokman and B. S. Schweigert, *J. Biol. Chem. 186:*153(1950).
35. L. M. Henderson and G. B. Ramasarma, *J. Biol. Chem. 181:*687 (1949).
36. L. V. Hankes and M. Urivetsky, *Arch. Biochem. 52:*484(1954).
37. D. M. Bonner and C. Yanofsky, *J. Nutr. 44:*603(1951).
38. L. M. Henderson, *J. Biol. Chem. 181:*677(1949).
39. L. Henderson and J. Hirsch, *J. Biol. Chem. 181:*667(1949).
40. Y. Nishizuka and O. Hayaishi, *J. Biol. Chem. 230:*3369(1963).
41. K. Iwai and H. Taguchi, in *Methods in Enzymology*, Vol. LXI (D. B. McCormick and L. G. Wright, eds.), Academic Press, New York, 1980, p. 96.

42. Y. Nishizuka and S. Nakamura, in *Methods in Enzymology*, Vol. XVII (M. Tabor and C. W. Tabor, eds.), Academic Press, New York, 1970, p. 491.
43. P. M. Packman and W. B. Jakoby, in *Methods in Enzymology*, Vol. XVII (M. Tabor and C. W. Tabor, eds.), Academic Press, New York, 1970, p. 501.
44. J. Preiss and P. Handler, *J. Biol. Chem. 233:*488(1958).
45. J. Preiss and P. Handler, *J. Biol. Chem. 233:*493(1958).
46. J. Imsande and P. Handler, *J. Biol. Chem. 236:*525(1961).
47. R. Seifert, M. Kittler, and H. Hilz, in *Current Aspects of Biochemical Energetics* (N. O. Kaplan and E. P. Kennedy, eds.), Academic Press, New York, 1966, p. 413.
48. J. M. Price and R. R. Brown, *J. Biol. Chem. 222:*835(1956).
49. I. Ohara, S-I. Otsuka, Y. Yugari, and S. Ariyoshi, *J. Nutr. 110:*634(1980).
50. I. Ohura, S-I. Otsuka, Y. Yugari, and S. Arigoshi, *J. Nutr. 110:*641(1980).
51. H. H. Loh and C. P. Berg, *J. Nutr. 101:*465(1971).
52. D. R. Celander and C. P. Berg, *J. Biol. Chem. 202:*339(1953).
53. R. R. Langer and C. P. Berg, *J. Biol. Chem. 214:*699(1955).
54. L. V. Hankes, R. R. Brown, J. Leklen, M. Schmaeler, and J. Jesseph, *J. Invest. Dermatol. 58:*85(1972).
55. K. C. Triebwasser, P. B. Swan, L. M. Henderson, and J. A. Budney, *J. Nutr. 106:*642(1976).
56. J. M. Price, R. R. Brown and N. Yess, *Adv. Clin. Chem. 000:* 000(0000).
57. B. Smith and D. J. Rockop, *N. Engl. J. Med. 267:*1338(1962).
58. R. E. Olson, D. Gursey, and J. W. Vester, *N. Engl. J. Med. 263:*1169(1960).
59. G. Curzon, G. Ettlinger, M. Cole, and J. Walsh, *Neurology 13:* 431(1963).
60. I. A. Mirsky, G. Perisutti, and R. Jinks, *Endocrinology 60:*318 (1957).
61. M. Winitz, P. Cuillino, J. P. Greenstein, and S. M. Birnbaum, *Arch. Biochem. Biophys. 64:*333(1964).
62. J. S. Wittman, *J. Nutr. 106:*631(1976).
63. B. C. Korbitz, J. M. Price, and R. R. Brown, *J. Nutr. 80:*55 (1963).
64. L. M. Henderson, I. M. Weinstock, and C. B. Ramasarma, *J. Biol. Chem. 189:*19(1951).
65. F. Rosen, J. W. Huff, and W. A. Perlzweig, *J. Nutr. 33:*561(1947).
66. B. S. Schweigert and P. B. Pearson, *J. Biol. Chem. 168:*555(1947).
67. J. K. Yeh and R. R. Brown, *J. Nutr. 107:*261(1977).
68. E. C. Miller and C. A. Baumann, *J. Biol. Chem. 157:*551(1945).
69. C. Schwartzmann and L. Strauss, *J. Nutr. 38:*000(1949).
70. H. E. Axelrod, A. F. Morgan, and S. Lepkovsky, *J. Biol. Chem. 160:*155(1945).

71. G. E. Cartwright, M. M. Wintrobe, P. Jones, M. Lauritsen, and S. Humphreys, *Bull. Johns Hopkins Hosp. 75:*35(1944).
72. I. D. Greenberg and J. F. Rinehart, *Proc. Soc. Exp. Biol. Med. 70:*20(1949).
73. Z. H. M. Vergjee, *Int. J. Biochem. 2:*711(1971).
74. N. Yess, J. M. Price, R. R. Brown, P. B. Swan, and H. Linkswiler, *J. Nutr. 84:*229(1964).
75. R. R. Brown, N. Yess, J. M. Price, H. Linkswiler, P. Swan, and L. V. Hankes, *J. Nutr. 87:*419(1965).
76. E. M. Baker, J. E. Canham, W. T. Nunes, H. E. Samberlich, and M. E. McDowell, *J. Clin. Nutr. 15:*59(1964).
77. I. T. Miller and H. Linkswiler, *J. Nutr. 93:*56(1967).
78. J. Kelsay, L. T. Miller, and H. Linkswiler, *J. Nutr. 94:*27(1968).
79. D. P. Rose and P. A. Toseland, *Metabolism 22:*165(1973).
80. H. K. Shin and H. M. Linkswiler, *J. Nutr. 104:*1348(1974).
81. J. I. Patterson, R. R. Brown, H. Linkswiler, and A. E. Harper *Am. J. Clin. Nutr. 33:*2157(1980).
82. R. W. Vilter, J. F. Mueller, H. S. Glazer, T. Jarrold, J. Abraham, C. Thompson, and V. R. Hawkins, *J. Lab. Clin. Med. 42:*335(1953).
83. D. P. Rose, R. Strong, P. W. Adamas, and P. E. Harding, *Clin. Sci. 42:*465(1972).
84. I. Nakagawa, S. Ohguri, A. Sasaki, M. Kajimoto, M. Sasaki, and T. Tahahashi, *J. Nutr. 105:*1241(1975).
85. W. E. Knox and M. M. Piras, *J. Biol. Chem. 242:*2959(1967).
86. W. W. Coon and E. Nagler, *Ann. N. Y. Acad. Sci. 166:*30(1969).
87. O. J. Park, L. M. Henderson, and P. B. Swan, *Proc. Soc. Envir. Biol. Med. 142:*1023(1973).
88. H. Sidransky, E. Verney, and C. N. Murty, *J. Nutr. 110:*2231 (1980).
89. J. M. Price, R. R. Brown, and M. Yess, *Actv. Metab. Dis. 2:* 159(1965).
90. D. P. Rose, *Clin. Sci. 31:*265(1966).
91. D. P. Rose, *Lancet 1:*230(1967).
92. D. P. Rose and F. McGinty, *Clin. Sci. 35:*1(1968).
93. A. L. Luhby, M. Brin, P. Gordon, P. Davis, M. Murphy, and H. Spiegel, *Am. J. Clin. Nutr. 24:*684(1971).
94. S. A. Price, D. P. Rose, and P. A. Toseland, *Am. J. Clin. Nutr. 25:*494(1972).
95. R. R. Brown, D. P. Rose, J. E. Lehler, H. Linkswiler, and R. Anand, *Am. J. Clin. Nutr. 28:*10(1975).
96. J. E. Lehler, R. R. Brown, D. P. Rose, H. Linkswiler, and R. A. Arend, *Am. J. Clin. Nutr. 281:*146(1975).
97. T. R. Bossé and E. A. Donald, *Am. J. Clin. Nutr. 32:*1015(1979).
98. E. A. Donald and T. R. Bossé, *Am. J. Clin. Nutr. 32:*1024(1979).
99. J. P. Biehl and R. W. Viller, *Proc. Soc. Exp. Biol. Med. 85:*389 (1954).

100. D. A. P. Evans, K. A. Mauley and V. A. McKusick, *Brit. Med. J. 2:*485(1960).
101. L. Levy, *Ann. N. Y. Acad. Sci. 166:*184(1969).
102. P. A. DiLorenzo, *Acta. Dermat. Venereol. 47:*318(1967).
103. P. S. Shankar, *J. Ind. Med. Soc. 54:*73(1955).
104. M. M. Wood, *Brit. J. Tuber. 49:*20(1955).
105. I. H. Porter, in *Heredity and Disease,* McGraw-Hill Book Co., New York, 1968, p. 246.
106. L. J. Zatman, N. O. Kaplan, and S. P. Colowick, *J. Biol. Chem. 200:*197(1953).
107. L. T. Zatman, N. O. Kaplan, S. P. Colowick and M. M. Ciotti, *J. Biol. Chem. 209:*467(1954).
108. H. Okamoto and O. Hayaishi, *Biochem. Biophys. Res. Commun. 29:*394(1967).
109. L. M. Henderson, R. S. Koski, and F. D'Angeli, *J. Biol. Chem. 212:*369(1955).
110. P. B. Jungueira and B. S. Schweigert, *J. Biol. Chem. 175:*535 (1948).
111. C. C. Porter, J. Clark and R. H. Silber, *Arch. Biochem. Biophys. 18:*339(1948).
112. M. Mason, *J. Biol. Chem. 201:*995(1953).
113. N. Sampitak and Q. Chayutimonkul, *Lancet 1:*836(1974).
114. M. Briggs and M. Briggs, *Lancet 1:*1234(1974).
115. R. C. Theuer, *J. Reprod. Med. 8:*13(1972).
116. W. S. Rosenthal, N. F. Adham, R. Lopez, and J. M. Cooperman, *Am. J. Clin. Nutr. 26:*858(1973).
117. D. A. Rose, *A Plague of Corn. The Social History of Pellagra,* Cornell University Press, Ithaca, N. Y., 1973.
118. L. F. Gray and L. J. Daniel, *J. Nutr. 103:*1764(1973).
119. J. E. Wilson and V. Vigneaud, *J. Biol. Chem. 184:*63(1950).
120. A. M. Asatoor, *Nature 203:*1382(1964).
121. I. A. Jaffe, *Ann. N. Y. Acad. Sci. 166:*57(1969).
122. L. E. Hollister, F. F. Moore, F. Forrest, and J. L. Bennett, *Am. J. Clin. Nutr. 19:*307(1966).
123. F. O. Brady, M. E. Monaco, H. J. Forman, C. Schutz, and P. Feigelson, *J. Biol. Chem. 247:*7915(1972).
124. G. Gopalan and K. S. J. Rao, *Vitamins and Hormones 33:*505(1975).
125. M. E. Shils, in *Modern Nutrition in Health and Disease,* (R. S. Goodhart and M. E. Shils, eds.), 6th Edition, Lea and Febiger, New York, 1980, p. 1153.
126. J. Jännes, V. V. E. Lepännen, and M. Oka, *Ann. Med. Exptl. Biol. Fenniae (Helsinki) 41:*115(1963).
127. E. Boyland and D. C. Williams, *Biochem. J. 64:*578(1956).
128. R. R. Brown, J. M. Price, G. Freidell, and S. W. Burner, *J. Natl. Cancer Inst. 43:*295(1969).
129. G. T. Bryan, *Am. J. Clin. Nutr. 24:*241(1971).

130. O. Yoshido, R. R. Brown, and G. T. Bryan, *Am. J. Clin. Nutr. 24:*248(1971).
131. R. R. Brown, J. M. Price, S. W. Burney, and G. H. Friedell, *Cancer Res. 30:*611(1970).
132. S. Gailani, G. Murphy, G. Kenny, A. Nussbaum, and P. Silvernail, *Cancer Res. 33:*1071(1973).
133. H. Wolf, H. R. Nielsen, and R. R. Brown, *Proc. Am. Assoc. Cancer Res. 17:*6(1976).
134. H. R. Neilsen and H. Wolf, *Third International Symposium on Detection and Prevention of Cancer,* Abstracts, p. 201 (1976).
135. H. R. Nielsen, H. Wolf, and R. R. Brown, *J. Natl. Cancer Inst. 59:*855(1977).
136. J. Roberts and H. T. Rosenfeld, *Proc. of AACR and ASCO 18:*31 (1977).
137. J. Roberts, F. A. Schmid, and H. J. Rosenfeld, *Cancer Treat. Rep. 63:*1045(1979).
138. M. M. Wiseman, N. Kalant, and M. M. Hoffman, *J. Lab. Clin. Med. 52:*27(1958).
139. M. Oka and V. V. E. Leppänen, *Acta. Med. Scand. 173:*361(1963).
140. Y. Kotake and S. Tani, *J. Biochem (Tokyo) 40:*295(1953).
141. D. A. Rosen, G. D. Moengwyn-Davies, B. Becker, H. H. Stone, and J. S. Friedenwald, *Proc. Soc. Exptl. Biol. Med. 88:*32(1965).
142. E. G. McDaniel, J. M. Hundley, and W. H. Sebrell, *J. Nutr. 59:* 407(1956).
143. A. M. Mehler, K. Yamo, and E. L. May, *Science 145:*817(1964).
144. N. Akarte and N. Shastri, *J. Nutr. Sci. Vitaminol. 22:*175(1976).
145. M. Ikeda, H. Tsuji, S. Nakamura, A. Idriyama, and O. Hayaishi, *J. Biol. Chem. 240:*1395(1965).
146. G. W. Evans and E. C. Johnson, *J. Nutr. 110:*1076(1980).
147. G. W. Evans, *Nutr. Rev. 38:*137(1980).
148. I. Krieger, *Nutr. Rev. 38:*148(1980).
149. L. S. Hurley and B. Lonnerdal, *J. Nutr. 110:*2536(1980).
150. C. Holt, L. S. Hurley, and B. Lonnerdal, *J. Nutr. 111:*2240(1981).
151. Y. Kotaka and T. Inada, *J. Biochem. 40:*291(1953).
152. Y. Kotake and E. Murakami, *Am. J. Clin. Nutr. 24:*826(1971).
153. J. M. Price, *University of Michigan Medical Bulletin 24:*461(1958).
154. R. S. Rivlin, in *The Thyroid A Fundamental and Clinical Treatise* (S. C. Werner and S. H. Lybar, eds.), 3rd Edition, Harper and Rowe, New York, 1971.
155. I. R. Payne, E. M. Walsh, and E. J. R. Wittenberg, *Am. J. Clin. Nutr. 27:*565(1974).
156. J. D. Stratigos and A. Katsambas, *Brit. J. Dermatol. 96:*99(1977).

3

NICOTINIC ACID BIOSYNTHESIS IN PLANTS AND MICROORGANISMS

The need for nicotinic acid as its amide in coenzymes is uniform in nature. Conversely all living organisms can theoretically act as a source of nicotinic acid. There are a number of species that have a requirement for preformed nicotinic acid, but the majority synthesize the compound from simple precursors without going through tryptophan biosynthesis and subsequent conversion. The conversion of tryptophan to nicotinic acid is confined to a few species outside vertebrates, giving various alternate fates of tryptophan depending on the organism tested (Fig. 1). Soda and Tanizawa named the alternate pathways quinoline and aromatic (1). Those microorganisms that have the ability to catabolize tryptophan to nicotinic acid frequently have a constitutive kynureninase, primarily for the path of tryptophan to niacin, and a tryptophan-inducible enzyme when tryptophan is metabolized to CO_2 and water via either pathway (2–4)

I. NICOTINIC ACID BIOSYNTHESIS IN FUNGI

Neurospora are able to convert tryptophan to niacin by the identical pathway that mammals use. The use of *Neurospora* mutants helped unravel the biosynthetic pathway in general, and therefore some discussion of the metabolism of this species is of interest. Tryptophan is converted to anthranilic acid (5) and hydroxyanthranilic acid (6). When the *Neurospora* is exposed to tryptophan, kynureninase levels are increased 600-fold (7). There are two kynureninases. The constitutive enzyme has favorable kinetic constants for 3-hydroxykynurenine (the K_m for 3-hydroxykynurenine is $3.7 \times 10^{-6} M$ and for kynurenine $8.4 \times 10^{-5} M$). For the inducible enzyme the affinity is about equal. The ratio of activity for 3-hydroxykynurenine to kynurenine is 3.16 for the constitutive and 1.92 for the inducible enzyme

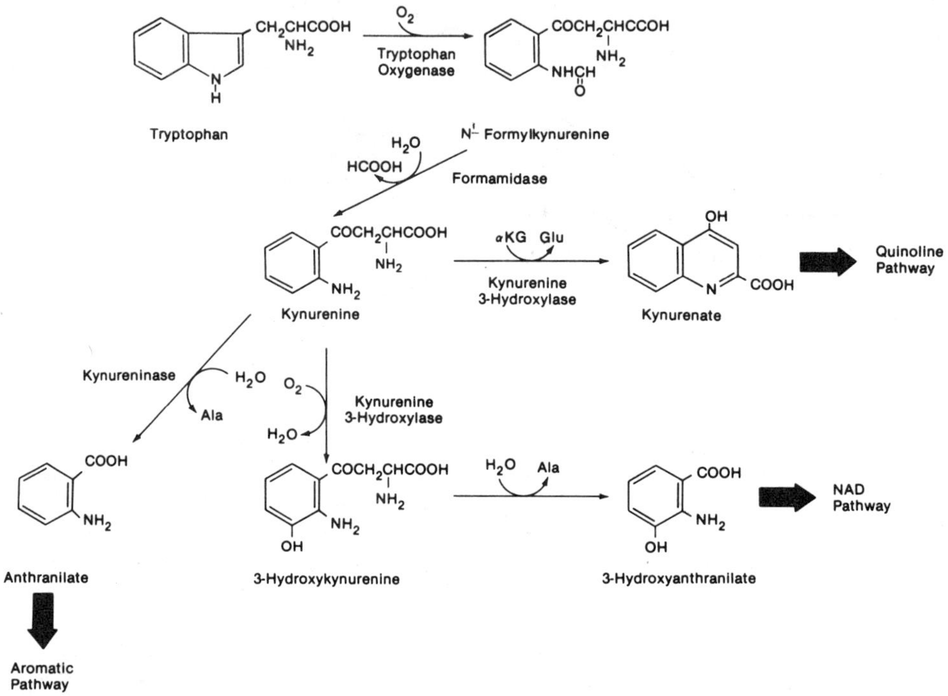

Figure 1 The major pathways of tryptophan degradation in nature. (Adapted from Ref. 1.)

(8). Therefore it is assumed that the constitutive enzyme is the enzyme responsible for biosynthesis on the nicotinamide adenine dinucleotide pathway (9). The dual pathway enzymes are found in many fungi and the ratio of constitutive to inducible enzyme varies widely (4). Some *Neurospora* strains have only the inducible enzyme. The pathway in yeasts varies under aerobic conditions. *Saccharomyces cerevisiae* converts tryptophan to niacin via hydroxykynurenine and quinolinic acid as in the mammalian pathway. However, under aerobic conditions, the more direct pathway from the amino acid glutamate is utilized (10).

II. NICOTINIC ACID BIOSYNTHESIS IN BACTERIA

Almost all bacteria can make niacin. As will be discussed in a later chapter, many of the *Lactobacilli* have fastidious growth requirements. They have been used often for the microbiological assay of nutrients, including tryptophan or niacin. However, that is an artificial emphasis because of nutritional science.

The majority of organisms do not use the conversion of tryptophan to quinolinic acid, but use simpler precursors for quinolinic acid. However, the pathway is not completely defined. In *Escherichia coli* the precursors dihydroxyacetone phosphate and aspartate are used to form quinolinic acid via unknown intermediates (11), but other microorganisms such as *Clostridium butylicum* have different pathways (12,13). Only *Xanthomonas pruni* among bacteria can convert tryptophan to nicotinic acid (14,15). All pathways utilize quinolinic acid as an intermediate.

III. NICOTINIC ACID BIOSYNTHESIS IN PROTOZOA

The information is sparse since detailed metabolic studies are only available on a few species. Most ciliates require preformed tryptophan and niacin for axenic growth (16). The ciliate organism *Tetrahymena pyriformis* has been studied in greatest detail, and it has an absolute requirement for niacin even when tryptophan is added (17). This organism is of special interest since it is often promoted to evaluate protein quality (18). The inability to execute this pathway makes direct translation of such data to human nutrition doubtful.

For most other protozoa, when defined media have been found to support growth, both tryptophan and niacin are added, and both are often thought to be essential. However, frequently no definitive experiments which attempt to determine their relative need have been done. *Acanthameba* has a simpler requirement and can grow on a tryptophan- and niacin-free medium though tryptophan stimulates growth (19), but nothing is known about metabolic interrelationships. The nutritional requirements of many other species which are important to veterinary or human medicine have been evaluated for nutritional requirements. These include: *Trypanosomes, Giardia, Leishmanii,* malaria parasites, and the protozoal fish parasites which are of great commercial importance. The photosynthetic protozoa *Ochromonas* is frequently used for the microbiological assay of nutrients, but niacin is not a requirement.

IV. NICOTINIC ACID BIOSYNTHESIS IN PLANTS

Like bacteria, plants synthesize pyridine derivatives, including nicotinic acid. However, unlike animals, plants synthesize from simpler precursors like bacteria (20,21). In tobacco, nicotine can be converted to nicotinic acid, but the pyridine ring of nicotine is synthesized de novo from simple precursors (22). However, details of the pathway in plants and bacteria are not worked out. Knowledge of the plant pathway would be of interest to human nutrition, since plant varieties rich in tryptophan and niacin might possibly be used in human staple diets in areas where pellagra is still endemic. Corn of high tryptophan and niacin content exists and has greater biological value than standard varieties (23).

REFERENCES

1. K. Soda and K. Tanizawa, *Adv. Enzymol. 49:*1(1979).
2. J. R. Turner and H. Druden, *Biochem. Biophys. Res. Comm. 42:* 698(1971).
3. F. H. Gaertner, K. W. Cole, and G. R. Welch, *J. Bacterol. 108:* 902(1971).
4. A. S. Shetty and F. H. Goertner, *J. Bacterol. 122:*235(1975).
5. F. A. Haskins and H. K. Mitchell, *Proc. Natl. Acad. Sci. 35:*500 (1949).
6. D. M. Bonner and C. Yanofsky, *J. Nutr. 44:*603(1951).
7. W. B. Jacoby and D. M. Bonner, *J. Biol. Chem. 205:*699(1953).
8. K. Tanizawa, T. Yamamoto, and K. Soda, *Biochemistry. 12:*2969 (1973).
9. F. M. Gaertner, K. W. Cole, and G. R. Welch, *J. Bacteriol. 122:* 235(1975).
10. F. Ahmed and A. G. Moat, *J. Biol. Chem. 241:*775(1966).
11. B. M. Steiner, J. T. Heard, Jr., and G. J. Tritz, *J. Bacteriol. 141:*989(1980).
12. N. Ogasawara, T. L. R. Chandler, R. K. Gholson, R. J. Rosser, and A. J. Andrioli, *Biochim. Biophys. Acta. 141:*199(1967).
13. T. A. Scott, E. Bellia, and M. Matney, *Eur. J. Biochem. 10:*318 (1969).
14. A. T. Brown and C. Wagner, *J. Bacteriol. 101:*456(1970).
15. D. Davis, L. M. Henderson, and D. Powell, *J. Biol. Chem. 189:* 543(1951).
16. R. G. Wilson and L. M. Henderson, *J. Bacteriol. 85:*221(1963).
17. G. G. Holz Jr., in *Biochemistry and Physiology of Protozoa* (S. P. P. Hunter, ed.), Academic Press, New York, 1964, p. 199.
18. D. L. Hill. *The Biochemistry and Physiology of Tetrahymena,* Academic Press, New York, 1972.
19. R. H. Mackler, in *Evaluation of Proteins for Humans* (C. E. Bodwell, ed.), AVI, Westport, Conn. 1975, pp. 55–67.
20. T. J. Byers, R. A. Akins, B. J. Maynard, R. A. Leftin, and S. M. Martin, *J. Protozool. 27:*216(1980).
21. L. A. Hadwiger, S. E. Badiei, G. R. Waller, and R. K. Gholson, *Biochem. Biophys. Res. Comm. 13:*466(1963).
22. R. F. Dawson, D. R. Christman and R. U. Byerrum, in *Methods in Enzymology,* Vol. XVIII (D. B. McCormick and L. D. Wright eds.), Academic Press, New York, 1971, pp. 90–113.
23. G. D. Griffith, T. Griffith, and R. U. Byerrum, *J. Biol. Chem. 235:*3536(1960).
24. G. C. Graham, R. C. Placko, and W. C. MacLean, Jr., *J. Nutr. 110:*1061(1980).

4

NICOTINIC ACID METABOLISM AND EXCRETION

I. NICOTINIC ACID EXCRETION AFTER PHARMACOLOGICAL DOSES

A careful distinction must be made between the physiological disposition of nicotinic acid in the state of suboptimal to normal nutrition, and the disposition of nicotinic acid when given in pharmacological doses. The pathway from nicotinic acid through the pyridine nucleotides and back to nicotinamide will not be detailed here, but will be discussed in Part II. However, it must be remembered that biosynthesized nicotinic acid does not result in free nicotinic acid but rather is utilized directly to form nicotinic acid nucleotides (Fig. 1). When excess nicotinic acid is administered, it is handled in part like benzoic acid and is converted into a series of "detoxification" products. Figure 2 indicates some of those substances. In fact, while metabolites of nicotinic acid have been known for a long time before the role of nicotinic acid as a vitamin was realized, they were indeed generally regarded as "detoxification products" rather than physiological metabolites. Nicotinuric acid was isolated as early as 1913 from the urine of dogs fed nicotinic acid (1). Later this compound was recognized as a normal metabolite in urine. However, it is usually present in small amounts compared to the total niacin. For example, in human urine nicotinuric acid is present in traces only (2).

Other products which can be found include the products of conjugation with glucoronic acid in rat urine (3). However, most excess nicotinic acid and nicotinamide given acutely is excreted unchanged in rats (4,5). Nicotinoyl glucuronide is also found in chicken excreta when more physiological amounts of niacin are given as $[C^{14}]$niacin (6), and in rat urine without extra administration of nicotinic acid (3). In birds, the various ornithine conjugates α-, δ-, and α,δ-nicotinoyl ornithine (6,7) are formed. This is analogous to the difference between birds and mammals in the metabolism of benzoic acid.

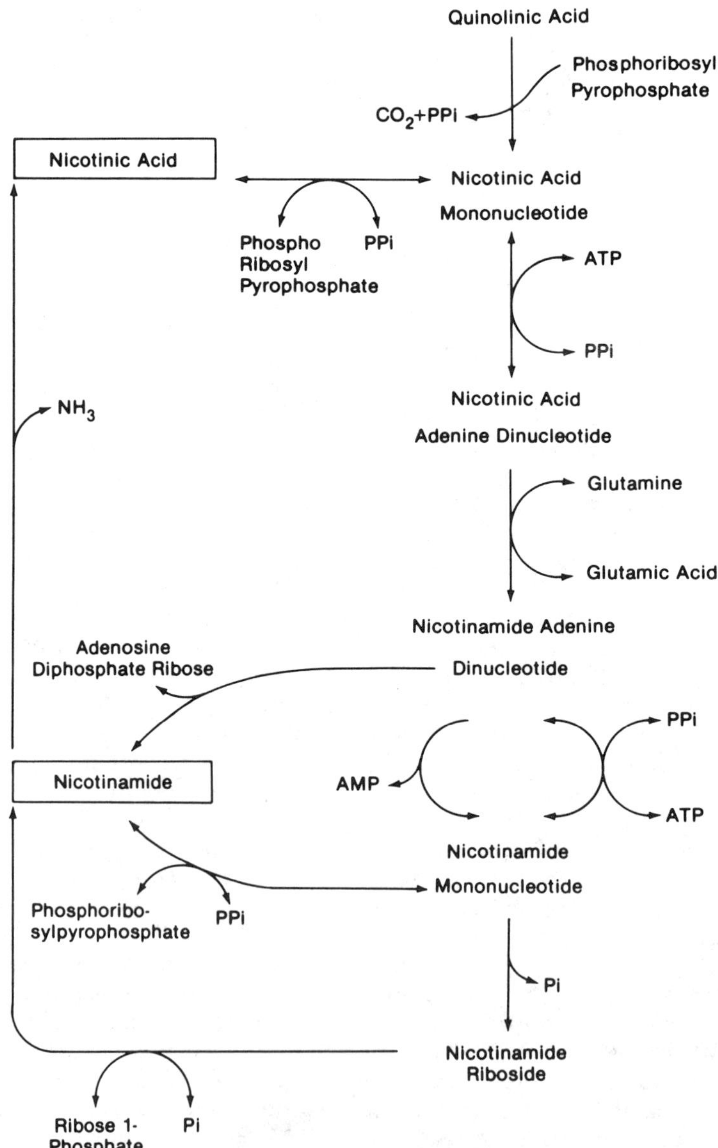

Figure 1 The interconversions of nicotinamide nucleotides, of free nicotinic acid, and of nicotinamide.

Nicotinamide is not actively deamidated in the mammalian organism. The deamidase has a low affinity, as evidenced by a very high K_m for nicotinamide. Deamidation and hydrolysis of a variety of pyridine carboxylic amides and esters are catalyzed (8).

II. METABOLISM OF PHYSIOLOGICAL AMOUNTS OF NICOTINIC ACID

When physiological turnover of nicotinic acid and nicotinamide is considered for balance studies, it is convenient to divide excretory products into those which support bacterial growth on a niacin-deficient medium, and those which are altered to such an extent that this nutrient property is lost. Through the property of microbiological growth support, it was ascertained that nicotinic acid (and its glucuronide, which is so labile that it is measured as free nicotinic acid) and nicotinamide are present in urine as such, and also as nicotinuric acid. Also, a small quantity of the niacin coenzymes is detected. Many reports have confirmed the presence of these substances: in the urine of rabbits (9); of humans (10); of dogs, rats, mice, and hamsters (11); again of rats (12); and of monkeys (13).

Those niacin products which do not support growth of microorganisms proved confusing to investigators for a long time. It was claimed that

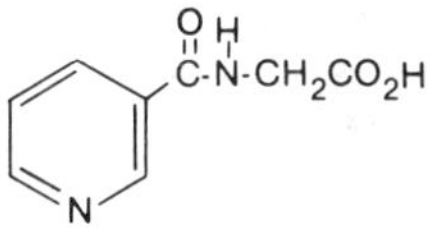

Nicotinuric acid

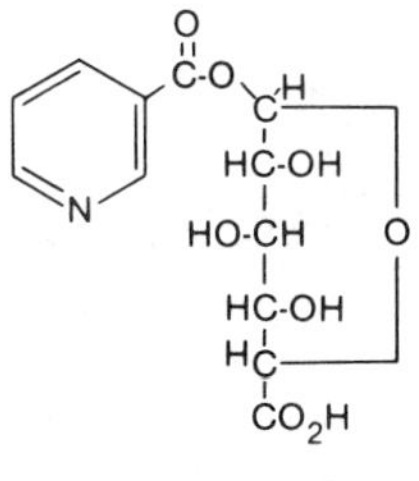

Nicotinoyl glucuronide

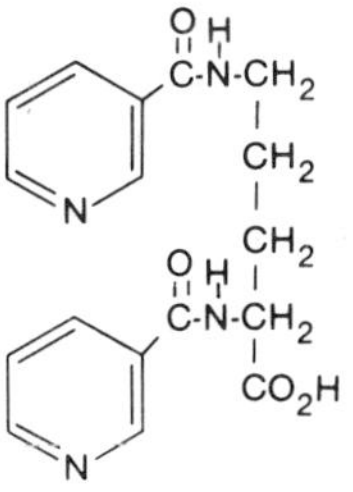

Di-Nicotinoyl Ornithine

Figure 2 Nicotinic acid detoxification products.

trigonelline was a normal excretory product of niacin, and the compound was actually isolated from urine (1). It is true that this base is a natural material (14), but there is now adequate evidence that the compound is not a niacin metabolite in mammals, since labeled nicotinic acid does not give rise to labeled trigonelline in several species investigated (11,12). Until 1940 the origin of trigonelline was unsettled. Melnick and co-workers (15) reported trigonelline to be a normal constituent of urine, although they noted that the amount increased after smoking (16). In that same year Perlzweig and his associates reported that the compound is not a normal metabolite of nicotinic acid, but that its presence in urine was attributable to the drinking of coffee (17). There probably is a trace in human urine, but it does not increase after niacin ingestion (10). N^1-methylnicotinamide was identified as a nicotinic acid and nicotinamide metabolite (18), and some of the trigonelline observed could arise from the decomposition of this material. One circumstance of niacin-related trigonelline excretion has been claimed: the bound form of nicotinic acid from wheat (niacytin) was given to rats and 30–40% of the nicotinic acid content was said to be converted to trigonelline and excreted in the urine, though none was found in the feces (19).

The above metabolites account for only about 40% of the niacin ingested. This discrepancy was largely solved by Knox and Grossman (20) with their discovery of N^1-methyl-6-pyridone-2-carboxylamide* in urine. This accounted for most of the remaining niacin in urine. To evaluate the possibility that the 2-pyridone might also exist in urine, Holman synthesized both pyridones (21) and devised determinations for both pyridones separately (10). But, as will be further discussed later, it is established that ingested niacin can be largely accounted for by the sum of N^1-methyl-6-pyridone-3-carboxylamide, N^1-methyl-nicotinamide, and nicotinic acid plus nicotinamide in the urine. No 2-pyridone is detectable. As an additional check, an analytical procedure for N^1-methyl-3-pyridone-3-carboxylic acid was devised (10), but this material is not a niacin metabolite either. These results were later confirmed with improved methods of determination (22). The old methods were so complicated that only one determination could be carried out in 36 hr.

It was not our intention to imply that Holman was the first to synthesize the pyridones. Huff had previously synthesized the 6-pyridone from trigonelline (23). Furthermore, the metabolite was an established, well-known compound, which Pechmann and Welsch (24) had prepared

---

*Alternate names for this compound are: N'-methyl-2-pyridone-5-carboxylic acid amide, which is the proper chemical designation; N^1-methylnicotinamide-6-pyridone is more commonly used, but constitutes a sort of chemical pleonasm. The most common designation found in the literature is "the 6-pyridone."

by an unequivocal, though tedious route. In 1954, Pullman and Colowick further improved the synthesis of both the 2- and 6-pyridones (25).

It was deemed necessary by other investigators to rule out certain possible metabolites, and hence Bandler established the absence of 3-amino-pyridine from urine (26), while Reddi and Kodicek established the absence in human urine of N^1-methylnicotinamide betaine (27).

Even though the metabolites of nicotinic acid in urine had already risen to an unprecedented number, other excretory products were yet to be discovered. Leifer and co-workers injected carboxyl-labeled nicotinic acid into rats, mice, dogs, and hamsters (11). The urine was chromatographed on paper and the spots detected by radioautographic techniques. The following metabolites were detected in all species: nicotinic acid, nicotinamide, nicotinuric acid, N^1-methylnicotinamide, the 6-pyridone, and one unidentified metabolite. Pei Hsing Lin and Johnson confirmed this, identifying these five metabolites plus coenzyme I by both bioautographic and radioautographic techniques (12,28). They also found one unidentified metabolite after the administration of nicotinic acid in rats, and three unidentified metabolites after the administration of nicotinamide. The unidentified metabolites had biological activity in the yeast *Torula cremoris*. Furthermore, both groups agreed that the expired CO_2 was heavily labeled, accounting for up to 13% of the injected dose (11).

The glucuronide was one of these metabolites and a substance claimed to be the 4-pyridone was likewise identified (29–31), as were hydroxylated metabolites of nicotinic acid or nicotinamide (32). The 4-pyridone was seen in rat (4,5,24), monkey (30), pig (31), sheep (31), and human (6) urine. The 4-pyridone does not increase in quantity in the urine of humans after a test load of nicotinic acid (30). In monkeys the 4-pyridone was only a minor metabolite of [^{14}C]nicotinic acid as was the 6-pyridone (30). In the rat, 6-hydroxynicotinamide and 6-hydroxynicotinic acid are minor products (0.6–1.2%) of injected nicotinamide or nicotinic acid. Since these products are also produced in germ-free animals they are true animal metabolites (34). When [^{14}C]-nicotinamide is injected, a major metabolite is nicotinamide *N*-oxide in mouse (33,35) and rat (4,5), in addition to the products already mentioned: N^1-methylnicotinamide, 6-pyridone, and 4-pyridone. There is indeed deamidation, since free nicotinic acid (4,5,33,35), 6-hydroxynicotinic acid (34), and nicotinic acid (35) are seen in the mouse and rat after injection of nicotinamide. Those data confirm the wide spectrum of quantitative niacin metabolites that one can find in the urine of varying species. Perlzweig studied the metabolism of the 6-pyridone in a variety of species—humans, dog, rat, pig, calf, sheep, goat, guinea pig, and rabbit. In man, dog, rabbit, or pig, most of the 6-pyridone is recovered quantitatively, while in rabbit, guinea pig, sheep, or goats, much of the 6-pyridone could not be accounted for (36). Figure 3 summarizes the metabolic transformations of nicotinic acid.

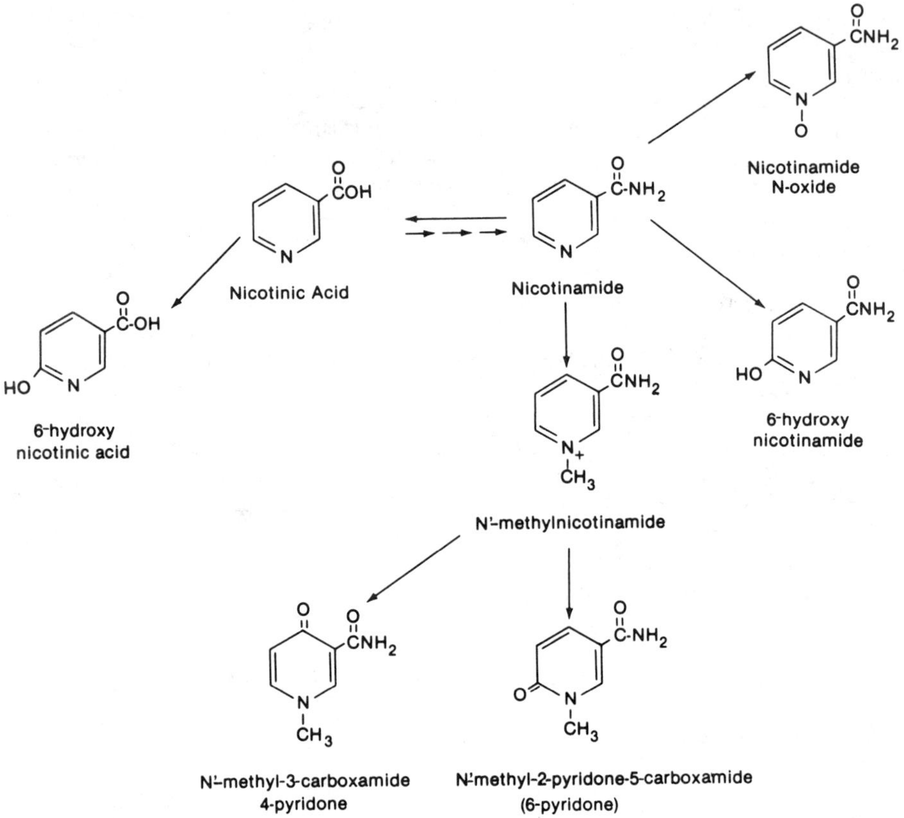

Nicotinamide N-oxide

Nicotinic Acid

Nicotinamide

6-hydroxy nicotinic acid

6-hydroxy nicotinamide

N¹-methylnicotinamide

N¹-methyl-3-carboxamide 4-pyridone

N¹-methyl-2-pyridone-5-carboxamide (6-pyridone)

Figure 3 The metabolism of nicotinic acid and nicotinamide in mammals.

III. NORMAL EXCRETION OF NICOTINIC ACID METABOLITES IN HUMANS

In man, however, the sum of N^1-methylnicotinamide and the 6-pyridone gives a satisfactory measure of niacin status (37). On a control diet providing 11.3 mg of niacin (and 885 mg of tryptophan), the average amount excreted by normal women balanced the intake when measured as N^1-methylnicotinamide and the 6-pyridone (38). When a normal diet for women was supplemented with nicotinic acid at 8.3, 12,5, 20.8, or 29.2 mg, again most supplemental nicotinic acid was recovered as N^1-methylnicotinamide and the 6-pyridone with some free nicotinic acid (39) and like metabolites.

REFERENCES

1. D. Ackerman, *Z. Biol. 59:*17(1913).
2. B. C. Johnson, T. S. Hamilton, and H. H. Mitchell, *J. Biol. Chem. 159:*231(1945).
3. J. van Eys, O. Touster, and W. J. Darby, *J. Biol. Chem. 217:* 287(1955).
4. B. Petrack, P. Greengaard, and H. Kalinsky, *J. Biol. Chem. 241:*2367(1966).
5. P. Greengaard, B. Petrack, and H. Kqlinsky, *J. Biol. Chem. 242:*152(1967).
6. M. L. Wu Chang and B. Conner-Johnson, *J. Biol. Chem. 226:*799 (1956).
7. W. T. Dann and L. W. Huff, *J. Biol. Chem. 168:*121(1947).
8. J. Kirchner, J. G. Watson, and S. Chaykin, *J. Biol. Chem. 241:* 953(1966).
9. D. Chattopadhay, N. C. Ghosh, H. Chattopadhyay, and S. Banerjee, *J. Biol. Chem. 201:*529(1953).
10. W. T. M. Holman and D. J. de Lange, *Nature 165:*604(1950).
11. E. Leifer, L. H. Roth, D. S. Hogness and M. H. Corso, *J. Biol. Chem. 190:*595(1951).
12. Pei Hsing Lin and B. C. Johnson, *J. Am. Chem. Soc. 75:*2974 (1953).
13. S. Banerjee and B. Basak, *J. Nutr. 55:*179(1956).
14. D. Ackerman, *Z. Physiol. Chem. 295:*1(1943).
15. D. Melnick, W. D. Robinson and H. Field, Jr., *J. Biol. Chem. 138:*131(1940).
16. D. Melnick, W. D. Robinson, and H. Field, Jr., *J. Biol. Chem. 136:*15(1940).
17. W. A. Perlzweig, E. D. Levy and H. P. Sarrett, *J. Biol. Chem. 136:*729(1940).
18. J. W. Huff and W. A. Perlzweig, *J. Biol. Chem. 120:*515(1970).
19. J. B. Mason and E. Kodicek, *Biochem. J. 120:*515(1970).
20. W. E. Knox and W. I. Grossman, *J. Biol. Chem. 168:*363(1947).
21. W. I. M. Holman and C. Wiegand, *Biochem. J. 43:*423(1948).
22. W. I. M. Holman, *Biochem. J. 56:*513(1954).
23. J. W. Huff, *J. Biol. Chem. 171:*639(1947).
24. H. von Pechmann and W. Welsch, *Ber. 17:*2384(1884).
25. M. E. Pullman and S. P. Colowick, *J. Biol. Chem. 206:*121(1954).
26. R. Bandier, *On nicotinic acid,* Ejner Kunksgaard, Copenhagen, 1940.
27. K. K. Reddi and E. Kodicek, *Biochem. J. 53:*286(1953).
28. B. C. Johnson and Pei Hsing Lin, *J. Am. Chem. Soc. 75:*2971 (1953).
29. M. L. WuChang and B. Connor-Johnson, *J. Biol. Chem. 234:*1817 (1959).

30. M. L. WuChang and B. Connor-Johnson, *J. Biol. Chem. 236:*2096 (1961).
31. M. L. WuChang and B. Connor-Johnson, *J. Nutr. 76:*512(1962).
32. Y. C. Lee, R. K. Sholson, and N. Raica, *J. Biol. Chem. 244:* 3277(1969).
33. S. Chaykin, M. Dagani, L. Johnson, and M. Samli, *J. Biol. Chem. 240:*932(1965).
34. Y. C. Lee, R. K. Gholson, and N. Raica, *J. Biol. Chem. 244:* 3277(1969).
35. V. Bonavita, S. A. Narrod, and N. O. Kaplan, *J. Biol. Chem. 236:*936(1961).
36. W. A. Perlzweig, F. Rosen, and P. B. Pearson, *J. Nutr. 40:*453 (1950).
37. J. G. Prinsloo, J. P. DuPlessis, H. Kinger, D. J. DeLange, and L. S. de Villier, *Am. J. Clin. Nutr. 21:*98(1968).
38. E. I. Frazier, M. E. Prather, and E. Hoene, *J. Nutr. 56:*501 (1955).
39. T. Nakagawa, T. Takahashi, T. Suzuki, and Y. Masana, *J. Nutr. 99:*325(1969).

5

ASSESSMENT OF THE ADEQUACY OF NIACIN NUTRITURE

The symptoms of frank pellagra have been discussed in Chap. 2. The sequence of development of symptoms is characteristic, but there are no specific physical findings that will allow the diagnosis of inadequate niacin nutriture. The early diagnosis rests entirely on the laboratory evaluation of the patient. The methodology available allows the analysis of precursors, the vitamin itself, coenzymatic forms, and metabolic products. Microbiological, biochemical, and chemical assays are all available.

The choice of methodology is heavily dependent on the need for which the data are to be acquired. Nutritional adequacy can be assessed for research purposes, for the evaluation of a single patient, or for the survey of populations. The questions asked are very different. The assessment of the nutritional state of an experimental animal or human under controlled dietary conditions allows balance studies to be performed, and demands accurate evaluation of quantity and function of the vitamin. When an individual patient is evaluated for laboratory corroboration of suspected inadequacy of niacin nutriture, a few specific tests may be selected that accurately reflect niacin nutriture status. Evaluation of the nutrition status of a population generates entirely different demands. Individuals are not necessarily identifiably linked to a given abnormal test result, and global statistical correlations are required for the answer that is sought.

The conversion of tryptophan to niacin is a complicating factor in assessment of nicotinic acid nutrition. It is generally assumed that 60 mg of tryptophan can be converted into 1 mg of nicotinic acid. The derivation of this equivalency will be discussed in Chap. 6. It should be clear, however, from the discussion in Chap. 2 that the conversion ratio is not necessarily constant. Nor is all niacin activity equivalent:

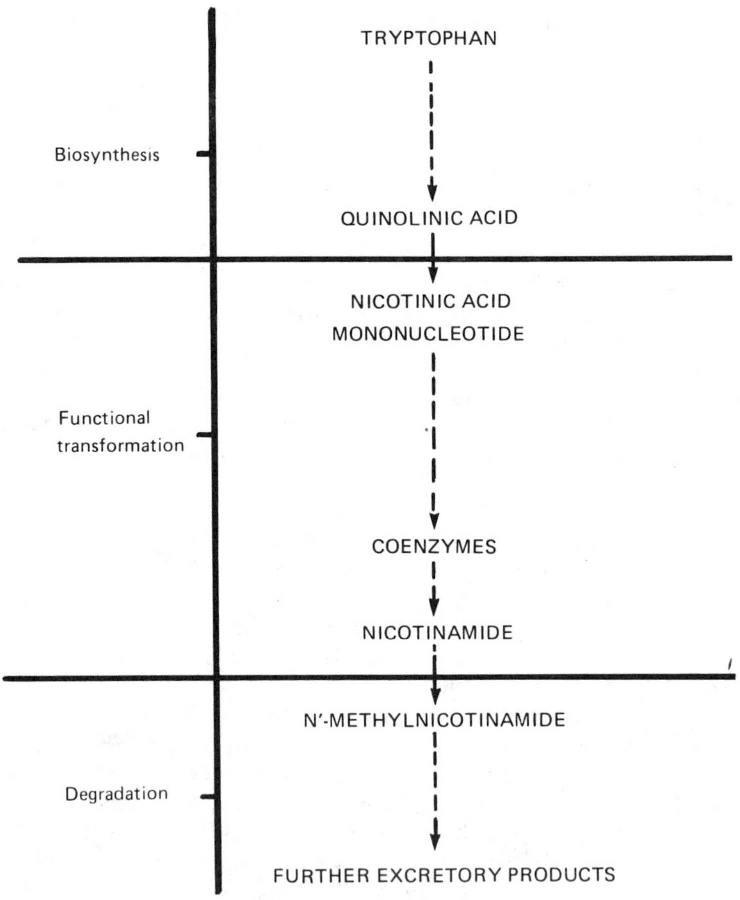

Figure 1 The subsets of the nutritional biochemistry of niacin.

there are bound forms of niacin, or niacin may be unavailable; and the distribution of niacin among coenzymatic or storage forms is variable. Unless such factors are taken into account, dietary surveys can be misleading, and interpretation of biochemical findings can be erroneous.

I. AVAILABLE METHODOLOGIES

A. Introduction

For purposes of organization, it is useful to divide the presentation in this section according to modes of analysis and aspects of niacin metabolism. Modes of analysis can be divided into chemical, biochemical, and microbiological. Chemical modes include optical and physical measurements made after chemical manipulation of the sample. Biochemical determinations are measurements in which the function of niacin or of relevant compounds is tested. Microbiological determinations are measurements in which the growth-promoting activity of niacin or related compounds in microorganisms is quantitated.

Aspects of niacin metabolism are most conveniently divided into three subsets (Fig. 1): Biosynthesis, functional transformations, and degradations. Biosynthesis comprises all steps from tryptophan to quinolinic acid; functional transformations, all steps from nicotinic acid mononucleotide or exogenous free niacin through pyridine nucleotides to endogenous nicotinamide; degradation comprises the transformation of nicotinic acid or nicotinamide to excretory products. The following reviews will make no attempt to describe all known methods of investigation. Rather, the principles, concepts, and application of methodology will be presented, illustrated by representative methods.

B. Chemical Methodology

Biosynthesis: In evaluating niacin nutriture the assumption that the tryptophan–niacin conversion is accurately reflected by the 60:1 ratio is frequently erroneous. In order to measure the flux through the pathway, the excretory products of the biosynthetic pathway often need to be analyzed. Since the total flux through the pathway is not as important as is the ratio of the transformation of α-amino-β-carboxymuconic acid-ε-semialdehyde to the uncyclized precursor of picolinic acid, α-aminomuconic-ε-semialdehyde, or to the spontaneous cyclization product, quinolinic acid, the determination of quinolinic acid has a special importance. Finally, quantitatively, tryptophan determinations are of importance to the assessment of dietary adequacy.

Tryptophan, when free in samples, can be readily converted to harman or norharman by reaction with acetaldehyde or formaldehyde (Fig. 2). This method, proposed originally by Denckla and Dewey (1), is

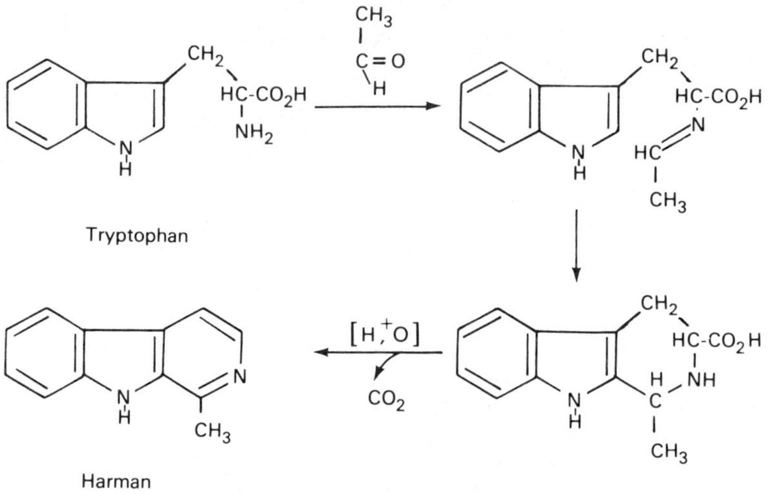

Tryptophan

Harman

Figure 2 The chemistry of the determination of tryptophan.

sensitive and accurate on a microscale; however, tryptophan is very photolabile and great care must be exercised in its analysis (2). Alternatively, tryptophan can be assayed in free form in biological fluids by methods generally applicable to amino acid analyses by column chromatography. Methods used include ion exchange chromatography, gas chromatography, and high-pressure liquid chromatography, often after suitable derivitization. The detailed discussion of such methodology would go too far afield here.

Tryptophan is destroyed in the usual acid hydrolysis of proteins and even to a degree with alkaline hydrolysis. Alkaline hydrolysis is usually employed when the tryptophan content of proteins needs to be determined, but the amino acid racemizes in that process. Alternatively, more biological methodology can be used, as will be discussed later. However, the chemical assay values exceed the biological availability in many cases as judged by rat growth assays (3).

The determination of substances in the pathway intermediate between tryptophan and niacin has been alluded to in Chap. 2. As already mentioned, such intermediates are not a measure of tryptophan conversion to niacin. Quinolinic acid, the spontaneous cyclization product of α-amino-β-carboxymuconic-ε-semialdehyde, is irreversibly formed and is the first obligatorily committed intermediate in the conversion of tryptophan to niacin.

Quinolinic acid can be determined as nicotinic acid after percholoric acid extraction, absorption and elution from charcoal, followed by decarboxylation by heat in glacial acetic acid (4). The nicotinic acid so formed is reacted with cyanogen bromide, which converts it to a glutaconaldehyde derivative which can react with *o*-toluidine to form a chromophore. This procedure is suitable for measurements in urine and liver. Pyridoxine and picolinic acid do not interfere.

This method appears to be a more reliable measure of nicotinic acid than microbiological assays which will be discussed later.

Functional transformation: In the conversion of quinolinic acid to nicotinic acid a quaternary ring nitrogen is generated. Because of this chemical transformation the pyridine ring is capable of additional reactions at the 4-position. This type of reaction was discovered by Meyer Meyerhof and co-workers when they tried to use the carbonyl reagents or bisulfite to study the stoichiometry of the reaction catalyzed by triose phosphate dehydrogenase (5). They observed the interaction of these reactants with nicotinamide-adenine dinucleotide (NAD), which resulted in a substance which had an ultraviolet absorption spectrum reminiscent of the reduced coenzyme (Fig. 3). The reaction between cyanide and the quaternary pyridinium compound is neither instantaneous nor instantaneously reversible. As will be discussed later, the reaction of the quaternary pyridine in NAD is similar to that in N^1-methylnicotinamide. N^1-methylnicotinamide was prepared by Karrer

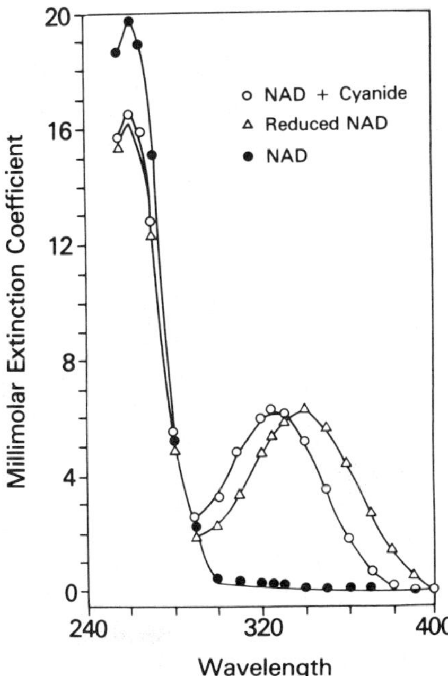

Figure 3 The spectrum of NAD, reduced NAD, and the NAD–cyanide complex.

and Warburg as a prototype for NAD long before it was known to be an excretory product of niacin metabolism (6). The actual chemical reaction is illustrated in Figure 4. The cyanide adducts between N^1-methylnicotinamide (7) and N^1-2,6-dichlorobenzylnicotinamide (8) have been isolated. Because of the ionization of the adduct, the positioning of the pyridine ring exchanges deuterium when the reaction is executed in deuterium oxide. By this method San Pietro could demonstrate that the reaction occurs at the 4-position (9). This is the site of reduction of NAD in catalytic hydride ion transfer (10). The cyanide reaction is the basis of the determination of the variety of nicotinamide-containing nucleotides and coenzymes (11,12). However, the method is not a microprocedure and therefore is more applicable to in vitro systems or chemical reactions. Also, the reaction between cyanide and nicotinic acid nucleotides is very slow to develop (13,14), while trigonelline (N^1-methylnicotinic acid) only forms the cyanide adduct in alcoholic solutions (14).

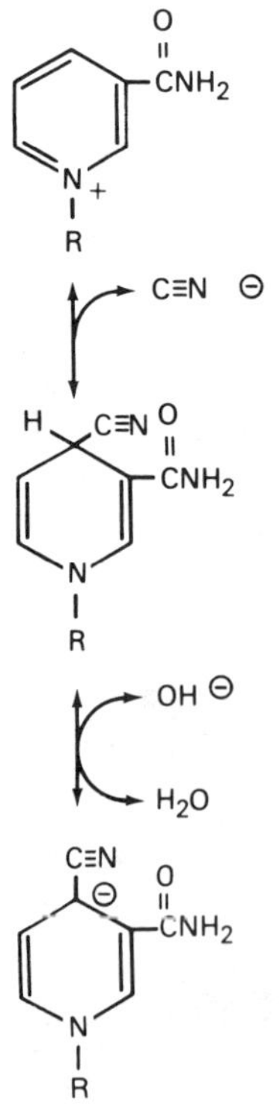

Figure 4 The chemical reaction between quaternary nicotinamide derivatives and cyanide.

More sensitive methods are based on the observations that many adducts at the 4-position of the pyridine ring generate stable compounds that are fluorescent. Of these the most frequently used are alkali and carbonyl compounds. It must be remembered, however, that numerous nucleophilic substances can generate such adducts or charge transfer complexes. Many reviews exist (15–20). Such adducts and complexes are not specific for nicotinamide derivatives but are formed with many pyridine derivatives that have been used as nicotinamide analogues. However, quaternary nicotinic acid derivatives do not yield such adducts as readily. Therefore, this approach to chemical determinations is primarily directed to nicotinamide derivatives.

Alkali itself generates an adductlike derivative (21,22). Dilute alkali tends to degrade nicotinamide nucleotides by splitting the pyridinium-ribose bond. In strong alkali a highly fluorescent material is formed (21). It is likely that in alkali, pseudobase formation occurs at all possible sites of the pyridine ring—the 2, 4-, or 6-position (19, 20,23). This can be deduced in part because oxidation of N^1-methylnicotinamide in alkali with ferricyanide gives both the 2- and 6-pyridones (24), while the 4- and 6-pyridones are natural oxidation products. The alkali adduct gives a higher fluorescent product which can be used for determination of NAD and derivatives, but the carbonyl adduct yields a more stable fluorescent compound. The adduct between quaternary nicotinamide and acetone is stabilized by the formation of dihydronapththyridine derivatives, which in acid can be oxidized to the highly fluorescent naphthyridines. Huff was the first to isolate these compounds and suggested that the structure for the acetone product with N^1-methylnicotinamide was 1,7-dimethyl-5-hydroxy-1,6-naphthyridine chloride. This would indicate an additional reaction at the 2-position. When acetone adds to the 4-position, the end product is 1,6-dimethyl-8-hydroxy-1,7-naphthyridine chloride (Fig. 5). These stabilized adducts are the basis for a fluorometric method for the determination of N^1-methylnicotinamide (25) or pyridine nucleotides (26). However, more reliable conversions seen with methyl ethyl ketone (2-butanone), and most derivatives are based on that reagent (27–29). It is of some interest that commercian NAD can contain sufficient tone to give, when dissolved, the acetone adduct of NAD, which through autoxidation will give the intermediate in the determination of NAD through the methods described (30). The structure of this substance is identical to the synthetically prepared adduct (31). It is therefore important to purify commercial NAD before it can be used as standard for any determination.

These methods do not differentiate among nicotinamide-adenine dinucleotide (NAD), nicotinamide-adenine dinucleotide phosphate (NADP), nicotinamide mononucleotide (NMN), nicotinamide riboside, or even N^1-methylnicotinamide. To determine these substances individually, it is necessary to separate the various pyridine nucleotides by chromatographic methods. Charcoal separation and direct spectrophotometric

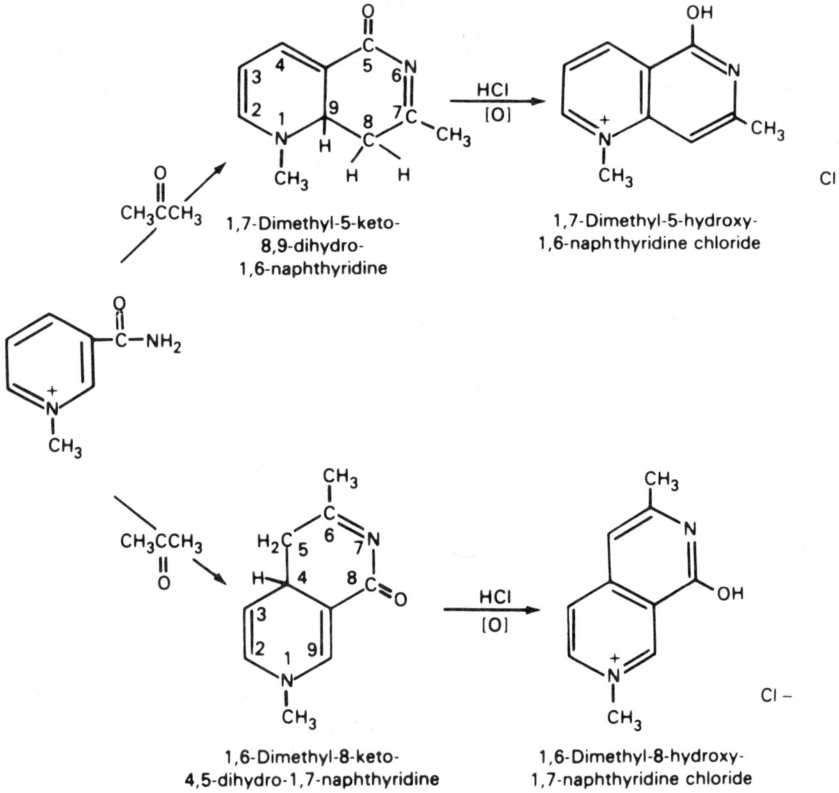

Figure 5 Chemistry of formation of acetone adduct with N^1-methylnicotinamide.

assay was the most commonly used early method for assaying pyridine nucleotides (32). One of the earliest methods using the adduct assay only separated the pyridine nucleotides and N^1-methylnicotinamide (33), but more modern methodology allows complete separation of all compounds of interest. Liquid chromatography on Dowex-1 formate columns allows complete separation of the large number of substances that might need to be determined separately. The details and principles have been described by Bernofsky (34). There are other chromatographic methods, which might be preferable under certain experimental conditions.

The identification of each substance is dependent on the change in spectral properties after treatment with potassium cyanide (35). The reduced coenzymes can be determined directly by their spectral properties (36).

The separate determination of the oxidized and reduced forms of the nicotinamide coenzymes depends on the extraction procedure for the coenzymes that allows them to remain stable. However, since NAD and NADP are alkali labile while NADH and NADPH are acid labile, separate extraction of two aliquots allows a more convenient determination of the coenzymes. The methodology has been described and reviewed for NAD by Pinder et al. (35) and for NADP by Burch (37). The actual determination of the coenzymes is usually done by a biochemical procedure amplifying the response to the presence of the nucleotides in the sample to be analyzed.

Degradation: The determination of the vitamin itself and of the excretory products derived from nicotinic acid and nicotinamide is the mainstay of nutritional evaluation. Methodologies for the determination of nicotinic acid derivatives fall in two broad classes: (1) those methods based on the reaction previously discussed as metabolic transformation reactions, in which the properties of quaternary pyridine derivatives are used, and (2) those methods by which the substance identity must be determined more directly.

Nicotinic acid and nicotinamide can be measured by the Koenig reaction, which involves the degradation of the pyridine ring utilizing cyanogen bromide. The open-ring olefinic dialdehyde can then react with an aromatic amine to form a colored product. This reaction was already described for the determination of quinolinic acid. The most widely used methods employ as the chromophore-generating base *p*-methylaminophenol sulfate (Metol) (38), or sulfanilic acid (39). Barbituric acid has also been used as chromogen, but that method is more widely used in pharmaceutical analysis (40). It has been said that the chromagen obtained with *p*-aminobenzoic acid does not extract in organic solvents when bound niacin in cereals is being measured. Therefore, a distinction between free and bound niacin could be made with this assay method (41). In all reactions, the color yield for nicotinamide is significantly lower than that for nicotinic acid, and acid hydrolysis is therefore used. However, because this method of hydrolysis is not as reproducible as one might wish, enzymatic deamination to nicotinic acid is recommended. A method based on that principle has been proposed by Fuller (42), utilizing chromatographically separated nicotinamide, deamidation by the nicotinamide deamidase from *Saccharamyces cerevisiae*, and determination of nicotinic acid by the method of Friedemann and Frazier (38). The sensitivity of these methods is in the range of 1–10 μg per sample.

More sensitive methods are frequently required. In principle, gas–liquid chromatography can be used for the determination of the vitamin and its derivatives, but it is not widely applied outside the pharmaceutical field (42). The sensitivity of the nicotinamide assay can be greatly increased by converting the vitamin to N^1-methylnicotinamide and utilizing the fluorescent properties of the adducts at the 4-position. A recent micromethod for N^1-methylnicotinamide and nicotinamide increased the sensitivity of the determination further by utilizing acetophenone as the carbonyl compound, resulting in a phenyl-substituted naphthyridine derivative. Nicotinamide is just converted to N^1-methylnicotinamide with methyliodide (44). Interference with pyridine nucleotides is avoided by analyzing unhemolyzed serum. The method is sensitive to picomole amounts.

The more common methodology for determination of N^1-methylnicotinamide is directed towards urine samples, since the nutritional state is usually evaluated through these samples. Najjar and Wood discovered a fluorescent substance in urine which resulted from niacin ingestion (45), which was later found to be a derivative of N^1-methylnicotinamide (46,47). The reaction between acetone and N^1-methylnicotinamide in alkali was reported (47), and the basic method of determination of N^1-methylnicotinamide was discovered. All methods are variants on the procedure described by Huff and Perlzweig (25) or by Carpenter and Kodicek (28). In nutritional studies the description of the procedure in the *Manual for Nutrition Surveys*, published by the Interdepartmental Committee on Nutrition for National Defense, is used (48). A more rapid determination using methyl ethyl ketone has been described by Pelletier and Campbell (49), and the ultramicro serum method has been described.

As already discussed, a variety of other metabolites of niacin has been described. The dominant products are the pyridones, especially the 6-pyridone (1-methyl-2-oxypyridine-5-carboxamide). The determination of this substance is far more difficult. The earlier methods based on chromatographic separation and direct ultraviolet absorption measurements were accurate but cumbersome. The most widely used method for urine was that of Price (50), but a method described by Rosen et al. is more reliable at very low concentrations (51).

More modern methodology of high-pressure liquid chromatography allows separation of the niacinamide metabolites on a Dowex-1 formate column. Even though many substances are not charged, the Dowex-1 column acts as a partition phase. Under the conditions described by Bernofsky (34), separation of N^1-methylnicotinamide, N^1-methylnicotinic acid (trigonelline), nicotinamide *N*-oxide, 1-methyl-4-pyridone-3-carboxamide, 1-methyl-3-carboxamide-6-pyridone, nicotinamide, 6-hydroxynicotinamide, and 4-hydroxynicotinamide can be accomplished with column elution in that order. This methodology can be used for both preparative and analytical purposes.

Individual substances can be assayed on the basis of their specific properties. For instance, nicotinamide *N*-oxide can be determined as nicotinamide after reduction with xanthine oxidase followed by measurement of nicotinamide through the cyanogen bromide reaction (54, 55). Alternatively an isotope dilution method can be used. However, as will be discussed later, the minor metabolites of niacin need not be determined when the goal is nutritional evaluation.

C. Biochemical Methods

Biochemical methods are those modes of analysis wherein the compounds are quantitated by their properties on isolated enzymes or enzyme systems. In principle, such methodology does not require catalytic properties. The extent of an irreversible reaction can be used to quantitate the substrate. However, this type of methodology is more often reserved for utilization of the catalytic properties of the compound to be quantitated because in that way ultramicro methodology can be devised. Because of this, the biochemical methods for niacin nutrition and biochemistry are primarily for the quantitation of the pyridine coenzymes. No directly relevant methodology exists in the biosynthetic pathway.

Functional transformation: The principle of magnifying the consequence of the presence of pyridine nucleotides through highly sensitive cycling assays was discovered and made practical by Lowry and co-workers (54). Amplification is up to 500-fold, but the assay suffers from being linear over only a small range of coenzyme concentrations. The principle is based on the sharing of the coenzyme by a pair of dehydrogenases by which the coenzyme alternatively oxidizes the substrate of one enzyme and reduces that of the other. After a period of cycling, if the latter enzyme reaction is irreversible, the product of the second enzyme can be measured, usually in the reverse direction with extra added pyridine nucleotide by the fluorescent properties of the reduced nucleotide. Because of the magnification process and the inherent sensitivity of the fluorometric methodology, exquisitely sensitive assays can be devised. These methods can be specific for NAD or NADP, by choosing the proper enzyme pair. While the method does not distinguish between the oxidized and reduced coenzyme present in the original sample, the sum can first be determined, and the amount of the reduced coenzyme can be separately determined after destruction of the oxidized coenzyme with alkali. When a weak alkali extract is used, the extract contains both oxidized and reduced coenzymes because the reduced forms are quite stable under those conditions, while the oxidized forms are stable for the duration of the extraction (55,56). If desired, the reduced form can be destroyed by acid. Ascorbic acid is added to avoid hemoglobin-catalyzed oxidation of the reduced coenzymes (55), and cysteine is added to stabilize the reduced coenzymes during heating (57). This methodology is far more sensitive than direct reduction without cycling, which has a sensitivity of only 3×20^{-6} *M* (58). The chemical methods utilizing the fluorescence of the alkali or methyl ethyl ketone have a sensitivity of about $10^{-7} M$, as does the direct determination of reduced pyridine nucleotides with fluorescence. The cycling methods have a sensitivity from 10^{-10} to 10^{-15} *M*. The cycling systems are varied. The original system for NAD and NADH was based on lactate dehydrogenase and

Table 1 Catalytic Assays for the Ultramicro Determination of Pyridine Nucleotides

Pyridine nucleotide assayed	Cycling enzymes	Product Assay	Sensitivity	Reference
NAD	(1) Alcohol dehydrogenase	Reduced phenazine methosulfate with 2,6-dichlorophenol	$10^{-7}M$	59,60
	(2) Phenazine methosulfate	Indophenol (spectrophotometric)		
NAD	(1) Lactate dehydrogenase	Pyruvate with excess lactate dehydrogenase and NADH (fluorometric)	$10^{-15}M$	54
	(2) Glutamate dehydrogenase			
NAD	(1) Alcohol dehydrogenase	Reduced phenazine methosulfate with molecular oxygen (polarographic)	$10^{-10}M$	61
	(2) Phenazine methosulfate			
NAD	(1) Glyceraldehyde 3-phosphate dehydrogenase (with arsenate)	Glutamate with excess glutamate dehydrogenase and NAD	$3 \times 10^{-13}M$	62

	(2) Glutamate dehydrogenase			
NADP	(1) Glucose 6-phosphate dehydrogenase	Reduced phenazine methosulfate with 2,6-dichlorophenol Indophenol (spectrophotometric)	$10^{-7}M$	59,60
	(2) Phenazine methosulfate			
NADP	(1) Glucose 6-phosphate dehydrogenase	Reduced phenazine methosulfate with molecular oxygen (polarographic)	$10^{-10}M$	61
	(2) Phenazine methosulfate			
NADP	(1) Glucose 6-phosphate dehydrogenase	6-Phosphogluconate with excess 6-phosphogluconate dehydrogenase and NAD (fluorometric)	$10^{-9}M$	55
	(2) Glutamate dehydrogenase			

glutamate dehydrogenase. The system could be varied and the reoxidation step could be chemical or enzymatic, and the ultimate assay could be spectrophotometric, polarographic, or fluorometric. Table 1 summarizes some of the available assay systems.

These methods are quite reproducible and surprisingly reliable. Because of their sensitivity they are applicable to very small samples of tissue. For that reason they have found more applicability in biochemical and histochemical research than in nutritional research. Much of the methodology has been reviewed and detailed by Pinder et al. (35).

D. Microbiological Assays

The microbiological techniques for the assays of vitamins and coenzymes have enjoyed great popularity. They have the advantage that multiple forms of a vitamin may be assayed simultaneously; thus vitamin activity can be assessed. For vitamins which have complex transformations, and which are inherently unstable, this is a great advantage. Folic acid derivatives are a good example. Vigorous hydrolysis of tissue or other samples will convert all niacin-containing coenzymes to nicotinic acid, so that such considerations do not pertain. However, the distinction between free and bound niacin in grain could in principle be made through microbiological assays. Quite apart from such biological considerations, the assays have remained popular because of their ready applicability to natural materials. Pearson reviewed the principles in some detail and described the microbiological assay for niacin as the paradigm (63).

Biosynthesis: Microbiological assays for tryptophan have been described (64,65). They are not in general use any more, since the current chromatographic methodology is quite accurate while extraction methods are about the same. Quinolinic acid has also been measured microbiologically (66), but the results are much higher than those obtained by the chemical technique, suggesting that the assay is nonspecific.

Functional transformation: There are no developed and standardized methods for the assay of nicotinic acid or nicotinamide ribosides. However, *Hemophilus parainfluenza* was found to require a growth fraction (growth factor "V") that could be satisfied with NAD or NADP (67). Furthermore, this requirement could be satisfied in two strains of *H. parainfluenza* and five strains of *H. influenza* with nicotinamide riboside (65). In principle one could use these organisms to determine the nicotinamide riboside content of samples.

Degradation: The most widespread use of microbiological assays in niacin nutriture has been in the assessment of niacin excretion or niacin content of natural products. The sample method described by Pearson, which is based on a method by Snell and Weight (64), has already been mentioned (63). Alternative methods exist (70). Some microorganisms can use conjugated nicotinic acid, while others cannot. In mammalian urine, nicotinuric acid is the primary conjugated form of nicotinic acid (the glucuronide is so unstable it is spontaneously converted to nicotinic acid). Johnson utilized the differential growth requirements of these microorganisms for an assay of nicotinuric acid (71). The assay of bound niacin in cereals has also been attempted by microbiological assays. This is usually based on the acid stability and alkaline lability of bound niacin, and the inability of most test organisms to utilize bound niacin (72–74).

II. PRACTICAL NUTRITIONAL ASSESSMENT

A. Individual Assessment

Whether a given patient has adequate niacin nutrition can be evaluated before overt symptoms and signs of niacin deficiency occur by measuring the levels of excreted niacin metabolites and the levels of pyridine nucleotides in accessible tissues. As has already been discussed, there is considerable variation in niacin metabolite excretion in animals of various species. In humans, excreted niacin metabolites in a healthy, well-nourished adult include 4–6 mg of N^1-methylnicotinamide and 40–50% as the 6-pyridone (76–80). Therefore, there is normally a ratio of about 1.3–2.0 between the 6-pyridone and N^1-methylnicotinamide (76, 78). A ratio of 1.0 or less has been considered a latent niacin deficiency (77,78). This is a promising approach to the evaluation of niacin nutriture in individual patients since N^1-methylnicotinamide does not fall to a minimal level until signs of pellagra begin to appear (81, 82). Thus some investigators have claimed that there is no relationship between N^1-methylnicotinamide and niacin intake (83), but this is not the case (84,85); rather, the relationship manifests itself late in the course of development of the deficiency. However, pyridone levels not only decrease much earlier, but may be absent weeks before pellagra is manifest (82,86,87). The ratio between N^1-methylnicotinamide and pyridone holds for all age groups (77,88).

Guidelines for normal excretion levels of N^1-methylnicotinamide have been generated for use in nutrition surveys (89) because the pyridone is not easily assayed (Table 1). The table shows an increased excretion of N^1-methylnicotinamide during pregnancy (90,91).

The level of functional niacin could be deduced from tissue levels of pyridine nucleotides. Both erythrocytes and leukocytes contain appreciable quantities. There has been some controversy about the usefulness of that determination. It has been claimed that the levels of pyridine nucleotides were not depressed in the blood cells of experimental subjects made niacin deficient (92,93) or of patients with pellagra (94). However, it is possible to depress the NAD and NADP content of red cells using an experimental diet low in tryptophan and niacin (87). In patients with pellagra the relative amounts of the various nicotinamide and nicotinic acid ribosides, nucleotides, and coenzymes are altered (95). However, the tissue concentration of coenzymes is clearly not useful as a measure of niacin deficiency.

Attempts have been made to do saturation tests. A load of niacinamide should give rise to marked increase in niacin metabolites. If it does not, the individual could be considered deficient. This test has also given conflicting results. For instance, 50 mg of niacinamide did not differentiate pellagrins from normal persons when a 24-hr excretion of N^1-methylnicotinamide was measured (96,97), while others felt that it could measure a chemical deficiency (98,99). The blood test is not generally useful.

B. Population Assessment

From the foregoing it must be clear that the evaluation of the nutritional state of individuals is far from satisfactory. The evaluation of the niacin nutriture in populations is even more complex. Urinary excretion and serum levels are the mainstay of nutritional assessment. Urinary excretion levels are quite useful in estimating dietary intakes in populations, but because of individual variations they are less useful when they are used on an individual basis. Large populations will minimize such differences. However, population surveys will also have to rely on "casual" urine specimens instead of timed, fasting specimens, or complete 24-hr collections. It is usual to relate the excretion to creatinine excretion. Plough and Consolazio have evaluated the relationship of vitamin excretion to creatinine excretion in 10 healthy normal male subjects (100). The N^1-methylnicotinamide/creatinine ratio varied considerably and more than the thiamine/creatinine or riboflavin/creatinine ratios. Furthermore, creatinine excretion in children differs from that in adults and therefore, the values vary with age as well. However, the determinations of the pyridones are all cumbersome so that the pyridone/N^1-methylnicotinamide ratio is not a practical measure in surveys. Table 2 shows the data derived from the experience of the Interdepartmental Committee on Nutrition for National Defense (101). In an extensive experience of nutritional surveys among countries that now would be considered part of the Third World, very few instances of low N^1-methylnicotinamide excretion were found (102).

Tryptophan intake and tryptophan excretion or quinolinic acid excretion do not correlate well (81,82). Therefore, even if the assays were practical in the survey setting they would be of little use in evaluating population intakes of niacin precursors.

Nicotinic acid excretion is very low. The amount does not relate to the nicotinic acid intake of either niacin or tryptophan (81,82). Serum levels of nicotinic acid likewise are not related to nutritional status. During the survey of the extreme malnutrition and starvation

Table 2 Urinary N' Methyl Nicotinamide Excretion

Level	Milligrams /6 hours	Milligrams /gram creatinine	Milligrams 24 hours
"Deficient"	<0.02	< 0.5	< 0.8
"Low"	0.2-0.59	0.5-1.59	0.8-2.3
"Acceptable"	0.6-1.5	1.6-4.2	2.4-6.3
"High"	> 1.6	> 4.3	> 6.4

Adapted from W. N. Pearson, J. Am. Med. Ass. 180, 49 (1962).

in the Western Netherlands after World War II, the nicotinic acid levels in blood were not different from normal (104).

C. Concluding Remarks

There still is endemic pellagra in the world, and many alcoholics are still severely malnourished. However, the actual diagnosis of the disease must remain primarily clinical. Biochemical evaluation of the pyridone/N^1-methylnicotinamide ratio is useful as corroboration. In spite of the still prevalent tryptophan—niacin malnutrition that is deduced clinically, survey data to date are not indicative of widespread niacin malnutrition, detectable by current methodology.

REFERENCES

1. W. D. Denckla and H. K. Dewey, *J. Lab. Clin. Med. 69*:160 (1967).
2. M. Kenney, R. F. Lambe, D. A. O'Kelley, and A. Darragh, *Clin. Chem. 26*:1511(1980).
3. P. Pongpaew and K. Guggenheim, *Nutr. Diet 10*:297(1968).
4. H. G. McDaniel, in *Methods in Enzymology, Vol. 66* (D. B. McCormick and L. D. Wright, eds.), Academic Press, New York, 1980, p. 91.
5. O. Meyerhof, P. Ohlmeyer, and W. Möble, *Biochem. Z. 297*:113 (1938).
6. P. Karrer and O. Warburg, *Biochem. Z. 285*:297(1936).
7. W. Marti, M. Viscontini, and P. Karrer, *Helv. Chim. Act. 39*: 1451(1956).
8. K. Wallenfells and H. Schuly, *Ann. Chem. 621*:86(1959).
9. A. San Pietro, *J. Biol. Chem. 217*:579(1955).
10. M. E. Pullman, A. San Pietro, and S. P. Colowick, *J. Biol. Chem. 206*:129(1954).

11. M. M. Ciotti and N. O. Kaplan, in *Methods in Enzymology 3*, (S. P. Colowick and N. O. Kaplan, eds.), Academic Press, New York, 1957, p. 890.
12. S. P. Colowick, N. O. Kaplan, and M. M. Ciotti, *J. Biol. Chem. 191*:447(1951).
13. M. Lamborg, F. E. Stolzenbach, and N. O. Kaplan, *J. Biol. Chem. 231*:685(1958).
14. M. Lamborg, as quoted in N. O. Kaplan, *Rec. Chem. Progr. 16*: 186(1955).
15. N. O. Kaplan, in *The Enzymes, Vol. 3*, (P. D. Boyer, H. Lardy, and K. Myrback, eds.), Academic Press, New York, 1960, p. 105.
16. N. O. Kaplan, *Record. Chem. Progr.* (Kresge-Hooker Scientific Library) *16*:176(1955).
17. E. M. Kosover, in *The Enzymes, Vol. 3*, (P. D. Boyer, H. Lardy, and K. Myrback, eds.), Academic Press, New York, 1960, p. 171.
18. G. M. Kosower, *Molecular Biochemistry*, McGraw-Hill, New York, 1962.
19. K. Wallenfells, in *Steric Course of Microbiological Reactions*, (G. E. W. Wolstenholme and C. M. O'Connor, eds.), Churchill, London, 1959, p. 10.
20. S. P. Colowick, J. van Eys, and J. M. Park, in *Biological Oxidations, Comprehensive Biochemistry, Vol. 14*, (M. Florkin and E. M. Stotz, eds.), Elsevier, New York, 1966, p. 1.
21. N. O. Kaplan, S. P. Colowick, and C. C. Barnes, *J. Biol. Chem. 191*:461(1951).
22. R. B. Martin and J. G. Hull, *J. Biol. Chem. 289*:1237(1964).
23. A. G. Anderson and G. Berkenhammer, *J. Am. Chem. Soc. 80*: 992(1958).
24. M. E. Pullman and S. P. Colowick, *J. Biol. Chem. 206*:126(1954).
25. J. W. Huff and W. A. Perlzweig, *J. Biol. Chem. 167*:157(1947).
26. N. Levitas, J. Robinson, F. Rosen, J. W. Huff, and W. A. Perlzweig, *J. Biol. Chem. 167*:169(1947).
27. H. B. Burch, C. A. Storvick, R. L. Bicknell, H. C. Kung, L. G. Alejo, W. A. Everhart, O. H. Lowry, C. G. King, and O. A. Bessey, *J. Biol. Chem. 212*:897(1955).
28. K. H. Carpenter and E. Kodicek, *Biochem. J. 46*:421(1950).
29. G. A. Goldsmith and O. N. Miller, in *The Vitamins, 2nd Edition, Volume VII* (P. Gyorgy and W. N. Pearson, eds.), Academic Press, New York, 1967, p. 137.
30. M. I. Dolin and K. B. Jacobson, *Biochem. Biophys. Res. Comm. 11*:102(1963).
31. R. M. Burton, A. San Pietro, and N. O. Kaplan, *Arch. Biochem. Biophys. 70*:87(1957).
32. P. Feigelson, J. N. Williams, Jr., and C. A. Elvehjem, *J. Biol. Chem. 185*:741(1950).

33. J. P. Kring and J. N. Williams, Jr., *J. Biol. Chem.* *207*:851(1954).
34. C. Bernofsky, in *Methods in Enzymology, Vol. 66* (D. B. McCormick and L. D. Wright, eds.), Academic Press, New York, 1980, p. 23.
35. S. Pinder, J. B. Clark, and A. L. Greenbaum, in *Methods in Enzymology, Vol. 66* (D. B. McCormick and L. D. Wright, eds.), Academic Press, New York, 1980, p. 23.
36. O. H. Lowry, N. R. Roberts, and J. J. Kapphahn, *J. Biol. Chem.* *224*:1047(1957).
37. H. B. Burch, in *Methods in Enzymology, Vol. 66* (D. B. McCormick and L. D. Wright, eds.), Academic Press, New York, 1980, p. 11.
38. F. E. Friedemann and E. J. Frazier, *Arch. Biochem.* *26*:361(1950).
39. Association of Official Agricultural Chemists, in *Official Methods of Analysis, 8th Ed.* (W. Horwitz, ed.), Assoc. Off. Agr. Chemists, Washington, D.C., 1955, p. 826.
40. O. Pelletier and J. A. Campbell, *J. Pharm. Sci.* *50*:926(1961).
41. M. L. Das and N. C. Ghosh, *Ind. J. Med. Res.* *45*:631(1957).
42. L. Fuller, in *Methods in Enzymology, Vol. 66*(D. B. McCormick and L. D. Wright, eds.), Academic Press, New York, 1980, p. 3.
43. A. J. Sheppard and A. R. Prosser, in *Methods in Enzymology, Vol. 18* (D. B. McCormick and L. D. Wright, eds.), Academic Press, New York, 1971, p. 17.
44. B. R. Clark, in *Methods in Enzymology, Vol. 66* (D. B. McCormick and L. D. Wright, eds.), Academic Press, New York, 1980, p. 5.
45. V. A. Najjar and N. W. Wood, *Proc. Soc. Exptl. Biol. Med.* *44*:386(1940).
46. J. W. Huff and W. A. Perlzweig, *J. Biol. Chem.* *150*:395(1943).
47. V. A. Najjar, V. White, and D. B. McNair-Scott, *Bull. Johns Hopkins Hosp.* *74*:378(1944).
48. Interdepartmental Committee on Nutrition for National Defense, *Manual for Nutrition Surveys*, U.S. Government Printing Office, Washington, D.C., 1957, p. 79.
49. O. Pelletier and J. A. Campbell, *Anal. Biochem.* *3*:60(1962).
50. J. M. Price, *J. Biol. Chem.* *211*:117(1954).
51. F. Rosen, W. A. Perlzweig, and J. G. Leder, *J. Biol. Chem.* *179*:157(1949).
52. M. Sparthan and W. S. Chaykin, *Anal. Biochem.* *31*:286(1969).
53. M. Sparthan and S. Chaykin, in *Methods in Enzymology, Vol. 18* (D. B. McCormick and L. D. Wright, eds.), Academic Press, New York, 1971, p. 46.
54. O. H. Lowry, J. V. Passoneau, D. W. Schultz, and M. Y. Rock, *J. Biol. Chem.* *236*:2746(1961).
55. H. B. Burch, M. E. Bradley, and O. H. Lowry, *J. Biol. Chem.* *242*:4546(1967).

56. F. M. Matschinskey, *J. Neurochem. 15*:643(1968).
57. O. H. Lowry, J. V. Passoneau, and M. K. Rock, *J. Biol. Chem. 236*:1756(1961).
58. E. Racker, in *Methods in Enzymology, Vol. 2* (S. P. Colowick and N. O. Kaplan, eds.), Academic Press, New York, 1954, p. 72.
59. M. A. Eichel and M. J. Spirtes, *Arch. Biochem. Biophys. 53*:309 (1954).
60. T. F. Slater and B. Sawyer, *Nature 193*:454(1962).
61. A. L. Greenbaum, J. B. Clark and P. McLean, *Biochem. J. 95*: 161(1965).
62. F. M. Matschinsky, in *Methods in Enzymology, Vol. 18* (D. B. McCormick and L. D. Wright, eds.), Academic Press, New York, 1971, pp. 3–11.
63. W. N. Pearson, in *The Vitamins, Vol. VII* (P. Györgi and W. N. Pearson, eds.), Academic Press, New York, 1964, p. 1.
64. R. D. Greene and A. Black, *J. Biol. Chem. 155*:1(1944).
65. I. T. Greenhut, B. S. Schweigert, and C. A. Elvehjem, *J. Biol. Chem. 165*:325(1946).
66. L. Henderson and G. Ramarsana, *J. Biol. Chem. 181*:687(1949).
67. A. Lwoff and M. Lwoff, *Proc. Roy. Soc. Lond., Sec. B, 122*: 3512(1937).
68. W. Gingrich and F. Schlenk, *J. Bacteriol. 47*:535(1944).
69. E. E. Snell and L. D. Wright, *J. Biol. Chem. 139*:675(1941).
70. W. A. Krehl, F. M. Strong, and C. A. Elvehjem, *Ind. Eng. Chem. Anal. Ed. 15*471(1943).
71. B. C. Johnson, *J. Biol. Chem. 159*:227(1945).
72. W. A. Krehl and F. M. Strong, *J. Biol. Chem. 156*:1(1944).
73. K. M. Clegg, E. Kodicek, and S. P. Mistry, *Biochem. J. 50*:326 (1952).
74. K. M. Clegg, *Brit. J. Nutr. 17*:325(1963).
75. W. I. M. Holman and D. J. DeLange, *Nature 166*:468(1950).
76. W. I. M. Holman and D. J. DeLange, *Nature 165*:604(1950).
77. J. P. Du Plessis, *Council for Scientific and Industrial Research Report No. 261,* National Nutrition Research Institute, Pretoria, South Africa, 1967.
78. D. J. DeLange and C. P. Joubert, *Am. J. Clin. Nutr. 15*:169 (1964).
79. C. P. Joubert and J. DeLange, *Proc. Nutr. Soc. South Africa 3*: 60(1962).
80. G. A. Goldsmith, O. N. Miller, W. Unglaub, and J. Gibbens, *Fed. Proc. 14*:434(1955).
81. M. K. Horwitt, C. C. Harvey, W. S. Rothwell, J. D. Cutler, and D. Haffron, *J. Nutr. Suppl. 1*:60(1956).
82. G. A. Goldsmith, H. P. Sarett, U. D. Register, and J. Gibbens, *J. Clin. Inv. 31*:533(1952).

83. O. Michelson and L. L. Erickson, *Proc. Soc. Exptl. Biol. Med. 588*:33(1945).
84. J. M. Ruffin, D. Cayer, and W. A. Perlzweig, *Gastroenterology 3*:340(1944).
85. F. Sargent, P. F. Robinson, R. E. Johnson, and M. Castiglione, *J. Clin. Inv. 23*:714(1944).
86. G. J. Walters, R. R. Brown, M. Kaikara, and J. M. Price, *J. Biol. Chem. 217*:489(1955).
87. V. M. Vivian, M. M. Chaloupka, and M. S. Reynolds, *J. Nutr. 66*:587(1958).
88. P. Careddu, L. Mainardi, G. Sacchetti, and L. T. Tenconi, *Acta Vitaminol. Milano, 19*:135(1965).
89. Interdepartmental Committee on Nutrition for National Defense, *Manual for Nutrition Surveys, 2nd. Ed.* U.S. Government Printing Office, Washington, D.C., 1963.
90. W. J. Darby, W. J. McGanity, M. P. Martin, G. Bridgforth, P. M. Densen, M. M. Kaser, P. J. Ogle, J. A. Newbill, A. Stockell, M. E. Ferguson, O. Touster, G. S. McClellan, C. Williams, and R. O. Cannon, *J. Nutr. 51*:565(1953).
91. M. E. Lojkin, A. W. Wertz, and C. G. Dietz, *J. Nutr. 46*:335 (1952).
92. J. R. Klein, W. A. Perlzweig, and P. Handler, *J. Biol. Chem. 145*:27(1942).
93. C. W. Carter and J. R. P. O'Brien, *Quart. J. Med. 14*:197(1945).
94. N. Raghuramula, S. G. Grikantia, B. S. Narasinga Rao, and C. Gopola, *Biochem. J. 96*:837(1965).
95. S. G. Srikantia, B. S. R. Narasinga, N. Raghuramulu, and C. Gopalen, *Am. J. Clin. Nutr. 21*:1306(1968).
96. F. T. Lossy, G. A. Goldsmith, and H. P. Sarett, *J. Nutr. 45*: 213(1951).
97. H. L. Rosenthal, G. A. Goldsmith, and H. P. Sarrett, *Proc. Soc. Exptl. Biol. Med. 84*:208(1953).
98. R. E. Johnson, F. Sargent, P. F. Robinson, and C. F. Consolazi, *War Medicine 7*:227(1945).
99. W. C. Unglaub and G. A. Goldsmith, in *Symposium Methods for Evaluation of Nutritional Status in Man* (M. Spector, M. Peterson, and T. E. Friedemann, eds.), National Research Council, Washington, D.C., 1954.
100. I. C. Plough and C. F. Consolazio, *J. Nutr. 69*:363(1959).
101. W. N. Pearson, *J. Am. Med. Assoc. 180*:49(1962).
102. W. N. Pearson, in *Nutrition, Vol. III* (G. H. Beaton and E. W. McHenry, eds.), Academic Press, New York, 1966, p. 265.
103. G. J. Gabuzda and C. S. Davidson, *Am. J. Clin. Nutr. 11*:502 (1962).
104. G. C. E. Burger, J. C. Drummond, and H. R. Standstead, *Malnutrition and Starvation in Western Netherlands*, The Hague General State Printing Office, Part I, 1948, p. 125.

6

NICOTINIC ACID NUTRITURE

Niacin is an essential nutrient, but as discussed in the previous chapter it is a complex nutritional requirement because the requirement can in many species be satisfied by adequate intake of the amino acid tryptophan. Furthermore, niacin occurs in multiple chemical forms in nature. This variability is seen within the usual components of an ordinary diet. There is a large body of literature that pertains to the dietary requirements for niacin activity in various species. This conspectus will concentrate on the human requirement. The requirements in animals will be discussed only in so far the data were derived with concern for the human physiology or because the data illustrate a pertinent point or principle.

Since the need for nicotinamide coenzymes is satisfied by a large variety of chemical entities, the discussion will be somewhat complex. This is the more true because the generation of nicotinic acid is only a minor, if not an incidental, metabolic function of tryptophan. The primary nutritional need for tryptophan is an amino acid in proteins. Therefore, a discussion of tryptophan requirements, independent of niacin nutriture, is germane in this context. It must be pointed out, however, that while the conversion of tryptophan to niacin is regulated, it is not regulated to the degree that niacin synthesis stops when tryptophan nutrition is marginal.

I. TRYPTOPHAN REQUIREMENTS

A. Tryptophan as Essential Amino Acid

The concept that some amino acids are indispensable while others are not is a relatively old notion. Blum noted that certain proteins, low in tyrosine and "indole-supplying" grains, were not able to support nitrogen balance in dogs. Specifically, casein-derived protein could, and fibrin-derived protein could not support nitrogen balance (1). The term "essential" amino acid is now regularly used. Willcock and Hopkins observed the growth-promoting effect of tryptophan when the protein zein was used as the major nitrogen source (2). It is of interest that zein is the principal protein in corn. These experiments were done in mice and confirmed for that species by Wheeler (3). Later Abderhalden reported the essentiality in dogs (4,5). However, convincing evidence for the need for tryptophan did not surface until the deficiency of zein in both lysine and tryptophan was recognized by Osborne and Mendel (6). The use of mixtures of amino acids in place of whole proteins showed that some of those amino acids were essential, but because not all amino acids were known, such experiments failed. It was hypothesized that peptide bonds were necessary for growth (7). However, with the discovery of threonine by Rose and associates (8,9), experiments with purified diets were made possible; from those experiments the modern concept of essential amino acids was generated. Through such experiments a list of indispensible and dispensible amino acids was proposed for the rat (10). Even at that time it was known that the "all or none" concept was clearly not valid: nonessential amino acids could spare essential ones, as exemplified by the pairs tyrosine/phenylalanine and cysteine/methionine or choline plus cysteine/methionine. When the experiments were further refined by using increasingly purified diets, it became clear that the essential/nonessential dichotomy is not absolute (11). The demonstration of tryptophan as an absolute dietary essential did not have to await the development of purified diets, however, because of the already known deficiency of tryptophan and lysine in zein, and the lability of tryptophan to acid hydrolysis of proteins allowed qualtitative destruction while sparing other amino acids to varying degrees.

Alkaline hydrolysis does preserve tryptophan but racemizes it. Therefore, the question of the availability of D-tryptophan had to be settled in order to interpret early experimental data. It is extremely instructive that such a simple question, approachable by well-established experimental procedures of the period with accepted nutritional toools, generated an ongoing controversy of conflicting data and interpretations. The reason is, again, that the data do not yield an all-or-none, yes/no answer. The early data suggested that DL-tryptophan was nearly as effective in promoting growth, and therefore, by implication, that D-tryptophan equaled L-tryptophan in biological activity (12). When the two isomers were resolved, it was concluded that D- and L-tryptophan and their acetylated forms were in fact equivalent for the growth of the rat (13,14). However, the conversion of the racemates to kynurenine was thought to not be equivalent (14). Rose accepted these data initially without question, and DL-tryptophan was therefore the usual component of purified diets. But when L-tryptophan was substituted, better growth was obtained, suggesting that D-tryptophan was inferior to the L-isomer, at least when pure diets were used (11,15). Similar findings for human nutrition are evident in the discussion on the tryptophan load test: data derived from DL-tryptophan might not be as readily interpretable because of lesser conversion to metabolites. The wide species variation in utilization was mentioned in that discussion. The human seems to use the D-isomer poorly, in that a considerable amount of a test dose is excreted unchanged (16), or gives rise to an "aberrant" metabolite (17,18). As will be discussed later in quantitative human studies the isomer is ineffective (19,20). Price and Brown also saw quantitative differences between D- and L-tryptophan metabolism (21), as did Hankes et al. (22). Not only is the D-isomer of tryptophan poorly converted to CO_2, but also D-isomers of kynurenine and of hydroxykynurenine are likewise poorly converted. In fact, less conversion to quinolinic acid was seen from D-tryptophan than from L-tryptophan. This was confirmedly the low excretion levels of N^1-methulnicotinamide and free nicotinic acid (22).

Acetyltryptophan is poorly utilized by man (20,21,23,24) even though the rat can use both acetyltryptophan and tryptophan equally well for growth (13,14). The poor utilization by the human can in part be ascribed to its poor absorption (23).

Similar evaluations were made of other amino acid derivatives. N^1-methyltryptophan was found to stimulate growth in rats maintained on a diet deficient in tryptophan (25), but it is not as efficient for rats as is tryptophan, and the racemic amino acid is substantially less active than is the L(+)-isomer, suggesting that the D(−)-isomer could not be utilized by the rat (26).

The fact that utilization of tryptophan by man varies depending on the isomer and derivative has no bearing on the indispensibility of tryptophan as an amino acid for the human. From the studies of Rose, a list of amino acids essential for man (27) was postulated (Table 1); tryptophan was one of them. There have been changes in the list since, especially with respect to histidine, which was thought to be essential in the rat but not in the human. However, the essentiality of tryptophan has been unquestioned. The need for tryptophan to maintain nitrogen balance was already demonstrated by Holt et al., who showed that tryptophan was required for nitrogen balance in young men on a diet in which the amino acids were supplied by acid-hydrolyzed casein (28). Shortly thereafter, the argument for tryptophan essentiality was strengthened by a study comparing the effect of feeding a protein hydrolysate obtained through enzymatic digestion, which would leave tryptophan intact, or through acid hydrolysis, which would

Table 1 Classification of Amino Acids with Respect to Their Role in Maintenance of Nitrogen Equilibrium in Normal Adult Man

Essential	Nonessential
Valine	Glycine
Leucine	Alanine
Isoleucine	Serine
Threonine	Cystine
Methionine	Tyrosine
Lysine	Aspartic acid
Phenylalanine	Glutamic acid
Tryptophan	Proline
	Hydroxypyroline
	Histidine
	Arginine
	Citrulline

Source: Ref. 27.

destroy it. Adding tryptophan aided nitrogen balance in the latter but not in the former (29). In his experiments, Rose used purified diets, which again confirmed the essential nature of tryptophan for man (27, 30).

B. The Quantitative Need for Tryptophan in Humans

The establishment of tryptophan as a dietary essential does not establish a priori the quantitative requirements. Furthermore, requirements do vary with age, sex, and physiological state (e.g., pregnancy and lactation) of the subjects.

The early quantitative studies were done primarily in young, healthy males and shortly thereafter in young, healthy females. Various methodologies were employed including: nitrogen balance studies, an evaluation of blood or urinary excretion levels, or the concept of nutrient saturation of body stores. Alternative approaches have included growth and nitrogen equilibrium measurements.

Holt et al. initiated the first quantitative studies, using their hydrolysate as a source of a tryptophan-free diet in two healthy adult male subjects. They tested urinary excretion patterns, the saturation concept, and arrived at a figure of 3.0–6.0 mg tryptophan/kg body weight as the requirement. However, up to 6.3–9.4 mg tryptophan/kg body weight were needed to restore nitrogen balance (31). A later study by Denko and Grundy utilized whole protein as the basis of the experimental diet, but they varied the quantity and quality of the protein to manipulate the tryptophan content. They found that 250 mg/day was sufficient to maintain nitrogen balance in seven healthy young men (32) and that body size had surprisingly little effect on the minimum requirement, a finding confirmed by Rose et al. (27). Rose discussed the experimental concept in detail. The limitations of the nitrogen balance method were clearly recognized, but with that approach the data suggested a requirement of 250 mg/day. The three carefully studied subjects had required 150, 150, and 250 mg/day to remain in balance. Fifteen other men were kept in balance on 200 mg/day. Rose and co-workers also recognized the possibility of a disparity between these minimal requirements and a *safe intake*. They therefore recommended a safe intake of 500 mg/day (30). Figure 1 shows a typical nitrogen balance curve for a subject deprived of or supplemented with tryptophan. Using serum levels of tryptophan to establish the requirement, Young and associates concluded that 3–5 mg tryptophan/kg body weight was appropriate, since in that range plasma levels responded to changes in dietary intake, while above 5 mg/kg, the body appeared saturated (33). Nitrogen balance studies, obtained simultaneously on the same subjects, suggested a slightly lower figure.

Studies on women were done by Leverton and her co-workers (34), and Fisher et al. (35). Leverton's studies showed a range of 82–120 mg tryptophan as the level at which some subjects failed to maintain nitrogen balance. Of 16 subjects, all remained in nitrogen balance at 157 mg/day, suggesting a requirement of 160 mg/day. When the total nitrogen supplied was diminished, the requirement of five young women dropped to 50 mg/day.

Children have different requirements during growth. Infants have been studied with purified diets. Albanese et al. (36) showed that the requirement in three 6- to 12-month-old infants was not less than 6 mg tryptophan/kg BW nor more than 59 mg tryptophan/kg BW. Albanese therefore recommended 30 mg L-tryptophan/kg/day. A later study of

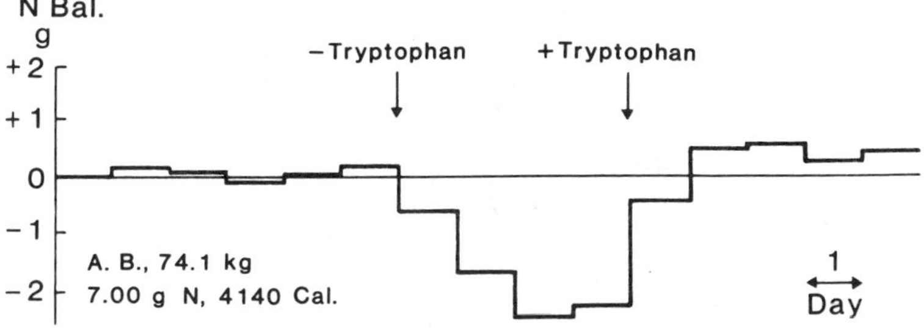

Figure 1 The role of arginine and tryptophan in the maintenance of nitrogen equilibrium in man. The basal diet was devoid of histidine and arginine throughout.

five infants suggested that 13 mg/kg was too low, but that all infants thrived on an intake of 16 mg tryptophan/kg/day (37).

In children of school age, a study from Japan suggested that for boys 120 mg tryptophan/day was adequate to maintain positive nitrogen balance (38) and that the diet, adequate for boys, was also adequate for girls (39).

The elderly may have slightly lower requirements than young adults (40). In four males and 20 females at an average age of 73 years, the requirement was two-thirds of that of young adults, or 2 mg/kg as judged by plasma levels. Separate data are available for pregnant or lactating women (41), though it has been reported that pregnant women excrete more tryptophan than do nonpregnant women (42). Table 2, which is adapted from Irwin and Hegsted (43), summarizes the information that is known to date. From such data the National Research Council generates recommended dietary allowances (RDA) which are generally set at 30% above those commonly reported to allow for individual variation (44). Table 3 shows the data for amino acids as published in the ninth edition of the Recommended Dietary Allowances, but which were taken from data published previously by the Food and Nutrition Board (45).

Table 2 Quantitative Tryptophan Requirement of Man

Subject	Criteria	Estimated requirement	Reference
Infants	N-retention, growth	12–40 mg/kg	17
	N-retention, growth	13–16 mg/kg	37
School children	N-retention	60–120 mg/day	38
Adult males	N-retention	6–9 mg/kg	31
	Urinary retention	3–6 mg/kg	31
	N-retention	150–250 mg/day	15
	N-retention	225 mg/day	20
	Plasma levels	3 mg/kg	33
	N-retention	2–2.6 mg/kg	33
Adult females	N-equilibrium	82–157 mg/day	34
	N-retention	50 mg/day	35
Elderly (male and female)	Plasma levels	2 mg/kg	40

Source: Adapted and expanded from Ref. 43.

Table 3 Estimated Amino Acid Requirements of Man

Amino acid	Requirement (mg/kg body weight/day)			Pattern for high-quality proteins (mg/g of protein)
	Infant (4-6 months)	Child (10-12 years)	Adult	
Histidine	33	?	?	17
Isoleucine	83	28	12	42
Leucine	135	42	16	70
Lysine	99	44	12	51
Total S-containing amino acids (cystine and methionine)	49	22	10	26
Total aromatic amino acids (phenylalanine and tyrosine)	141	22	16	73
Threonine	68	28	8	35
Tryptophan	21	4	3	11
Valine	92	25	14	48

[a]Two grams per kilogram of body weight per day of protein of the quality listed in column 4 would meet the amino acid requirement of the infant.

Source: Ref. 44.

Problems of amino acid imbalances, by which the apparent quantitative requirement of one amino acid may affect the requirement for another, are evidenced by the discrepancy between the results of Leverton et al. (34) and Fischer et al. (35). However, varying the source and preparation of amino acids did not change the observation that approximately 250 mg/day of tryptophan is likely to be adequate for adults (46,47). Women on a self-chosen diet seem to take in adequate quantities of tryptophan (48), and when in nitrogen balance, their tryptophan intake was 440–1280 mg/day.

II. THE REQUIREMENTS FOR TRYPTOPHAN PLUS NIACIN

A. Humans

Once the conversion of tryptophan to niacin was understood, it became essential to estimate the efficiency of conversion of tryptophan to niacin. Tryptophan was able to reverse the symptoms of pellagra in humans (49–51), but such observations were not sufficiently quantitative to allow estimates of conversion. Two sets of data have been used. One can measure in normal men the conversion to niacin metabolites (N^1-methylnicotinamide and N^1-methyl-2-pyridone-5-carboxamide, the 6-pyridone) from a test dose of tryptophan and compare that to metabolite levels following a test dose of niacinamide. This was done, and with the assumption that D-tryptophan was not converted to niacin metabolites, the data averaged to a conversion of 55.8 mg tryptophan equivalent to 1 mg of nicotinamide, with a range of 33.7–86.3 mg tryptophan (52–54). Another approach, used by Horwitt et al. (55), involved maintaining subjects for a long period of time on diets deficient in both tryptophan and niacin. Several groups were studied: one group received 100 mg of L-tryptophan, a second group received 10 mg of niacin, and the remaining group was unsupplemented. Using the amount of N^1-methylnicotinamide that was excreted as the basis for comparison, 52–65 mg tryptophan were found to be equivalent to 1 mg niacin. When 200 mg tryptophan was fed to the deficient subjects in the form of the protein lactalbumen, the conversion figure was 60 mg tryptophan equivalent to 1 mg niacin. Later data of Vivian seemed to confirm this equivalence (56) and the 60:1 ratio is usually cited (44). Different conversion factors have been reported. When there is calorie restriction, the excretion of N^1-methylnicotinamide increases two- to five-fold, so that the 60:1 ratio may not hold under such conditions (57). When patients are deficient, the excretion of the 6-pyridone will not increase until the blood cells are resaturated with NAD, suggesting that experiments should not be of too short duration (58). Nakayawa and co-workers found conversions from as low as 122:1, when the tryptophan plus niacin level was low, to 74:1, as amounts of tryptophan and niacin were increased (59). When the dietary tryptophan level was rapidly alternated between 250 mg/day and 25–30 mg/day, the excretion of niacin, N^1-methylnicotinamide, and the 6-pyridone did not change rapidly, suggesting that the efficiency of conversion may vary according to the dietary condition (60). It does appear that when a limited amount of tryptophan is provided in a diet low in niacin, the amino acid is metabolized preferably to maintain nitrogen balance and used for niacin synthesis only after nitrogen balance is achieved (56, 61). Other conversion factors have been claimed (62), but any differences in the conversion rate from tryptophan to niacin do not pertain to the steady state diet as one would ordinarily encounter in dietary calculations.

Knowing the conversion rate and knowing the tryptophan requirement so that subjects could be kept in nitrogen balance, it became possible to establish the niacin requirement for man. The niacin requirement could be defined as that amount of niacin needed if no conversion from tryptophan were to occur. If tryptophan were ingested, then for each 60 mg of tryptophan it was assumed that 1 mg of niacin was obtained. That basic assumption implies that the conversion rate is independent of tryptophan intake. That assumption is true as long as amino acid imbalances do not change the niacin conversion rate and the patient is in nitrogen balance.

The majority of experiments on niacin requirements in humans originate with Goldsmith and her associates and with Horwitt et al. Goldsmith has executed a number of long-term studies on experimental niacin deficiency. Her studies include diets with corn or without corn as the major "background" staple. The studies by Horwitt et al. at Elgin State Hospital specifically avoided corn as a variable.

Niacin deficiency was induced by Goldsmith in 15 subjects in 19 long-term studies; most were females (12 of 15 subjects). In 15 of the 19 experiments a range of niacin deficiency was induced, from mild to frank pellagra (63–66). The diets that induced pellagra furnished 3.4–5.4 mg niacin and 151–207 mg tryptophan per day. Therefore, the niacin equivalent of the deficient diets ranged from 5.9 to 8.8 mg/day, while the subjects who did not develop pellagra had intakes of 7.4–10.6 mg/day. These studies showed that lime treatment of corn did not change the results significantly (65), so that such data could be compared with results from other studies. Corn diets supplemented with niacin with a base of 200 mg tryptophan produced significant increase in the excretion of niacin metabolites to about 8–10 mg/day or 11–13 niacin equivalents (63).

Horwitt employed diets not containing corn, though zein was used to offer enough protein without tryptophan. No subject developed pellagra on a niacin intake of 5.2–7.0 mg/day with a tryptophan intake of 238–318 mg/day, a niacin equivalent of 9.2–12.3 mg/day (55). Horwitt also recalculated the data of Goldberger when he induced pellagra. He supplied 6.7 mg niacin/day with an estimated total tryptophan of 330 mg/day, for a niacin equivalent of 12.2 mg/day (55). The data are tabulated in Table 4. The discrepancy between the data of Goldsmith and

Table 4 Niacin and Tryptophan Intake in Relation to Experimental Production of Niacin Deficiency

Investigator	Number of subjects	Niacin deficiency	Niacin in diet	Tryptophan in diet	Niacin[a] equivalent in diet	Niacin equivalent /1000 cal	Niacin equivalent /kg body weight
			mg	mg	mg	mg	mg
Goldsmith	10	Severe	4.1–5.4	151–197	6.6–8.6	3.7–4.9[b]	0.11–0.18[c]
Goldsmith	5	Mild	3.4–5.4	152–207	5.9–8.8	4.3–4.9[b]	0.12–0.20
Goldsmith	4	None	4.2–5.4	193–264	7.4–10.6	4.0–5.4	0.13–0.19[d]
Horwitt	15	None	5.2–7.0	238–318	9.2–12.3	4.4	0.15–0.21
Goldberger	11	6/11	6.7	330[e]	12.2	4.1	0.14–0.22

[a]Niacin equivalent equals dietary niacin plus 1/60 of dietary tryptophan.

[b]Subjects with ratios of 4.4 or above received less than 2000 calories (1600-1900).

[c]Only one subject received more than 0.14 mg/kg.

[d]Only one subject received less than 0.16 mg/kg.

[e]Estimated value.

Source: Ref. 45.

Horwitt, which suggested a minimum daily requirement of approximately 9 mg/day, and the earlier data of Goldberger, which suggested a minimum daily requirement in excess of 12 mg/day, was thought to be due to a difference in caloric intake. Goldberger supplied 3000 calories/day, Horwitt between 2000 and 2800 calories/day, and Goldsmith between 2150 and 2300 calories/day. Goldsmith therefore believed that the niacin requirement should probably be calculated per 1000 calories (53). (Table 4). The minimum requirement should then be about 4.5 mg/1000 calories. A more reproducible estimate related to body weight (50,59, 65,66). A figure of 0.15 mg/kg appeared to be the borderline between deficiency and sufficiency (Table 4). Horwitt (55) and Goldsmith (53) also attempted to estimate retrospectively the niacin and tryptophan content of other older experiments. All data were in rough agreement with the estimates in their direct experiments.

In infants, a niacin equivalent of 6 mg tryptophan/1000 calories in a casein niacin-free formula was considered adequate while 4 mg tryptophan/1000 calories was inadequate, as judged by excretion levels of N^1-methylnicotinamide (67). When human milk is used as a gauge, 8 niacin equivalents/1000 calories must be adequate, two-thirds of which comes from tryptophan (68).

The average American diet supplies 16–34 mg niacin equivalents; 500–1000 mg as tryptophan and 8–17 mg as niacin proper (44). The recommended allowances for niacin are summarized in Table 5. They are expressed as a total amount, relative to body weight.

The estimates for pregnancy and lactation are based on assumptions that the increased excretion of tryptophan metabolites in pregnancy (42) might indicate an increased efficiency in conversion of tryptophan to niacin. In other words, more tryptophan is needed per se, but a higher niacin equivalent is present. However, an opposing idea is that more methylation of nicotinamide to N^1-methylnicotinamide occurs, and therefore more niacin is needed (68,69). During lactation there is a loss of about 1.6 mg of preformed niacin per 850 ml (42). The actual increases recommended are based more on the increased caloric demand than a known increased requirement. In pregnancy the recommended increased energy intake is 300 calories daily or 2 niacin equivalents. In lactation the recommended increase in caloric intake is 500 kcal, thus suggesting an increase in 3.3 mg niacin equivalents. This coupled with the 2.6 mg daily loss in milk suggests an increase of 5 mg niacin equivalents.

B. Animals

Knowledge of the actual vitamin requirements for domestic animals is of great commercial importance. A large body of literature has documented the needs for various species. The availability of the D- and L-isomers of tryptophan in various species has been briefly reviewed. However, by far the major portion of tryptophan in a natural diet is the L-isomer form, so that in practical nutrition (in contrast to experimental nutrition) only the L-isomer needs to be considered.

Table 5 Recommended Daily Dietary Allowances[a]

	Age (years)	Weight (kg)	Height (cm)	Niacin (mg NE)[b]
Infants	0.0–0.5	6	60	6
	0.0–1.0	9	71	8
Children	1–3	13	90	9
	4–6	20	112	11
	7–10	28	132	16
Males	11–14	45	157	18
	15–18	66	176	18
	19–22	70	177	19
	23–50	70	178	18
	51+	70	178	16
Females	11–14	46	157	15
	15–18	55	163	14
	19–22	55	163	14
	23–50	55	163	13
	51+			
Pregnancy	–	–	–	+2
Lactating	–	–	–	+5

[a]The allowances are intended to provide for individual variations among most normal persons as they live in the United States under usual environmental stresses. Diets should be based on a variety of common foods in order to provide other nutrients for which human requirements have been less well defined.

[b]NE (niacin equivalent) is equal to 1 mg niacin or 60 mg of dietary tryptophan.

Source: Ref. 44.

It is important to realize that certain species cannot use tryptophan as a source of niacin. The best known example is the cat (70), but the mink, too, cannot utilize tryptophan as niacin precursor (71). As discussed earlier, this is to a large degree due to the ratio of the rate of 3-hydroxyanthranilic acid conversion to α-amino-β-carboxymuconic-ε-semialdehyde. If that latter conversion is too rapid, not enough intermediate accumulates to allow the spontaneous conversion to quinolinic acid (72).

As expected, the range of niacin requirement is a continuum. In poultry the range is very wide, from 27 mg/kg of diet for growing chickens, to 55 mg/kg for ducks, and 70 mg/kg for turkeys and pheasants (73). Again the activity of the picolinic acid carboxylase seems to correlate with the dietary need (74). When chickens were bred for high or low nicotinic acid requirement, which is feasible by feeding diets low in nicotinic acid (74), the picolinic acid formation correlated with the nutritional requirement (74,76,77).

The rat is the primary animal for which experimental data are obtained in order to understand mammalian nutrition in general and human nutrition in particular. It is fortuitous that the rat has similar niacin equivalence for L-tryptophan and that therefore data can be translated to humans.

III. AMINO ACID IMBALANCE AND NIACIN NUTRITURE

Amino acid imbalance is a term commonly used to designate a relative deficiency of one essential amino acid when one or more amino acids are in excess (78). This concept arose out of early experiments in which attempts were made to utilize gelatin as a major protein source when supplemented with various amino acids (79). These experiments were performed before the discovery of threonine (8), but even using proteins now known to contain threonine, it was still impossible to supplement gelatin to such a degree that satisfactory growth could be obtained (80). In this context the germane examples of amino acid imbalance are those in which excess additions of certain amino acids depress growth in rats while tryptophan and/or niacin can restore growth. Data derived from rat nutrition are legion but the relevance to the human situation to date is unclear. The phenomenon of amino acid imbalance is, in fact, readily provoked in a large number of species, including bacteria (81).

While amino acid imbalances can be created in a variety of ways, thereby generating relative deficiencies for a number of amino acids, the most frequently cited and investigated problem is the effect of amino acid imbalance on the tryptophan requirement, and thus the niacin requirement. The earliest studies in this area introduced the experimental setting of a niacin-free, 9% casein diet supplemented with gelatin, acid-hydrolyzed casein, or specific amino acids. The supplements generated a growth inhibition which could be prevented by adding niacin to the diet. Elvehjem and associates therefore claimed that such diet modification increased the requirements for niacin (82–84). The specific amino acids most markedly affecting the niacin–tryptophan requirements were found to be threonine and cystine/methionine (84). Singal and co-workers obtained the same results for threonine (85,86). The effects of amino acid additions to a 9% casein diet are observed with as low a supplementation as 2% casein hydrolysate; that effect could be overcome by niacin, but the effects of niacin and tryptophan are to a degree separate. At

12% casein hydrolysate, tryptophan *and* niacin both are required (87, 88) (Table 6). At 18% casein, threonine, cystine, and tryptophan are adequate. The effect of threonine and cysteine is also seen to a degree with phenylalanine (89). The mouse does not satisfy its requirement for niacin on a 9% casein diet at all (90), so that the phenomenon is not general across all species.

These effects of relative tryptophan and niacin deficiency on growth are only part of the phenomenon. If a 9% casein diet supplemented with niacin is continued, fatty livers develop in rats (91–93). This effect is primarily mediated by methionine, and is counteracted by threonine and to a lesser extent glycine (94). Using cereals as the sole source of protein, similar effects on fat content of liver are noted with rice but not with corn or wheat, suggesting that natural products do not always have the effect which is predicted from their amino acid composition (95,96).

Another amino acid imbalance can be precipitated by adding excessive leucine at 3% to a 9% casein diet. Significant growth retardation is observed in rats, which is not seen on an 18% casein diet. The active amino acid in casein is isoleucine. DL-isoleucine (0.4%) will counteract approximately 1.5% of L-leucine (81). When the leucine intake is raised even further, isoleucine alone will not counteract leucine, and valine is

Table 6 Effect of Tryptophan Imbalance Produced by Casein Hydrolysate Addition to 9% Casein Diet[a]

Diet supplements				
Casein hydrolysate (%)	Niacin (mg/kg)	DL-Tryptophan (%)	Average gain 4 weeks (g/rat)	Food/g gain (g)
–	–	–	21	7.38
–	20	–	73	3.85
2	–	–	14	9.75
2	20	–	90	3.15
2	20	0.05	104	3.08
12	20	–	47	4.62
12	20	0.10	81	3.41

[a]All diets contained 0.2% choline chloride, 0.30% L-cystine, 1.0% corn oil, complete mineral and vitamin supplements (except niacin, folacin, and vitamin B_{12}), and sucrose to 100%.
Source: Ref. 62.

also required (97). Valine and isoleucine actually inhibit leucine utilization (97). Again the effect seen in a pure amino acid diet may not be reproduced using cereals. Zein, the corn protein, has a leucine/isoleucine ratio that should not fire antagonism, but a definite growth response is seen in rats when isoleucine is added to a zein-based diet. This has been explained on the basis of unavailability of isoleucine in zein (98).

The toxicity of leucine is relevant to niacin nutriture because the excess leucine in corn has been postulated as an etiological factor in pellegra by Gopalan et al. (99). As already discussed, these claims are not uniformly accepted. While there is significant evidence for the effect of leucine and isoleucine in the experimental setting, the extrapolation to the etiology of pellagra is still tenuous. A recent balance study in humans showed that there was no consistent effect of L-leucine or vitamin B_6 supplements on the excretion of any of the metabolites measured (99).

IV. DIETARY DISTRIBUTION

The distribution of niacin in food is as wide as are the ubiquitous cofactors in which it is found. Most food tables indicate the niacin equivalents; thus, niacin represents the sum of mg niacin + 1/60 X mg tryptophan. For most dietetic purposes, this calculation is adequate, but for more detailed studies, it is inadequate since the presence of bound niacin and amino acid imbalances may lower the actual availability of niacin. There are multiple food tables available, and there are now computerized data bases.

The content of various elemental diets and proprietary formulas is beginning to be a major concern in the nutriture of the very sick. There are a large number of formulations, and their indications for use vary according to the nutritional status of the patient (Table 7) (100). All formulas contain adequate protein (tryptophan) and niacin. To date, no pellagra-like symptoms have been recorded with their use in adequate quantities. The following two tables which summarize the composition of these special formulas are derived from the diet manual of the M.D. Anderson Hospital (101). Table 8 supplies data on elemental diets and proprietary formulas.

Intravenous hyperalimentation is now frequently used to satisfy the nutritional requirements of sick patients. This has become an accepted and sophisticated method of nutritional support and rehabilitation (102). Here the tryptophan content of the various amino acid sources varies widely. As an example, the solutions used at the M. D. Anderson Hospital range in their tryptophan/leucine ratio from 1:12.7 (50:636) for Aminosol to 1:3.5 (152:526) for Travasol (Table 9) (103). However, pellagra-like symptoms have rarely been described in patients to date.

Table 7 Therapeutic Placement of Formulas for Enteral Hyperalimentation

General	Milk Base	Lactose free	Chemically defined	Elemental	Parenteral
Blenderized	Meritine	Ensure	Precision LR	Vivonex	TPN
Compleat B	Nutri 1000	Ensure Plus	Precision HN	Vivonex HN	Peripheral Systems
Formula 2	Carnation Instant Breakfast	Isocal	Isotonic	Vipep	
Vitaneed		Nutri 1000 LF	Flexical		
		Sustacal	Vital		
		Citrotein			
		Osmolite			
		Renu			
		Magnacal			
Functional gut			Afunctional gut		

Table 8A Composition of Proprietary Supplements: Intact Protein, Milk-Containing Products

Product/100 cc	Sustacal	Sustagen + Water	C.I.B.*[d] + Milk	Compleat-B	Formula 2	Nutri-1000
Protein (g)	6.0	10.4	5.8	4.3	3.8	4.0
Protein source	Casein, soy isolate	nonfat dry milk, Ca caseinate	milk, soy isolate, nonfat dry milk, Na caseinate	beef, casein	nonfat dry milk, beef, casein, albumin	skim milk
Fat (g)	2.3	1.6	3.1	4.3	4.0	5.5
Fat source	soy oil	milk fat	milk fat	corn oil, beef fat	corn oil, beef fat, egg yolk	corn oil
CHO (g)	13.8	29.6	13.5	12.8	12.3	10.1
CHO source	corn syrup, solids, sucrose, lactose	corn syrup solids, sucrose, lactose glucose	corn syrup solids, sucrose, lactose	maltodextrins, fruits, sucrose, veg., orange juice, lactose	sucrose, lactose, wheat flour, orange juice, vegetables	dextrin-maltose sucrose, lactose, dextrose
Kcal/100 cc	100	173	120	107	100	106
Lactose (g)	1.67	8.6	NAPH[e]	2.4	3.74	5.01
mOsm/kg H_2O	625	1200	2000	4057490	4357510	2.3/52.8
Na mEq/mg	4.0/92.5	5.2/120	4.0/93.1	7.4/170	2.7/63	3.8/147.8
K mEa/mg	5.3/205.6	8.1/315.6	7.0/273.5	3.6/140	4.9/191	3.8/147.8
Ca MeQ/mg	5.0/100	15.8/315.6	7.8/156.5	3.4/67	5.5/110	6.0/120.5
P mEq/mg	5.9/91.7	15.2/235.6	9.8/148.5	9.8/150	6.1/95	6.1/95.1
Residue level	low	low	low	medium	medium	low

Table 8A (Continued)

Product/100 cc	Sustacal	Sustagen + Water	C.I.B.*[d] + milk	Compleat-B	Formula 2	Nutri-1000
Volume need to meet 100% RDAs for protein, vitamins, and minerals	1080 cc	960 cc	1373 cc	1600 cc	2000 cc	1920 cc
Manufacturer	Mead Johnson	Mead Johnson	Carnation	Doyle	Cutter	Cutter
General comment	a,c	a,c	a*	c	b,c	a,c

[a]Recommended for oral feeding due to good patient acceptance.

[b]Not recommended for oral feeding due to poor patient acceptance.

[c]Recommended for tube feeding.

[d]Carnation Instant Breakfast.

[e]NAPH; Not available, presumed high.

*Flavor packets available for increased palatability.

Source: Composition data calculated from manufacturer's product information based on standard dilution: vanilla-flavored supplements used.

Table 8B Intact Protein, Lactose-Free

Product/100 cc	Ensure*	Ensure Plus**	Isocal	Osmolite
Protein (g)	3.7	5.5	3.4	3.7
Protein source	Casein, soy isolate	Na & Ca caseinates,	Casein, soy isolate	Na & Ca caseinates,
Fat (g)	3.7	5.3	4.4	3.8
Fat source	corn oil	corn oil	soy oil, MCT oil	MCT oil, corn oil, soy oil
CHO (g)	14.5	19.7	13.0	14.3
CHO source	corn syrup solids, sucrose	corn syrup solids, sucrose	corn syrup solids	corn syrup solids (glucose polymers)
Kcal/100 cc	100	150	100	106
Lactose (g)	0	0	0	0
mOsm/kg H_2O	450	600	350	300
Na mEq/mg	3.2/74	4.6/106	2.3/52.1	2.4/54.2
K mEq/mg	3.3/126.8	4.9/190	3.3/130	2.2/87.5
Ca mEq/mg	2.6/52.8	3.2/63	3.1/62.5	2.7/54.2
P mEq/mg	3.4/52.8	4.0/63	3.4/52.1	3.5/54.2
Residue level	low	low	low	low
Volume needed to meet 100% RDAs for protein, vitamins, and minerals	1920 cc	1540 cc	1920 cc	2000 cc

Table 8B (Continued)

Product/100 cc	Ensure*	Ensure Plus**	Isocal	Osmolite
Manufacturer	Ross	Ross	Mead Johnson	Ross
General comment	a,c	c	b,c	b,c

[a]Recommended for oral feeding due to good patient acceptance.

[b]Not recommended for oral feeding due to poor patient acceptance.

[c]Recommended for tube feeding due to good patient tolerance.

*Flavor packets available for increased palatability.

**Patient must be gradually increased to full strength tube feeding due to hypertonicity of solution.

Source: Composition data calculated from manufacturer's product information.

Table 8C Defined Formula Diets: Products Containing Amino Acids, Hydrolyzed Protein, or Protein Isolates

Product/100 cc	Flexical*	Vivonex*	Vivonex HN*	Precision HN**	Precision LR**	Precision Isot.+	Vital++
Protein (g)	2.24	2.1	4.2	4.4	2.63	2.9	4.2
Protein source	amino acids, casein hydrolysate	amino acids	amino acids	egg albumin	egg albumin	egg albumin	hydrolyzed soy, whey, meat, free amino acids
Fat (g)	3.4	0.1	0.1	.05	.08	3.0	1.03
Fat source	soy oil, MCT oil, soy lecithin	safflower oil	safflower oil	soybean oil, mono- and diglycerides	soybean oil, mono- and diglycerides	soybean oil, mono- and diglycerides	sunflower oil
CHO (g)	15.4	22.6	21.0	21.8	24.9	14.4	18.5
CHO source	corn syrup solids, modified tapioca starch, citrate	glucose oligosaccharides	glucose, glucose oligosaccharides	maltodextrins, sucrose	maltodextrins, sucrose	maltodextrins, sucrose, glucose oligosaccharides	glucose oligosaccharides polysaccharides
Kcal/100 cc	100	100	100	100	110	100	100
Lactose (g)	0	0	0	0	0	0	0
mOsm/kg H_2O	723	500	850	557	525	300	450
Na mEq/mg	1.6/36	3.7/86	3.4/77	4.4/100	3.0/70	3.5/80	2.0/45
K mEq/mg	3.2/124	3.0/117	1.8/70	2.3/90	2.2/87	2.5/96	3.5/137
Ca mEq/mg	3.0/60	2.2/44	1.3/27	2.6/40	3.7/58	4.1/64	5.0/78.4
Residue level	low	low	low	low	low	low	low

Table 8C (Continued)

Product/100 cc	Flexical*	Vivonex*	Vivonex HN*	Precision HN**	Precision LR**	Precision Isot.+	Vital++
Volume needed to meet 100% RDAs for protein, vitamins, and minerals	2000 cc	1800 cc	3000 cc	2850 cc	1710 cc	1560 cc	1500 cc
Manufacturer	Mead Johnson	Eaton	Eaton	Doyle	Doyle	Doyle	Ross
General comment	b,c	b,c	b,c	a,c	a,c	a,c	a,c

[a]Recommended for oral feeding due to good patient acceptance.

[b]Not recommended for oral feeding due to poor patient acceptance.

[c]Recommended for tube feeding due to good patient tolerance.

*Unflavored, product flavors available.

**Citrus flavor.

+Vanilla flavor.

++Banana flavor.

Source: Composition data calculated from manufacturer's product information based on standard dilution.

Table 8D Supplements[a]

Product/ 100 cc	Polycose	MCT oil	Pedialyte	Citrolein	Delmark Eggnog	Controlyte	Sustagen + Milk	Susta Protein
Protein (g)	0	0	0	4.0	6.0	trace	5.6	30
Protein source	–	–	–	egg albumin	nonfat day milk, egg albumin	–	nonfat dry milk, Ca caseinate, milk	82% hydrolyzed collagen, 18% free amino acids
Fat (g)	0	93.3	0	.17	3.6	9.6	3.8	0
Fat source	–	MCT oil	–	mono- and diglycerides, soybean oil	cottonseed oil, soybean oil, egg yolk	soybean oil	milk fat	–
CHO (g)	50	0	5.0	12.0	14.8	28.6	20.4	0
CHO source	modified corn starch	–	dextrose	maltodextrins, sucrose, glucose	maltodextrins, lactose, sucrose	maltodextrins	sucrose, lactose, corn syrup solids, glucose	–
Kcal/ 100 cc	200	775	20	65.3	116	200	98.2	120
Lactose (g)	0	0	0	0	NAPH[b]	0	NAPH[b]	0
mOsm/ kg H_2O	570	NA[b]	80	500	NA[b]	570	NA[b]	NA[b]

Table 8D (Continued)

Product/ 100 cc	Polycose	MCT oil	Pedialyte	Citrolein	Delmark Eggnog	Controlyte	Sustagen + Milk	Susta Protein
Na mEq/ mg	2.7/62	0	3.0	2.9/68	4.0/91.6	.26/6.0	3.2/74	trace
K mEa/mg	⩽ 1 mEq	0	2.0	1.7/68	6.7/262.5	.08/3.2	5.2/203	trace
Ca mEq/mg	⩽ 1 mEq	0	0.4	5.2/104	10.2/204	.08/1.6	9.0/179	trace
P mEa/mg	⩽ 1 mEq	0	0	6.74/104	10.5/162.5	.10/1.6	9.0/139	trace
Residue level	low	low	low	low	NA[b]	low	NA	low
Type of supplement	Caloric supplement	Caloric supplement	Caloric and electrolyte supplement	Protein minearl, and vitamin supplement	Protein and calorie supplement	Low pro- and electrolyte; high calorie supplement	Protein and calorie supplement	Liquid protein supplement
Manufacturer	Ross	Mead Johnson	Ross	Doyle	Delmark	Doyle	Mead J ohnson	Mead J ohnson

[a]Protein quality consistent with ideal A. A. pattern.

[b]NAPH, Not available, presumed high; NA, Not available.

Source: Composition data calculated from manufacturer's product information based on standard dilution; vanilla flavor used.

Table 9 Comparison of IVH Solutions Stocked at M. D. Anderson

	Aminosol 5% (700 ml)	Travasol 8.5% with Electrocytes (500 ml)	Travasol 8.5% without Electrocytes (500 ml)	Aminosyn 7% (500 ml)	Aminosyn 10% (500 ml)	Freamine III 8.5% (500 ml)	Nephramine (250 ml)
Protein source	Fibrin	Soybean	Soybean	Soybean	Soybean	Soybean	Soybean
Gram utilizable nitrogen	3.9	7.15	7.15	5.5	7.86	6.25	1.5
Calorie/nitrogen ratio (final dilution)	194:1	119:1	119:1	155:5	108:1	120:1	815:1
Essential amino acids (mg/100 ml)							
Leucine	636	526	526	660	940	770	880
Phenylalanine	100	526	526	310	440	480	880
Methionine	100	492	492	280	400	450	880
Lysine	400	492	492	520	720	620	640
Isoleucine	218	406	406	510	720	590	560
Valine	163	390	390	560	800	560	650
Threonine	232	356	356	370	520	340	400
Tryptophan	50	152	152	120	166	130	200
Semiessential amino acids (mg/100 ml)							
Histidine	116	372	372	210	300	240	–
Arginine	290	880	880	690	980	810	–

Table 9 (Continued)

	Aminosol 5% (750 ml)	Travasol 8.5% with Electrocytes (500 ml)	Travasol 8.5% without Electrocytes (500 ml)	Aminosyn 7% (500 ml)	Aminosyn 10% (500 ml)	Freamine III 8.5% (500 ml)	Nephramine (250 ml)
Electrolytes (mgEq in final dilution)							
Na	7	35	–	–	–	5	1.5
K	1.2	30	–	2.7	2.7	–	–
Ca	–	–	–	–	–	–	–
Mg	–	5	–	–	–	–	–
Cl	–	35	17	–	–	N2	–
HPO_4 (Phosphate)	–	30	–	–	10	10	–
Acetate	–	65	26	–	–	37	–

V. ALTERNATIVE SOURCES OF NICOTINIC ACID ACTIVITY

The niacin requirement of humans and most other animals can be satisfied with many substances other than nicotinic acid and nicotinamide. In microorganisms, the number of sources seems even greater, though it is possible that higher organisms are similar to microbes in this regard, but just have not been challenged.

In animals the compounds mentioned previously, which are biosynthetic precursors of niacin, are active in replacing or generating niacin in the whole animal: tryptophan, N^{α}-formylkynurenine, kynurenine, 3-hydroxykynurenine, 3-hydroxyanthranilic acid, and quinolinic acid. Compound substances and coenzymes are likewise active: nicotinamide riboside, nicotinamide mononucleotide, nicotinamide-adenine dinucleotide (NAD), and nicotinamide-adenine dinucleotide phosphate (NADP). Certain other pyridine derivatives are converted in vivo to nicotinic acid activity. It was shown early that 3-acetylpyridine, which is a niacin antagonist, is converted by the dog into N^1-methylnicotinamide (104). Furthermore, erythrocytes, liver, and kidney rapidly metabolize the analogue (105). When injected into mice, it generates increased quantities of NAD almost to the degree of nicotinamide itself (106).

As early as 1938, β-picoline (3-methylpyridine) was found to be partially effective in healing black tongue in dogs (107). In the duckling, β-picoline, pyridine-3-carbinol, and pyridine-3-aldehyde are active as niacin (108). β-Picoline, pyridine-3-carbinol, and pyridine-3-aldehyde are all active in generating increased NAD levels but not quite to the degree as acetylpyridine and pyridine-3-methylcarbinol (109). Other substances, active as inducers of NAD are the monomethyl, monoethyl and dimethyl amides of nicotinic acid. Coramine (the *N*-diethylamide of nicotinic acid) was too toxic to induce NAD formation (110). Some caution must be exercised in assuming that these data showing an NAD increase automatically reflect a conversion of the pyridine derivative to nicotinic acid since nicotinamide is an inducer of glucocorticoid inducible enzymes, including tryptophan pyrrolase (110). The doses of nicotinamide, and thus by analogy, of the pyridine derivatives required to achieve the NAD biosynthesis are nonphysiologic (106,109). However, the growth of ducklings on these substances implies potential conversion. Administration of one structural analogue of nicotinic acid, nicotinyl hydroxamic acid, leads to increased NAD in the liver after injection (111), probably through its conversion to nicotinic acid, since a mitochondrial enzyme system exists which can reduce nicotinic hydroxamic acid and the nicotinic hydroxamic acid-adenine dinucleotide to nicotinamide or NAD (112). Since the NAD analogue can readily be formed in vitro (113), whether the pathway of hydroxamate conversion to nicotinamide is direct or occurs at the NAD level, is not certain. There are far more substances that can replace niacin for bacteria than for mammals. Again, the N-substituted nicotinic amides and esters of nicotinic acid are active. These substances were synthesized by Badgett et al. (114,115). Thionicotinamide was also found to be active.

Table 10 Activity of Some Nicotinic Acid Derivatives and Related Compounds

Compound	Activity[a]
Nicotinic acid	++++
Methyl nicotinate	++++
Ethyl nicotinate	++
Propyl nicotinate	+
Butyl nicotinate	+
Quinolinic acid (pyridine-2,3-dicarboxylic acid)	+,0[b]
Nicotinuric acid (nicotinylglycine)	+++
Nicotinamide	++++
Nicotinic acid *N*-methyl amide	+++
Nicotinic acid *N*-ethyl amide	++,+,0[c]
Nicotinic acid *N*-diethyl amide (coramine)	+,0[d]
Thionicotinamide	++
Nicotinamidoxime	++++[e]
Thiazole-5-carboxylic acid	+
Thiazole-5-carboxylamide	+
Guvacin	+
Piperidine 3-carboxylic acid	+

Inactive compounds
(with possibly a few exceptions, as noted)

Isonicotinic acid (pyridine-4-carboxylic acid)	Nipecotic acid[k]
Picolinic acid (pyridine-2-carboxylic acid)	Arecoline
2-Aminonicotinic acid	Pyridine
6-Methylnicotinic acid	Nicotine[l]
Nicotinamide methochloride	Cinchomeronic acid (3,4-pyridine-dicarboxylic acid)
Nicotinamide methiodide[f]	Dinicotinic acid (pyridine-3,5-dicarboxylic acid)
Pyridine-3-sulfonic acid[g]	2,4,6-Trimethylpyridine-3,5-dicarboxylic acid
Pyridine-3-nitrite (nicotininonitrile)[h]	2-Methyl-3-hydroxy-4,5-bis (hydroxymethyl)pyridine (vitamin B_6)
3-Aminopyridine	Pyrazine monocarboxylic acid
3-Methylpyridine (β-picoline)	Pyrazine-2,3-dicarboxylic acid
4-Methylpyridine (γ-picoline)	Arecaidine
Ethylnicotinoacetate	
Trigonelline[j]	

Table 10 (Continued)

---

[a]The number of + signs represents roughly the comparative activity: ++++ = fully active as judged by nicotinic acid or nicotinamide, + = about 1% or less of full activity, only slightly active. The organisms tested were *Staphylococcus*, *Shigella*, *Proteus* and *Lactobacillus* plus those mentioned in the footnotes below.

[b]Quinolinic acid supported growth of 3 dysentery bacilli at 10^{-4} *M* but not at 10^{-5} *M*. It did not support growth of *Staphylococcus aureus*. Reported active for *Proteus* at 10^{-4} *M* but found inactive for a large collection of *Proteus* cultures. It showed some activity for *Proteus* and *L. plantarum* in amounts of 0.02–0.2 μg per ml. For *Candida pseudotropicalis* 2512 it was reported inactive.

[c]Light growth or negative results were reported for *L. plantarum* and *Proteus*.

[d]Coramine did not support growth of *S. aureus*. It supported growth of *Proteus* and dysentery bacilli at levels of 10^{-5} and 10^{-4} *M*, respectively, but not at lower levels. For *L. plantarum*, coramine showed 0.01 to 0.03% of the activity of nicotinic acid. For *C. pseudotropicalis* 2512 it was inactive.

[e]Nicotinamidoxime was reported quite active for both *Proteus* and *Staphylococcus aureus*, but the corresponding derivative of isonicotinic acid was inactive.

[f]Nicotinamide methiodide, in 10^{-4} to 10^{-6} *M* concentrations was inactive for dysentery bacilli but was found slightly active, at about 10^{-4} *M*, for *Proteus vulgaris* and *L. plantarum*.

[g]This compound was reported to support growth of *Proteus* at 10^{-4} *M*. It supported light growth of *Proteus* and *L. plantarum*. Some reported activity but others found it inactive.

[h]Pyridine-3-nitrile was reported to support growth of *Proteus* at 10^{-4} *M*. Others found that it supported good growth of both *P. vulgaris* and *S. aureus*, but the nitrile of isonicotinic acid showed no growth-promoting activity. Nicotinonitrile was quite active for *C. pseudotropicalis* 2512.

[i]3-Acetylpyridine substituted to some extent for some strains of dysentery bacilli which required nicotinic acid.

[j]Trigonelline possessed quite high activity as a nicotinic acid substitute for *C. pseudotropicalis* 2512.

[k]Nipecotic acid has usually shown little or no activity for dysentery bacilli. However, it was reported to show full activity for *L. plantarum*.

[l]Nicotine did not support growth of *S. aureus* at 10^{-5} *M* but for *P. vulgaris* there was growth-promoting activity at 1.6×10^{-5} *M* or higher concentrations.

A tabulation of this spectrum of activity, adopted from Koser (116), is shown in Table 10, with the addition of one substance not included by Koser: guvacine (1,2,5,6-tetrahydronicotinic acid) could replace nicotinic acid, but arecaidine (N^1-methyl guvacine) is inactive in microorganisms tested (117).

As can be seen from the table, claims have even been made for non-pyridine substances, such as pyridine sulfonic acid, 3-piperidinecarboxylic acid (118), and thiazole 5-carboxylic acid (119). The latter was investigated extensively during the development of sulfathiazole as an antimicrobial. Thiazole 5-carboxylic acid and its amide can act as an antimetabolite for *Itaphylococci* (119). On the other hand, under many circumstances the thiazole analogue of niacin can act as a niacin-sparing or possibly even as a niacin-replacing factor. Thiazole 5-carboxylic acid or its amide was said to replace niacin for *Shigella dysenteriae* (118). It presumably will also, in low concentrations, replace niacin for *Staphylococcus* (119), as will the amide. Finally, thiazole 5-carboxylic acid will reverse the inhibition of pyridine 3-sulfonic acid for a *Lactobacillus* (120), while the amide was said to have the same effect for a *Proteus* species (121). Whether all these actions can be traced to a niacin-sparing action or an actual niacin-replacing effect has never been decided by experiment, but our current knowledge of the mode of nicotinic acid action makes these claims for vitamin action seem rather dubious. Therefore, great care in extrapolation from bacterial data to higher animals must be exercised. While bacteria can synthesize nicotinic acid from primitive precursors, they do so only reluctantly when no preformed niacin is available (122). The action of pyridine derivatives could in fact merely represent an enzyme induction phenomenon in the auxotrophs rather than replacement of the vitamin.

REFERENCES

1. L. Blum, *J. Physiol. Chem. 30:*15(1900). Cited by: J. P. Greenstein and M. Winitz, *Chemistry of Amino Acids, Vol. I,* Wiley, New York, 1961.
2. E. G. Willcock and F. G. Hopkins, *J. Physiol. 35:*88(1906).
3. R. Wheeler, *J. Exp. Zool. 15:*209(1913).
4. E. Abderhalden, *Z. Physiol. Chem. 77:*22(1912).
5. E. Abderhalden, *Z. Physiol. Chem. 83:*449(1913).
6. T. B. Osborne and L. B. Mendel, *J. Biol. Chem. 17:*325(1914).
7. J. F. McClendon, *Proc. Soc. Exp. Biol. Med. 28:*915(1930–1931).
8. R. H. McCoy, C. E. Meyer, and W. C. Rose, *J. Biol. Chem. 112:*238(1935–1936).
9. C. E. Meyer and W. C. Rose, *J. Biol. Chem. 165:*721(1936).
10. W. C. Rose, *Physiol. Rev. 18:*109(1938).
11. W. C. Rose, J. Oesterling, and M. Womack, *J. Biol. Chem. 176:*753(1948).

12. C. P. Berg and M. Potgieter, *J. Biol. Chem. 94:*661(1931–1932).
13. V. du Vigneaud, R. R. Sealock, and C. van Etten, *J. Biol. Chem. 98:*565(1932).
14. C. P. Berg, *J. Biol. Chem. 101:*373(1934).
15. M. Womack and W. C. Rose, *J. Biol. Chem. 171:*37(1947).
16. R. R. Langner and C. P. Berg, *J. Biol. Chem. 214:*699(1955).
17. A. A. Albanese, J. E. Frankston, and V. Irby, *J. Biol. Chem. 100:*31(1945).
18. A. A. Albanese and J. E. Frankston, *J. Biol. Chem. 155:*101(1944).
19. W. C. Rose, G. F. Lamberg, and M. J. Coon, *J. Biol. Chem. 211:* 815(1954).
20. H. R. Baldin and C. P. Berg, *J. Nutr. 39:*203(1949).
21. J. M. Price and R. R. Brown, *J. Biol. Chem. 222:*835(1956).
22. L. V. Hankes, R. R. Brown, T. Leklem, M. Schmaeler, and T. Jesseph, *J. Inv. Derm. 58:*85(1971).
23. R. R. Langner and C. M. Volkmann, *J. Biol. Chem. 213:*433(1955).
24. W. C. Rose, M. J. Coon, G. F. Lambert, and E. E. Howe, *J. Biol. Chem. 213:*(1955).
25. W. G. Gordon, and R. W. Jackson, *J. Biol. Chem. 110:*151(1935).
26. W. G. Gordon, *J. Biol. Chem. 129:*309(1939).
27. W. C. Rose, *Fed. Proc. 8:*546(1949).
28. L. E. Holt, Jr., A. A. Albanese, J. E. Brumback, Jr., C. Kajdi, and D. M. Wangerin, *Proc. Soc. Exp. Biol. Med. 48:*726(1941).
29. W. M. Cox, A. J. Mueller and D. Fickas, *Proc. Soc. Exp. Biol. Med. 51:*303(1942).
30. W. C. Rose, W. J. Haines, and D. T. Warner, *J. Biol. Chem. 206:* 421(1954).
31. L. E. Holt, Jr., A. A. Albanese, J. E. Frankston, and V. Irby, *Bull. Johns Hopkins Hosp. 75:*353(1944).
32. C. W. Denko and W. E. Grundy, *J. Lab. Clin. Med. 34:*839(1949).
33. V. R. Young, M. A. Hussein, E. Murray, and N. S. Scrimshaw, *J. Nutr. 101:*45(1971).
34. R. M. Leverton, N. Johnson, L. Pazur, and J. Ellison, *J. Nutr. 58:*219(1956).
35. H. Fisher, M. K. Brush, and P. Griminger, *Am. J. Clin. Nutr. 22:*1190(1969).
36. A. A. Albanese, L. E. Holt, Jr., V. Irby, S. E. Snyderman, and M. Lein, *Bull. Johns Hopkins Hosp. 80:*158(1947).
37. S. E. Snyderman, A. Boyer, S. V. Phansalkar, and L. E. Holt, Jr., *Am. J. Dis. Child. 102:*163(1961).
38. I. Nakagawa, T. Takahashi, T. Suzuki, and K. Kobayashi, *J. Nutr. 80:*305(1963).
39. I. Nakagawa, T. Takahaski, T. Suzuki, and K. Kobayashi, *J. Nutr. 86:*333(1965).
40. K. Tonsirin, V. R. Young, M. Miller, and N. S. Scrimshaw, *J. Nutr. 103:*1220(1973).
41. M. I. Irwin and D. M. Hegsted, *J. Nutr. 101:*539(1971).

42. A. M. Wertz, M. B. Derby, P. K. Ruttenberg, and G. P. French, *J. Nutr. 68:*583(1959).
43. M. I. Irwin and D. M. Hegsted, *J. Nutr. 101:*539(1971).
44. Food and Nutrition Board. Committee on Dietary Allowances, *Recommended Dietary Allowances,* 9th Revised Ed., Natl. Acad. Sci., Washington, D.C., (1980).
45. National Research Council. Food and Nutrition Board, *Improvement of Protein Nutriture,* Natl. Acad. Sci., Washington, D. C., 1975.
46. H. E. Clark, P. Meyers, K. Goyal, and J. Rinehart, *Am. J. Clin. Nutr. 18:*91(1966).
47. H. E. Clark, W. H. Moon, J. L. Malzer, D. F. Birt, and R. L. Pang, *J. Nutr. 104:*121(1974).
48. M. S. Reynolds, M. F. Futrell, and C. A. Baumann, *J. Am. Diet Ass. 29:*359(1957).
49. H. P. Sarett and G. A. Goldsmith, *J. Biol. Chem. 182:*679(1950).
50. R. W. Vilter, J. F. Mueller, and W. B. Beam, *J. Lab. Clin. Med. 34:*409(1949).
51. W. B. Bean, M. Franklin, and K. Parum, *J. Lab. Clin. Med. 38:*167(1951).
52. G. A. Goldsmith, O. N. Miller, and W. G. Unglaub, *Fed. Proc. 15:*553(1956).
53. G. A. Goldsmith, *Am. J. Clin. Nutr. 6:*479(1958).
54. G. A. Goldsmith, O. N. Miller, and W. G. Unglaub, *Nutr. 73:*172(1961).
55. M. K. Horwitt, C. C. Harvey, W. S. Rothwell, J. L. Cutler, and D. Haffron, *J. Nutr. 60* (Suppl. 1):43(1956).
56. V. M. Vivian, .. *Nutr. 82:*395(1964).
57. C. F. Consolazio, H. L. Johnson, H. J. Krzywick, and N. F. Witt, *Am. J. Clin. Nutr. 25:*572(1972).
58. R. R. Brown, W. M. Vivian, M. S. Reynolds, and J. M. Price, *J. Nutr. 66:*599(1958).
59. I. Nakagawa, T. Takahashi, T. Suzuki, and Y. Masana, *J. Nutr. 99:*325(1969).
60. I. Nakagawa, T. Takahashi, A. Sasaki, M. Kajimoto, and T. Suzuki, *J. Nutr. 103:*1195(1973).
61. V. M. Vivian, M. M. Chaloupka, and M. S. Reynolds, *J. Nutr. 66:*587(1958).
62. J. I. Patterson, R. R. Brown, H. Linkswiler, and A. E. Harper, *Am. J. Clin. Nutr. 33:*2157(1980).
63. G. A. Goldsmith, H. P. Sarett, U. D. Register, and T, Gibbens, *J. Clin. Invest. 31:*533(1952).
64. G. A. Goldsmith, H. L. Rosenthal, J. Gibbens, and W. G. Unglaub, *J. Nutr. 56:*371(1955).
65. G. A. Goldsmith, J. Gibbens, W. G. Unglaub, and O. N. Miller, *Am. J. Clin. Nutr. 4:*151(1956).
66. G. A. Goldsmith, *Am. J. Diet. Assn. 32:*312(1956).

67. L. E. Holt, Jr., *Arch. Dis. Child. 31:*427(1956).
68. K. U. G. Toverd, G. Stearns, and I. G. Macy, in *National Research Council Bull. 123:* Nat. Acad. Sci., Washington, D. C., 1950, p. 174.
69. W. J. Darby, W. J. McGannity, M. P. Martin, E. Bridgforth, P. M. Densen, M. M. Kaser, P. J. Ogle, J. A. Newbill, A. Stockell, M. E. Ferguson, O. Touster, G. S. McClellan, C. Williams, and R. O. Cannon, *J. Nutr. 51:*565(1953).
70. A. C. DaSilva, R. Fried, and R. C. DeAngelis, *J. Nutr. 46:*399 (1952).
71. National Research Council, *Nutrient Requirements of Domestic Animals, No. 7. Nutrient Requirements of Mink and Foxes.* Nat. Acad. Sci., Washington, D. C., 1968.
72. L. M. Henderson and P. B. Swan, in *Methods in Enzymology, Vol. XVIII* (D. B. McCormick and L. D. Wright, eds.), Academic Press, New York, 1962, p. 1974.
73. National Research Council, *Nutrient Requirements of Domestic Animals, No. 1. Nutrient Requirements of Poultry, 6th Rev. Ed.* Nat. Acad. Sci., Washington, D. C., 1971.
74. R. N. DiLorenzo, Ph.D. Thesis, Cornell University, Ithaca, New York, 1972.
75. M. C. Nesheim, *Science 22:*290(1966).
76. R. N. DiLorenzo and M. C. Nesheim, *Fed. Proc. 31:*2903(1972).
77. M. C. Nesheim, in *The Effect of Genetic Variance on Nutritional Requirements of Animals,* Nat. Acad. Sci., Washington, D. C., 1975, p. 47.
78. W. D. Salmon, *Am. J. Clin. Nutr. 6:*487(1958).
79. R. W. Jackson, B. E. Sommer, and W. C. Rose, *J. Biol. Chem. 80:*167(1928).
80. S. W. Hier, C. E. Graham, and D. Videin, *Proc. Soc. Expt. Biol. Med. 56:*187(1944).
81. A. E. Harper, D. A. Benton, and C. A. Elvehjem, *Arch. Biochem. Biophys. 57:*1(1955).
82. W. A. Krehl, L. M. Henderson, J. de la Huerga, and C. A. Elvehjem, *J. Biol. Chem. 166*:531(1946).
83. L. M. Henderson, T. Deodhar, W. A. Krehl, and C. A. Elvehjem, *J. Biol. Chem. 170*:261(1947).
84. L. V. Hankes, L. M. Henderson, W. L. Brickson, and C. A. Elvehjem, *J. Biol. Chem. 179*:873(1948).
85. S. A. Singal, V. P. Sydenstricker, and J. M. Littlejohn, *J. Biol. Chem. 171:*203(1947).
86. S. A. Singal, V. P. Sydenstricker, and J. M. Littlejohn, *J. Biol. Chem. 176:*1063(1948).
87. W. D. Salmon, *Arch. Biochem. Biophys. 51:*30(1954).
88. M. E. Sauberlich and W. D. Salmon, *J. Biol. Chem. 214:*463(1955).
89. L. V. Hankes, L. M. Henderson, and C. A. Elvehjem, *J. Biol. Chem. 180:*1027(1949).

90. N. N. Pearson, J. S. Valenzuela, and J. van Eys, *J. Nutr. 66:* 277(1966).
91. S. A. Singal, S. J. Hazan, V. P. Syndenstricker, and J. M. Littlejohn, *J. Biol. Chem. 200:* 867(1953).
92. A. E. Harper, W. J. Monson, D. A. Benton, and C. A. Elvehjem, *J. Nutr. 50:* 383(1953).
93. A. E. Harper, W. J. Munson, D. A. Benton, M. E. Winje, and C. A. Elvehjem, *J. Biol. Chem. 206:* 151(1952).
94. A. E. Harper, D. A. Benton, M. E. Winje, and C. A. Elvehjem, *J. Biol. Chem. 209:* 159(1954).
95. A. E. Harper, M. E. Winje, D. A. Benton, and C. A. Elvehjem, *J. Nutr. 54:* 155(1954).
96. C. A. Elvehjem, *Fed. Proc. 15:* 965(1956).
97. D. A. Benton, A. E. Harper, H. E. Spivey, and C. A. Elvehjem, *Arch. Biochem. Biophys. 60:* 147(1956).
98. D. A. Benton, A. E. Harper, and C. A. Elvehjem, *Arch. Biochem. Biophys. 57:* 13(1955).
99. G. Gopalan and K. S. J. Rao, *Vitamins and Hormones 33:* 505(1970).
100. J. I. Patterson, R. R. Brown, H. Linkswiler, and A. Entharper, *Am. J. Clin. Nutr. 33:* 2157(1980).
101. W. P. Steffee and S. H. Krey, in *Enteral Hyperalimentation of the Cancer Patient* (G. R. Newell and N. M. Ellison, eds.), Raven Press, New York, 1981, p. 367.
102. R. K. Schamburg and J. J. Wollard, in *Diet Manual,* The University of Texas System Cancer Center, M. D. Anderson Hospital and Tumor Institute, Houston, TX. 1980.
103. J. P. Grast, in *Handbook of Total Parenteral Nutrition,* W. B. Saunders, Philadelphia, 1980.
104. W. T. Beher, W. M. Holliday, and O. H. Gaebler, *J. Biol. Chem. 198:* 573(1952).
105. W. T. Beher, S. P. Marfey, W. L. Anthony, and O. H. Gaebler, *J. Biol. Chem. 205:* 521(1953).
106. N. O. Kaplan, A. Goldin, S. R. Humphreys, and F. E. Stolzenbach, *J. Biol. Chem. 226:* 365(1957).
107. D. W. Woolley, F. M. Strong, R. J. Madden, and C. A. Elvehjem, *J. Biol. Chem. 124:* 715(1938).
108. R. van Reen and F. E. Stolzenbach, *J. Biol. Chem. 226:* 373(1957).
109. R. L. Blake, S. L. Blake, H. H. Lok, and E. Kun, *Mol. Pharm. 3:* 412(1967).
110. N. O. Kaplan, A. Goldin, S. R. Humphrey, M. M. Ciotti, and F. E. Stolzenbach, *J. Biol. Chem. 219:* 287(1956).
111. B. M. Anderson. Ph.D. Thesis. The Johns Hopkins University, Baltimore, 1958.
112. P. F. Hirsch and N. O. Kaplan, *J. Biol. Chem. 236:* 926(1961).
113. B. M. Anderson and N. O. Kaplan, *J. Biol. Chem. 234:* 1226(1959).
114. C. O. Badgett, R. C. Provost, Jr., C. L. Ogg, and C. F. Woodward, *J. Am. Chem. Soc. 67:* 1135(1945).

115. C. O. Badgett, and C. F. Woodward, *J. Am. Chem. Soc. 69:*2907 (1947).
116. S. A. Koser in *Vitamin Requirements of Bacteria and Yeast,* Charles C. Thomas, Springfield, Illinois, 1971, pp. 124–125.
117. H. von Euler, B. Hoegberg, P. Karrer, H. Salomon, and H. Ruckstuhl, *Helv. Chim. Act. 27:*382(1944).
118. S. A. Koser, A. Dorfman, and F. Saunders, *Proc. Soc. Expt. Biol. Med., 43:*391(1940).
119. H. Erlenmeyer, H. Bloch, and H. Kiefer, *Helv. Chim. Act. 25:* 1066(1942).
120. B. E. Moeller, and L. Birkhofer, *Ber. 75B:*1118(1942).
121. H. Erlenmeyer, and W. Wuergler, *Helv. Chim. Act. 25:*249(1942).
122. J. McLaren, D. T. C. Ngo, and B. M. Olivera, *J. Biol. Chem. 248:*5144(1973).

7
NICOTINIC ACID ANTAGONISTS

Because of the simple structure of nicotinic acid and nicotinamide, numerous substituted pyridines and substituted heterocyclic compounds have been tested for their biological effect on nicotinic acid and nicotinamide nutriture or metabolism. The discovery that the nicotinamide present in nicotinamide-adenine dinucleotide could be exchanged with a related pyridine, imidazole, thiazole, or thiadizole compound to generate the corresponding base–adenine–dinucleotide analogue of NAD has resulted in a voluminous literature on niacin analogues. No coenzyme has been modified so frequently to generate (1) coenzyme analogues inhibitory to NAD- or NADP-requiring reactions; (2) coenzyme analogues that can substitute for NAD or NADP in reactions, but with altered redox potential and enzyme binding properties; (3) coenzyme analogues that serve as starting material for the biochemical synthesis of ribosides or ribonucleotides. Some of this literature will be reviewed in the section on nicotinic acid as coenzyme (Part II). In this section only the direct action of nicotinic acid analogues on enzymes and biological systems will be reviewed. Our interest is here primarily antivitamin action in whole organisms and animals. Some of the compounds discussed in detail are shown in Figure 1. A few compounds with antiniacin action have been seriously considered as antineoplastic agents.

I. NICOTINIC ACID ANTAGONISTS FOR BACTERIAL GROWTH

Among the earliest attempts to develop nicotinic acid antagonists were studies of structural analogues and their effect on bacteria. However, substances with antinicotinic acid activity for bacteria may not have antivitamin action in animals or vice versa.

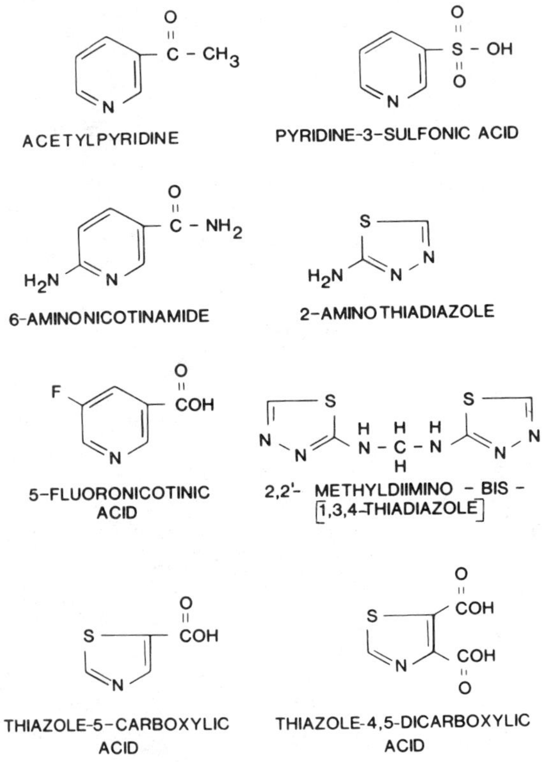

ACETYLPYRIDINE

PYRIDINE-3-SULFONIC ACID

6-AMINONICOTINAMIDE

2-AMINOTHIADIAZOLE

5-FLUORONICOTINIC ACID

2,2'- METHYLDIIMINO - BIS - [1,3,4-THIADIAZOLE]

THIAZOLE-5-CARBOXYLIC ACID

THIAZOLE-4,5-DICARBOXYLIC ACID

Figure 1 The structure of active niacin antagonists.

In bacteria, the most extensively studied substances are acetylpyridine, pyridine-3-sulfonic acid, 6-aminonicotinamide, 5-fluoronicotinic acid, and thiazole-5-carboxylic acid.

Acetylpyridine has no activity as a niacin antagonist in microorganisms even though it is one of the most potent nicotinic acid antagonists in animals (1).

A more generally active nicotinic acid antagonist is pyridine-3-sulfonic acid. This acts as a growth inhibitor for several species of microorganisms (2–10), but it is of relatively low potency. Susceptible organisms include *Proteus* (6), *Staphylococcus* (10), *Escherichia coli* (2,7), and *Streptobacterium plantarum* (4,5). However in the latter case the action can be reversed by manganous and cupric ions rather than just by nicotinic acid or nicotinamide (5,6). Pyridine-3-sulfonic acid has even been said to have slight growth-promoting activity when substituted for niacin (11–13).

Thiazole-5-carboxylic acid can act as an antagonist to nicotinic acid for *Staphylococci*, as can its amide (10). However, as discussed in Chapter 6, Sec. V, claims for its vitamin action have been more frequent.

Halogen-substituted nicotinic acids acted as effective competitive antagonists to nicotinic acid in a series of microorganisms tested (14). In all species, the order of activity was 6-fluoronicotinic acid < 5-bromonicotinic acid < 5-chloronicotinic acid < 5-fluoronicotinic acid. A series of 43 compounds was tested by Streightoff (15); 5-fluoronicotinic acid was the most active substance found (Table 1). A somewhat unusual substance said to inhibit growth of *Proteus* species, reversible by nicotinamide, is 1,3-diphenyl-1,3-propanedione (16).

Isonicotinic acid hydrazide (INH) may in part act as a niacin antagonist in the nutrition of *Mycobacterium tuberculosis*. This widely used antituberculosis drug is a potent inhibotor of nicotinamide-adenine dinucleotide hydrolase of beef spleen (17). The isolation of the INH analogue of nicotinamide-adenine dinucleotide, synthetized by NADase (18), in fact opened up the whole field of NAD analogue biosynthesis. Since INH can precipitate pellagra (19,20), it was thought to be a pure niacin antagonist. However, the previously discussed antipyridoxine (vitamin B_6) effects of INH and the necessity for adequate vitamin B_6 nutriture for optimal conversion of tryptophan to niacin makes this conclusion less certain. The relationship is very complex between INH and niacin.

Nevertheless, because of the great clinical usefulness of INH, a multitude of analogues have been synthesized and tested for activity. The work by Beyerman et al. (21) exemplifies this. One group of substances has been of special interest: the thionicotinamides and isonicotinamide derivatives. Some have significant antituberculosis activity with little antibacterial activity (22). 2-Ethyl-4-pyridinecarbothioamide (ethioniamide) is, however, very hepatotoxic in animals and is no longer in general clinical use (23,24). Whether this substance acted as an antiniacin is unknown.

Quinolinic acid is an effective substitute for nicotinic acid in many species, especially in *Pseudomonas* (25). While thiazole-5-carboxylic acid inhibits niacin action or allegedly replaces it, thiazole-4,5-dicarboxylic acid has no such effect on the growth of several microorganisms, including a *Pseudomonas* sp. (nor does it act as a niacin antagonist in animals) (26).

Not only do several antagonists for nicotinic acid exist in microbiology, nicotinic acid itself can be inhibitory to the growth of several species. In one *Saccharomyces* species, low concentrations of nicotinic acid can be inhibitory (27), while in the common *S. cerevisiae* nicotinic acid augments the inhibitory action of adenine or hypoxanthine (28).

In large amounts, nicotinamide can inhibit several strains of *Mycobacterium* (29), while *Neisseria* (30) and *Leptospira* (31) are inhibited by much smaller amounts of niacin in the medium (30,31). A substantial number of other organisms are inhibited by large amounts of nico-

Table 1 Inhibition of Bacterial Growth by 5-Fluoronicotinic Acid and Related Compounds and Its Reversal by Nicotinic Acid

	First series					
	Streptococcus		*Staphylococcus*		*Escherichia coli*	
Compound	Inhibition (%)	Reversal	Inhibition (%)	Reversal	Inhibition (%)	Reversal
5-Bromonicotinamide	0		0		100	Yes
5-Bromonicotinic acid	0		0		30	
5-Chloronicotinic acid	0		0		35	
5,6-Dichloronicotinic acid	100	Yes	0		35	
5-Fluoronicotinamide (5-FNAM)	90	Yes	80	Yes	100	Yes
5-Fluoronicotinic acid (5-FNA)	90	Yes	0		80	Yes
5-Fluoronicotinic acid, sodium salt					95	Yes
5-Iodonicotinamide	80		0		50	Yes
5-Iodonicotinic acid	0		0		40	
Methyl-5-bromonicotinate	100	Yes	100	Yes	100	
Methyl-5-fluoronicotinate	100	Yes	50	Yes	100	Yes

Compound	Second series					
	Streptococcus		*Lactobacillus plantarum*		*Escherichia coli*	
	Inhibition (%)	Reversal	Inhibition (%)	Reversal	Inhibition (%)	Reversal
5-Aminonicotinamide HCl			0			
4-Aminonicotinic acid	70					
5-Aminonicotinic acid	0		0		0	
5-Aminonicotinic acid HCl	40 0		0		0	
6-Aminonicotinic acid	100 40	Yes	0	Yes	100	Yes
5,6-Diaminonicotinic acid			0			
5-Fluoro-3-acetylpyridine HCl			0			
5-Fluoro-3-cyanopyridine	100	Yes	0	Yes	100 50	Yes
5-Fluoro-N^1-dimethylamino-methylnicotinamide	100 50	Yes	100 50		100	Yes
5-Fluoro-*N*,*N*-dimethyl-nicotinamide	50					
6-Fluoronicotinamide	0		0		0	
5-Fluoronicotinamide *N*-oxide	100					

Table 1 (Continued)

Compound	Second series					
	Streptococcus		*Lactobacillus plantarum*		*Escherichia coli*	
	Inhibition (%)	Reversal	Inhibition (%)	Reversal	Inhibition (%)	Reversal)
5-Fluoronicotinic acid (5-FNA)	100		70 60	Yes	100 60	Yes
6-Fluoronicotinic acid	0					
5-Fluoronicotinic acid hydrazide	100 50	Yes	0		100 40	Yes
5-Fluoronicotinic acid *N*-oxide	90					
5-Fluoronicotinhydroxamic acid	90	Yes	60	Yes	100 40	Yes
5-Fluoronicotinuric acid	100 60		0		25	
5-Fluoroquinolinic acid, monosodium salt	30					
5-Fluorothionicotinamide	100	No	0		0	
5-Hydroxynicotinic acid	50					
5-Methylnicotinamide	0					

4-Methylnicotinic acid	80 0		0		
5-Methylnicotinic acid	0		0	50	
5-Methylthionicotinamide	100	No	0		
Nicotinic hydroxamic acid	0		0	100	No
4-Nitronicotinic acid *N*-oxide	100		0	100	
5-Phenylnicotinic acid	0		0	30	
5-Pyrimidinecarboxylic acid	0		0	0	
5-Thiazolecarboxamide	0		0	100 70	Yes
5-Thiazolecarboxylic acid	100	No	0	30	
Thioisonicotinamide	100 75	No	0	0	
Thionicotinamide	100	No	0	0	

Source: Ref. 15.

tinic acid or nicotinamide. The inhibitory action is reversible by yeast extract, which in turn can be replaced by hydrolyzed casein and a vitamin mixture (32). This effect has been attributed to nutrient imbalance, but meta-chelating effects, or actual inhibition of dehydrogenases by the pyridine derivatives, could also be an explanation. The required concentrations of between 0.01 and 0.2 *M* are far in excess of potential clinical usefulness.

II. NICOTINIC ACID ANTAGONISTS IN ANIMALS

A. The Effects In Vitro

While the discussion will be largely limited to the in vitro effect, it is difficult to gain an understanding of the in vitro effects unless the in vivo effects are understood. Nicotinic acid as a nutrient exerts its physiologic and metabolic role entirely through nicotinamide-adenine dinucleotide (NAD) and nicotinamide-adenine dinucleotide phosphate (NADP). These substances are coenzymes for multiple dehydrogenases. Many pyridine derivatives act as real but weak antagonists to that coenzymatic role. It is unlikely that such effects are responsible for the in vivo action in animals, though it might help explain some antibacterial action, especially that observed with nicotinamide itself.

A second role of NAD is its action as substrate for the polymerization of adenosine diphosphate-ribose with release of nicotinamide. In addition NAD (and NADP) are substrates for nicotinamide-adenine dinucleotide glycohydrolase (NADase). It is here where pyridine derivatives could exert a great deal of their inhibitory action. An extraordinary number of substances have been studied in vitro to investigate their effects on the action of NADase, but only a few have been studied in vivo to determine the relevance of this reaction to their antiniacin action. Finally, pyridine and related derivatives can substitute for nicotinic acid or nicotinamide in other reactions in which those participate, usually at the mononucleotide or riboside level.

Pyridine bases inhibit dehydrogenases. In the past pyridine-3-sulfonic acid, nicotinic acid, and nicotinamide were found to inhibit glucose-6-phosphate and lactate dehydrogenase activities (33,34). The best studied example is yeast alcohol dehydrogenase (35). The inhibition of this enzyme is proportional to the pK_a of the ring nitrogen, except for pyridine-3-sulfonic acid (Fig. 2). The relationship is truly competitive with NAD. N-methylpyridine derivatives are also competitive inhibitors of NAD, and the series here show an inhibitory action that is greater the stronger the electronegativity of the side chain (Table 2). Imidazoles are also inhibitory to dehydrogenases, and, while it is possible that the action is analogous to that of pyridine derivatives, additional reactions between the quaternized nicotinamide in NAD and the imidazole are a more likely explanation (36).

The discovery of the effect of pyridine derivatives on NADase through the study of INH has already been mentioned. Not all NADases are equally sensitive to an inhibitory effect of INH; thus the concept of INH-sensitive and -insensitive enzymes was generated (18). In both cases, an exchange reaction occurs:

Nicotinamide-adenine dinucleotide + base $\rightleftarrows$ base-adenine dinucleotide phosphate _ nicotinamide.

This reaction competes with the normal reaction:

Nicotinamide-adenine dinucleotide + H_2O → nicotinamide + adenosine diphosphate ribose.

If either the new base-adenine dinucleotide is an inhibitor of NAD hydrolysis, or a free equilibrium between NAD and the NAD analogue is established, the base will appear inhibitory. If the analogue is formed ans is further hydrolyzed, the base will not appear to inhibit (37). The prototype of an INH-sensitive enzyme is the beef spleen NADase, while the prototype of an insensitive enzyme is pig brain NADase. The basic difference is illustrated in Figure 3, in which the effects of the INH analogue Marsilid (isonicotinic acid 2-isopropylhydrazide) on the activity of beef spleen and pig brain NADase are compared.

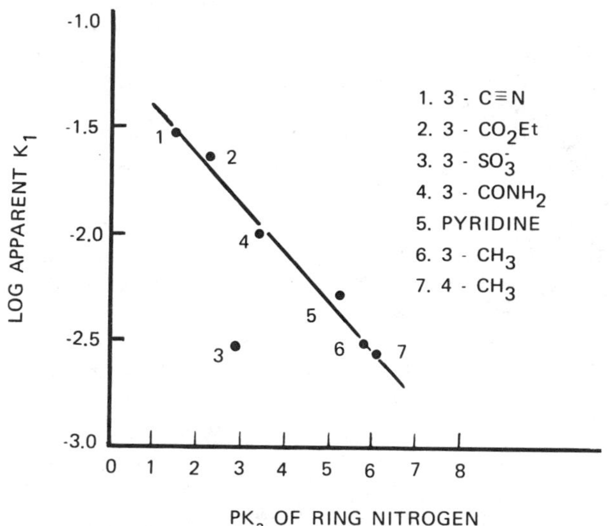

Figure 2 An examination of the inhibition of dehydrogenase activity by various pyridine bases.

Table 2 Comparison of Inhibition by Pyridine Derivatives[a]

		Concentration required for 50% inhibition		
		Pyridine derivatives		
Substituent	pK_a pyridine base	Total base (molar)	Pyridinium ion (molar)	N^1-methyl-pyridinium derivatives (molar)
4-CH_3	6.11	0.020	1.3×10^{-5}	–
3-CH_3	5.82	0.040	1.3×10^{-5}	5.8×10^{-3}
None	5.27	0.070	6.6×10^{-6}	5.5×10^{-3}
3-$CONH_2$	3.40	0.23	2.9×10^{-7}	5.0×10^{-3}
3-$COCH_3$	3.39	–	–	4.0×10^{-3}
3-CHO	3.37	–	–	4.2×10^{-3}
3-$CO_2C_2H_5$	2.24	0.30	2.6×10^{-8}	3.6×10^{-3}
3-CN	1.45	0.60	8.5×10^{-9}	3.2×10^{-3}
3-SO_3	2.9	0.013	5.1×10^{-9}	–

[a]The figures are based on a reaction mixture consisting of NAD 1.3×10^{-4} *M*, ethanol 0.5 *M*, Tris buffer 0.05 *M*, pH 9.3.

The NADase from *Neurospora* is said to be unaffected by INH and other bases and not to exchange bases to form analogues, but only to hydrolyze NAD rather specifically. However, it was once claimed that ergothioneine could impart to the *Neurospora* NADase the capacity to exchange radioactive nicotinamide into NAD (39). Ergothioneine was thought to be a cofactor for erythrocyte nicotinamide riboside phosphorylase (39). It was found to be required to impart nicotinamide "sensitivity" to both enzymes.

A very wide range of substances has been tested for their action on "sensitive" and "insensitive" NADases. Even pyridine is a potential reaction participant. A variety of corresponding analogues of NAD have been made since by chemical modification as well as enzymatic exchange. While these analogues are readily made in vitro, there is only sporadic evidence that such analogues are synthesized in vivo after administration of the nicotinic acid analogue, and even less evidence as to whether the formation of such analogues is indeed the mechanism of action of the antiniacin effect. The properties of these enzymes and of analogues will be discussed in Part II. There are a number of nicotinic acid analogues that inhibit nicotinic acid incorporation into NAD in platelets (41). These data were obtained to evaluate the mechanism of action of salicylates on platelet function before current concepts of aspirin action were known. A large number of substances have inhibitory action

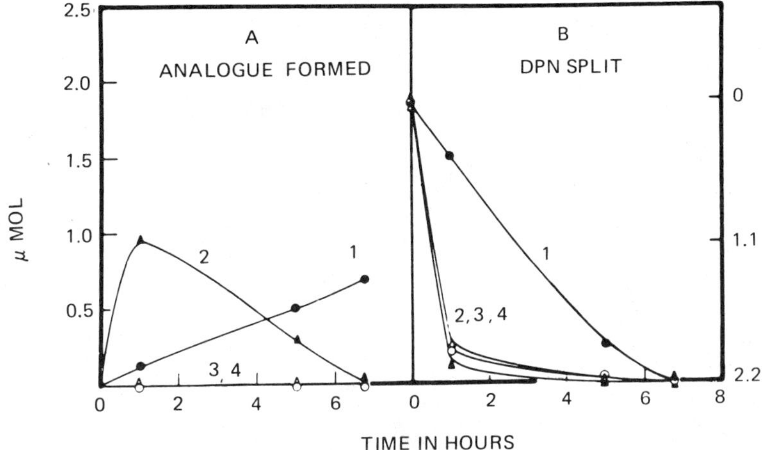

Figure 3 Comparison of the effects of pig brain and beef spleen NADases. Reaction mixtures contained 2.2 μmoles of NAD, 0.02 *M* Marsilid, 0.05 *M* potassium phosphate buffer, pH 7.5, 6 units each of pig brain and beef spleen NADase, and H_2O to a final volume of 0.6 cc with incubation at 37°C. Aliquots of 0.06 cc were removed for alcohol dehydrogenase assay, and for addition to 0.1 *M* NaOH to determine the formation of the Marsilid analogue of NAD. Curve 1, represents beef spleen + Marsilid + NAD; Curve 2, pig brain + Marsilid + NAD; Curve 3, pig brain + NAD; and Curve 4, beef spleen + NAD.

(Table 3). 2-Hydroxynicotinic acid is a competitive inhibitor of nicotinic acid phosphoribosyltransferase.

B. Nicotinic Acid Antagonists In Vivo

Among the various potential nicotinic acid antagonists only a few have been studied in some detail: acetylpyridine, 6-aminonicotinamide, and 2-amino-1,3,4-thiadiazole and derivatives. Another compound, pyridine-3-sulfonic acid, has been reported to have antiniacin activity in protein-deficient rats (42), but not in dogs (43) or mice (44). A large series of substances have been screened by Humphreys et al. (45) for toxicity and reversibility by nicotinamide, and some have been screened for antitumor activity. The three substances reviewed in the following discussion were studied by these investigators.

Acetylpyridine: Acetylpyridine is toxic, and gives rise to nicotinic acid deficiency signs in mice (1), dogs (9), and rats and chickens (46).

Table 3 Effect of Various Nicotinic Acid Analogs (1 m*M*) on the Isotopic Accumulation by the Human Platelet when Incubated for 1 hr with [7-^{14}C] Nicotinic Acid (10 μ*M*)[a]

Compound	Inhibition (%)
2-Hydroxynicotinic acid	90
2-Pyridylcarbinol	86
4-Hydroxynicotinic acid	74
2-Fluoronicotinic acid	71
3,5-Pyridyldicarboxylic acid	58
4-Chloronicotinic acid	26
5-Hydroxynicotinic acid	26
7-Chloronicotinic acid	24
2,3-Pyridyldicarboxylic acid	23
3-Pyridylcarbinol	22
Pyridine	21
5-Chloronicotinic acid	19
Nicotinuric acid	18
Nicotinic acid *N*-oxide	17
Ethyl-2-hydroxynicotinate	16
2-Hydroxynicotinamide	8
Nicotinamide	0

[a]Results of representative experiments are shown.
Source: Ref. 41.

The action can be prevented by the administration of nicotinic acid or nicotinamide before or simultaneously with acetylpyridine administration, but the toxic effects cannot be reversed (1). Extra tryptophan in the diet will also prevent the niacin deficiency symptoms in mice (47). The symptoms are mostly neurological and there are histological abnormalities demonstrable in the hippocampus (48–50). The 3-acetylpyridine analogue of NAD is readily formed in vitro, utilizing pig brain NADase (51). This analogue has been widely used in experimental and analytical enzymology because the redox potential favors substrate oxidation by two orders of magnitude. It can substitute for NAD in many dehydrogenase actions and can therefore, when formed in vivo, generate major metabolic disturbances. Kaplan and co-workers have shown that the analogue is indeed formed in vivo in tumor tissue (52,53).

Acetylpyridine is converted to nicotinic acid, as mentioned in Chapter 6, but the substance has primarily antiniacin toxicity. Yet even though the analogue is readily detectable in tumor tissue, it is not seen in substantial amounts in the normal brain (52).

6-Aminonicotinamide: 6-Amino-nicotinamide was developed as a potential inhibitor of amine acetylation (54,55) but was found to be too toxic in rabbits. The toxicity was similar to that seen in mice after acetylpyridine injection. Therefore, Woolley suggested it might be a nicotinic acid antagonist (56). It was shown that nicotinic acid, nicotinamide, or tryptophan could prevent the toxicity (57,58). 6-Aminonicotinamide is incorporated into the 6-aminonicotinamide-adenine dinucleotide in vivo in rats bearing the Walker-256 Sarcoma (58) and in mice which are tumor bearing (56,60). As observed for acetylpyridine, tumor tissues showed more NAD analogue formation than did normal tissue. In fact, the actual amount of NAD is not greatly affected by the 6-aminonicotinamide (61,62). 6-Aminonicotinamide does have central nervous system toxicity. The lesions in rats are said to be similar to those seen in pellagra (63), and these changes are more pronounced the older the animals (64). When injected into rats, a parkinsonian-like muscular rigidity develops (65), which seems to correlate with impaired dopamine function (66,67). In contrast to 3-acetylpyridine which causes primarily hopythalamic damage, 6-aminonicotinamide is associated with cerebellum and spinal cord damage (67). Disorders that are either hypo- or hypermotoric are produced with intracerebral injection of 6-aminonicotinamide. These correlate with the RNA levels in intracerebral neurons (68). Axial transport is impaired after subarachnoid injection of 6-aminonicotinamide (69).

There are major disturbances in carbohydrate metabolism which occur following administration of 6-aminonicotinamide. When the drug is intraventricularly administered, a centrally mediated hypoglycemia ensues (70).

6-Aminonicotinamide has been proposed as an antineoplastic agent. It was found to be active in a wide variety of experimental tumor models (59,71–73). Clinical trials have been held (74–76), but the drug was found to be too toxic for use (74,75). In a mouse model a poly-L-lysine polymer linked to a succinylated derivative of 6-aminonicotinamide was thought to have a wide therapeutic margin, while still antagonistic to niacin (77). Topical 6-aminonicotinamide has been used successfully for the treatment of psoriasis, using systemic niacin to counteract generalized toxicity (78).

6-Dimethylaminonicotinamide is borderline active as an antineoplastic agent in Lewis lung carcinoma in the form of L-1210 leukemia, but inactive against several other model tumors (79). *N*-methyl-6-amino-nicotinamide is inactive. A wide variety of 6-subsituted nicotinamide derivatives have been synthesized; the activity has varied, as has toxicity. Toxicity seemed to be correlated with the basicity of the substituent (80). A series of other nicotinic acid analogues were tested as antineoplastic agents. Pyrazine analogues are active against Erhlich ascites cells, but this action has not been further investigated (81).

2-Amino-1, 3, 4-thiadiazole: 2-Amino-1,3,4-thiadiazole is the parent compound of a series of derivatives that have marked antitumor activity in animals. In addition to this compound, the acetylamino- and ethylaminothiadiazoles have been shown to be particularly active (82) as has the compound 2,2'-methyldiiminobis[1,3,4-thiadiazole]. The antitumor effect can be reversed by nicotinamide or nicotinic acid (83–87) and the immunosuppressive activity is likewise counteracted by niacin (87).

The drug has been given to humans (88) and has shown to increase uric acid excretion and serum uric acid levels. This was thought to be the result of increased de novo synthesis (89,90), rather than nucleotide degradation. This effect on uric acid synthesis is also blocked by nicotinamide, and interestingly, also by 6-aminonicotinamide and 3-acetylpyridine (91). The antileukemic action of the thiadiazoles also is inhibited by 6-aminonicotinamide, 5-fluoronicotinamide, and 3-acetylpyridine (54,92). The aminothiadiazole and ethylamino-1,3,4-thiadiazole can form the corresponding base-adenine dinucleotide analogues of NAD in vitro, but no evidence of analogue formation could be obtained in vivo (86,93). 2-Ethylamino-1,3,4-thiadiazole actually stimulates the synthesis of NAD from nicotinamide at low doses of nicotinamide (90,91). 2-Amino-1,3,4-thiadiazole has a marked inhibitory effect on the conversion of inosinic acid to guanylic acid (94) which is reversed by niacin (94). 2,2'-Methylenediiminobis[1,3,4-thiadiazole] has a similar effect (95).

It appears that in L-1210 cells, a 2-amino-1,3,4-thiadiazole mononucleotide is formed (93). Nicotinamide prevents the formation of this mononucleotide. 2-Amino-1,3,4-thiadiazole-adenine dinucleotide and 2-amino-1,3,4-thiadiazole mononucleotide are potent inhibitors of inosinic acid (IMP) dehydrogenase (Fig. 4) and therefore this is the probable site of action of this antitumor agent. However, since there is no evidence that NAD analogues are formed in vivo, the mononucleotide, which is formed in vivo, is the actual inhibitor. Its synthesis could well be effected in a reaction with phosphoribosyl pyrophosphate, analogous to the entry of nicotinic acid or nicotinamide into the metabolic sequence (93). Alternatively, a nucleoside phosphorylase could generate the thiadiazole riboside which then could be converted to the mononucleotide. Nelson et al. (93) pointed out that this is analogous to the action of 1,2,4-triazole-3-carboxyamide riboside which is also an IMP dehydrogenase inhibitor.

The pharmacology of this group of compounds is still under investigation. Reports on its distribution and metabolism in dogs (96–100), mice (98,99), and monkeys (97–99) have been published, and a pharmacokinetic model for these compounds in mice, dogs and monkeys have been proposed (99). The drug has been tested in humans as mentioned, and in a recent review it was stated that phase I trials are on the way (101).

Paraquat: Paraquat is a herbicide with the structure 2,2'-dimethyl-4-, 4'-bipyridinium dichloride. It is highly toxic to mammals (102,103). Surprisingly, in rats niacin reduces paraquat toxicity, as it does in *E. coli* (104).

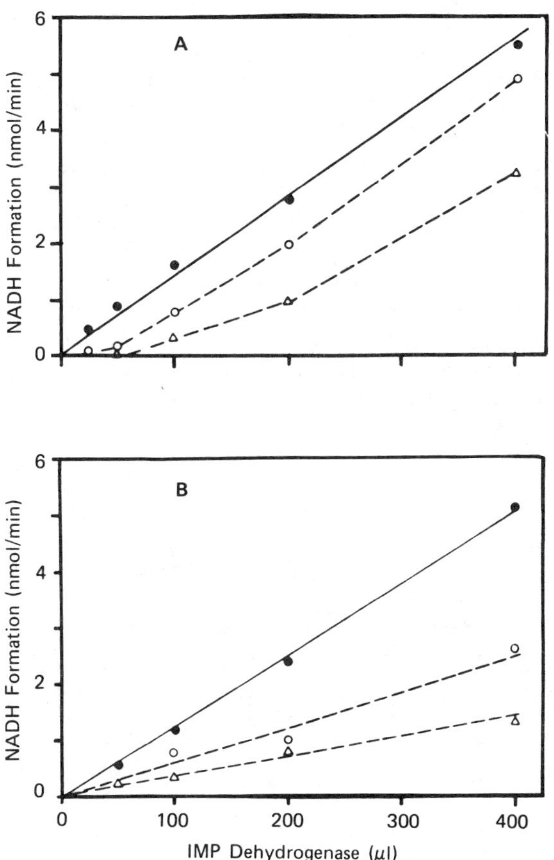

Figure 4 Ackermann-Potter plots of the inhibition of inosine 5'-monophosphate (IMP) dehydrogenase by derivatives of aminothiadiazole. Upper, inhibition by the NAD analog. o, 36 μ*M*; Δ, 72 μ*M*. Lower inhibition by thiadiazole mononucleotides. o, 0.05 μ*M*; Δ, 0.1 μ*M*. Upper and lower, •, control. The reaction was initiated by the addition of IMP to a final concentration of 200 μ*M* following a preincubation period of 5 min. The concentrations of other components were: 300 μ*M* NAD; 100 μ*M* KCl; 50 m*M* Tris-Cl, pH 7.4. The protein concentration of the IMP dehydrogenase preparation was approximately 10 mg/ml.

Teratogenic effects of nicotinamide antagonists: Pyridine derivatives, administered to pregnant animals, have powerful teratogenic effects, and likewise, their administration in chicken embryos produces major defects. It has been theorized that this indicates a specific role for nicotinamide in the developing embryo, but that is probably an unwarranted oversimplification.

In animals, a dominant finding is the development of cleft palate in rats (105–108) or mice (109–112). Embryonic mechanisms are being evaluated, but no biochemical basis has as yet been postulated (107, 108). There is no genetic correlation between different teratogens in different strains of mice (86). In the cited papers, many associated defects besides cleft palate have been reported. Eye defects have also been noted (113).

In developing chickens 6-aminonicotinamide and 3-acetylpyridine are both strong teratogens, especially causing defective limb development (113). It is in this system that the extrapolations to the mechanisms of differentiation mediated through niacin were primarily proposed, using both 3-acetylpyridine and 6-aminonicotinamide as models (115–122). Specific tissues such as cartilage were thought to be affected by 6-aminonicotinamide (123,124) while other authors thought that 3-acetylpyridine primarily affected muscle development (118). However, the effect appears not to be directed to specific tissues, but is rather more general. The neurological toxicity of 3-acetylpyridine in the embryo may mediate some of the effects (125). The 6-aminonicotinamide also affects many different tissues (126). 2-Amino-1,3,4-thiadiazole is also teratogenic in rats. The effect correlates with decreased DNA synthesis and is preventable by simultaneous nicotinamide administration (127). Interestingly, the drug also inhibits polyadenosine diphosphate ribose polymerase from baby hamster kidney cells (128).

III. CONCLUSIONS

Nicotinic acid antagonists as clinically useful compounds have not been as well investigated as we might expect. Attempts to generate alkylated derivatives of nicotinic acid have failed to yield successful antineoplastic agents (129), thus another blind alley precipitated by nonrational drug development.

The failure to exploit the therapeutic potential in nicotinic acid derivatives is unfortunate. One interesting attempt has been made utilizing 5-methylnicotinamide. This compound is a strong inhibitor of polyadenosine diphosphate ribose biosynthesis (127), and while not toxic itself, is a potentiator of streptozotocin action (130). There is still a reasonable hope that aminothiadiazole derivatives might be retested clinically as antineoplastic agents. In spite of the enormous investigative activity focused on modification of the NAD molecule at the nicotinamide sites, little else has been found useful thus far. In fact, the thought that these substances ought to act via incorporation at the

NAD level has dominated thinking to such an extent that the reassertion by Nelson et al. that incorporation into metabolically active pathways might follow nicotinic acid or nicotinamide reactions other than NADase is both obvious and revolutionary (93,94). It suggests the necessity for reinterpretation of data on other pyridine derivatives.

REFERENCES

1. D. W. Woolley, *J. Biol. Chem. 157*:455(1945).
2. H. McIlwain, *Brit. J. Expt. Path, 21*:136(1940).
3. J. Matti, F. Niti, M. Morel, and A. Lwoff, *Ann. Inst. Pasteur 67*:240(1941).
4. E. F. Moller and L. Birkofer, *Ber. 75*:1108(1942).
5. E. F. Moller and L. Birkofer, *Ber. 75*:1118(1942).
6. H. Eflenmeyer and W. Wurgler, *Helv. Chim. Act. 25:*249(1942).
7. H. McIlwain, *Nature 146:*653(1940).
8. D. W. Woolley, F. M. Strong, R. J. Madden, and C. A. Elvehjem, *J. Biol. Chem. 129:*715(1938).
9. D. W. Woolley and A. G. C. White, *Proc. Soc. Expt. Biol. Med. 52:*106(1943).
10. H. Erlenmeyer, H. Block, and H. Kiefer, *Helv. Chim. Act. 25:*1066(1942).
11. A. Lowff and A. Querido, *Compt. Rend Soc. Biol. 30:*1569(1939).
12. P. Ellinger, G. Frankel, and M. M. Abdel-Kader, *Biochem. J. 41:*559(1947).
13. P. Pitsch, B. Land, and M. Nakamura, *J. Bacteriol. 86:*159(1963).
14. D. E. Hughes, *Biochem J. 57:*485(1954).
15. F. Streightoff, *J. Bacteriol. 85:*42(1963).
16. Y. Raoul, *Bull. Soc. Chim. Biol. 30:*896(1948).
17. L. J. Zatman, N. O. Kaplan, S. P. Colowich, and M. M. Ciotti, *J. Biol. Chem. 209:*453(1954).
18. L. J. Zatman, N. O. Kaplan, S. P. Colowich, and M. M. Ciotti, *J. Biol. Chem. 209:*467(1954).
19. R. B. McConnel and H. D. Cheeham, *Lancet 1*(263):959(1952).
20. P. A. DiLorenzo, *Act. Dermat. Venereol. 47:*318(1967).
21. H. C. Beyerman, J. S. Bontkekoe, W. J. van der Burg, and W. L. C. van Veer, *Rec. Trav. Chim. Pays Bas 73:*109(1954).
22. F. Bonati, in *III International Congress of Chemotherapy* (H. P. Kuemmerle and P. Preziosi, eds.), George Thieme Verlag, Stuttgart, 1964, pp. 183–185.
23. K. H. Göggel, K. Hübner, H. Jungbluth, and K. L. Radenbach, in *III International Congress of Chemotherapy*, (H. P. Kuemmerle and P. Preziosi, eds.), George Thieme Verlag, Stuttgart, 1964, pp. 189–204.
24. H. J. Weinstine, W. Y. Hallett, and A. S. Sarauw, *Am. Rev. Resp. Dis. 86:*576(1962).

25. W. B. Jakoby, A. Schatz, S. H. Hatner, and M. M. Weber, *J. Gen. Microbiol. 6:*278(1952).
26. J. van Eys, Ph.D. Thesis, Vanderbilt University, 1955.
27. W. D. Braekean and G. Boge, *Nature 198:*585(1963).
28. J. W. Moulder and D. D. Woods, *J. Gen. Microbiol. 9:*IV(1953).
29. T. T. Hole, *Am. Rev. Resp. Dis. 86:*94(1962).
30. E. P. Casman, *J. Bact. 53:*561(1947).
31. M. D. Rosenfeld and M. R. Green, *J. Bact. 42:*165(1947).
32. S. A. Koser and G. J. Kasai, *J. Bact. 53:*743(1947).
33. H. von Euler and B. Skarzynski, *Arch. Gen. Mineral Geol. 16A:* (9):000(1943).
34. H. von Euler, *Ber. 75B:*1876(1942).
35. J. van Eys and N. O. Kaplan, *Biochim. Biophys. Acta 23:*574 (1957).
36. J. van Eys, *J. Biol. Chem. 233:*1203(1958).
37. N. O. Kaplan, M. M. Ciotti, J. van Eys, and R. M. Burton, *J. Biol. Chem. 234:*134(1959).
38. L. Grossman and N. O. Kaplan, *J. Biol. Chem. 231:*727(1958).
39. L. Grossman and N. O. Kaplan, *J. Biol. Chem. 231:*717(1958).
40. S. P. Colowick, J. van Eys, and J. M. Park, in *Comprehensive Biochemistry* (M. Florkin and E. M. Stots, eds.), Elsevier Publishing House, 1960.
41. Z. N. Gaut, in *Platelets and Thrombosis* (S. Sherry and A. Scriabine, eds.), University Park Press, Baltimore, 1972, p. 177.
42. W. A. Krehl, L. M. Henderson, J. de la Huenga, and C. A. Elvehjem, *J. Biol. Chem. 166:*531(1946).
43. O. H. Gaebler and E. V. Herman, *Fed. Proc. 6:*254(1947).
44. D. W. Woolley and A. G. C. White, *Proc. Soc. Expt. Biol. Med. 52:*106(1943).
45. S. R. Humphreys, J. M. Vendetti, C. J. Kline, A. Goldin, and N. O. Kaplan, *Cancer Res. [Supple.] Cancer Chemotherapy Screening Data, XV, 22:*683(1962).
46. W. Landauer, *J. Expt. Zool. 136:*509(1943).
47. D. W. Woolley, *J. Biol. Chem. 162:*179(1946).
48. R. E. Coggeshall and P. D. MacLean, *Proc. Soc. Expt. Biol. Med. 98:*687(1958).
49. S. P. Higgs, *Am. J. Path. 31:*189(1955).
50. R. E. Coggeshall and P. D. MacLean, *Proc. Soc. Expt. Biol. Med. 52:*106(1943).
51. N. O. Kaplan and M. M. Ciotti, *J. Biol. Chem. 221:*823(1956).
52. N. O. Kaplan, in *The Neurochemistry of Nucleotides and Aminoacids* (R. O. Brady and D. B. Towers, eds.) Wiley, New York, 1960, p. 70.
53. N. O. Kaplan, A. Golden, S. R. Humphrey, M. M. Ciotti, and J. M. Vendetti, *Science 120:*437(1954).
54. W. J. Johnson, *Nature 174:*744(1954).
55. W. J. Johnson, *Can. J. Biochem. Physiol. 33:*107(1957).

56. D. W. Woolley, *J. Biol. Chem. 157:*455(1945).
57. W. J. Johnson and J. D. McColl, *Science 122:*834(1955).
58. W. J. Johnson and J. D. McColl, *Fed. Proc. 15:*284(1956).
59. L. S. Dietrich, I. M. Friedland, and L. A. Kaplan, *J. Biol. Chem. 233:*964(1958).
60. L. S. Dietrich, L. A. Kaplan, I. M. Friedland, and D. S. Martin, *Cancer Res. 18:*1272(1958).
61. D. M. Shapiro, L. S. Dietrich, and M. E. Shihls, *Cancer Res. 17:*600(1957).
62. N. O. Kaplan, A. Golden, S. R. Humphreys, and F. E. Stokenbach, *J. Biol. Chem. 226:*365(1957).
63. N. Brita, S. Oyanagi, T. Tshii, and Y. Tumiyama, *Acta Neuropathol. 44:*111(1978).
64. N. Horita, T. Tshii, and Y. Tumiyama, *Acta Neuro Pathol. 49:*19 (1980).
65. D. Loos, K. Halbhubner, W. Kehr, and H. Herken, *Neurosciences 4:*667(1979).
66. W. Kehr, K. Halbhubner, D. Loos, and H. Herken, *Naunyn Schmiedeberg's Arch. Pharmacol. 304:*317(1978).
67. N. O. Kaplan, *Cancer Enzymology. Proceedings of the Miami Winter Symposia Papanicolaou Cancer Research Institute Vol. 12,* Miami, Florida, 1976, pp. 201–224.
68. E. Knoll-Kohler, F. Wojnorowicz, and H. J. Sarkandar, *Exp. Brain Res. 38:*173(1980).
69. R. J. Boegman and E. X. Albugerque, *J. Neurobiol. 11*:283(1980).
70. A. Baba, T. Baba, F. Matsuda, and H. Iwata, *J. Nutr. Sci. Vitaminol. 24:*429(1978).
71. S. L. Halliday, A. Sloboda, L. W. Will and J. J. Oleson, *Fed. Proc. 16:*190(1957).
72. J. D. McColl, W. B. Rice and V. W. Adamkiewicz, *Can. J. Biochem. Physiol. 35:*795(1957).
73. D. S. Martin, L. S. Dietrich, and R. A. Fugmann, *Proc. Soc. Expt. Biol. Med. 103:*58(1960).
74. C. P. Peilis, S. Kofman, M. Sky-Peck, and S. G. Taylor, III, *Cancer 14:*644(1961).
75. F. P. Herta, S. G. Weissman, H. G. Thompson, Jr., G. Hyman, and D. S. Martin, *Cancer Res. 21:*31(1961).
76. J. K. Wyatt and L. N. McAnich, *Can. Med. Ass. J. 84:*309(1961).
77. L. I. Arnold, A. Dagan, S. D. Simon, and N. O. Kaplan, *Proc. Am. Ass. Cancer Res. 19*:114(1978).
78. H. S. Zacherin, *Arch. Dermatol. 114:*1632(1978).
79. L. R. Lewis and C. C. Cheng, *J. Med. Chem. 15:*849(1972).
80. W. C. Ross, *Biochem. Pharm. 16:*675(1967).
81. K. Yoshida, Y. Takatsuka, and T. Fuku, *Mie Med. J. 16:*207 (1967).
82. J. J. Oleson, A. Sloboda, W. P. Tray, S. L. Halliday, M. J.

Landers, R. B. Angler, K. Serb, K. Cyr, and J. M. Williams, *J. Am. Chem. Soc. 77:*6713(1955).
83. T. J. Oleson, *Fed. Proc. 15:*372(1956).
84. D. M. Shapiro, M. E. Shotts, R. A. Fugman, and I. M. Friedland, *Cancer Res. 17:*29(1957).
85. M. M. Ciotti, N. O. Kaplan, A. Goldin, and J. M. Vendetti, *Proc. Am. Ass. Cancer Res. 2:*287(1957).
86. M. M. Ciotti, S. R. Humphreys, J. M. Vendetto, N. O. Kaplan, and A. Goldin, *Cancer Res. 20:*1195(1960).
87. T. Matsumoto, K. Ootsu, and Y. Okada, *Cancer Chem. Rep. 58* (Part I):331(1974).
88. I. M. Krakoff and G. B. Magill, *Proc. Soc. Expt. Biol. Med. 91:* 470(1956).
89. I. M. Krakoff and M. E. Balis, *J. Clin. Invest. 38:*907(1959).
90. J. E. Seegmiller, A. I. Grarel, and L. Liddle, *Nature 183:*1463 (1959).
91. I. M. Krakoff, C. R. Lacon and D. A. Karnofsky, *Nature 184:* 1805(1959).
92. M. F. Oettgen, J. A. Reppert, V. Coley, and H. J. Burdenal, *Cancer Res. 20:*1597(1960).
93. J. A. Nelson, L. M. Rose, and L. L. Bennett, Jr., *Cancer Res. 37:*182(1977).
94. J. A. Nelson, L. M. Rose, and L. L. Bennett, Jr., *Cancer Res. 36:* 1375(1976).
95. K. Tsukamoto, M. Suno, K. Igarashi, Y. Kozai, and Y. Sugino, *Cancer Res. 35:*2631(1975).
96. K. Lu, J. P. Chang, and Ti Li Loo, *Proc. AACR and ASCO 18:* 169(1977).
97. S. M. El Dareer and K. Tilley, *Proc. AACR and ASCO 18:*41(1977).
98. S. M. El Dareer, K. F. Tilley, and D. L. Hill, *Cancer Treatment Rep. 62:*75(1978).
99. F. G. King and R. L. Dedrick, *Cancer Treatment Rep. 63:*1939 (1979).
100. K. Lu and Ti Li Loo, *Cancer Chemother. Pharmacol. 4:*275(1980).
101. D. L. Hill, *Cancer Chemother. Pharmacol. 4:*215(1980).
102. S. Campbell, *Clin. Toxicol. 1:*245(1968).
103. O. R. Brown, M. Heitkamp, and C. S. Song, *Science 212:*1510 (1981).
104. M. A. Heitkamp, Thesis University of Missouri, Columbia, Missouri 1981.
105. T. G. Chamberlain and M. M. Nelson, *J. Expt. Zool. 153:*285 (1963).
106. J. G. Chamberlain, *Anal. ROC. 156:*31(1966).
107. V. M. Diewert, *Teratol. 13:*113(1979).
108. V. M. Diewert, *Teratol. 19*(2):213(1979).
109. L. Pinsky and F. Fraser, *Biol. Neonatol. 1:*106(1958).
110. D. R. Pollard and F. Frazer, *Teratol. 7:*267(1973).

111. R. Leinek, and M. Peterlea, *Cleft Palate J. 14:*266(1977).
112. F. G. Biddle and F. C. Frazer, *Teratol. 19:*207(1979).
113. S. Sander et al., *Morphol. Embryol. (Bucur) 24:*311(1978).
114. W. Landauer, *J. Exp. Zool. 136:*509(1957).
115. A. I. Caplan, *Expl. Cell Res. 62:*341(1970).
116. A. I. Caplan, *J. Exp. Zool. 178:*351(1971).
117. A. I. Caplan, *Devel. Biol. 28:*71(1972).
118. A. I. Caplan, *J. Exp. Zool. 180:*351(1972).
119. A. I. Caplan and J. Koutroupas, *J. Embryol. Exp. Morph. 29:* 571(1973).
120. A. I. Caplan, in *Vertebrate Limb and Somite Morphogenesis. Brit. Soc. Dev. Biol. 3rd symposium* (M. Balls and J. R. Hinchliffe, eds.) Cambridge University Press, 1977.
121. D. McMahon, *Science 185:*1012(1974).
122. D. Schubert and M. Laorbiere, *Proc. Nat. Acad. Sci. 73:*1989 (1976).
123. R. E. Seegmiller, D. O. Overman, and N. R. Meredith, *Devel. Biol. 28:*555(1972).
124. R. E. Seegmiller, *Devel. Biol. 58:*164(1977).
125. J. McLachlan, M. Bateman, and L. Wolpert, *Nature 264:*267(1976).
126. J. C. McLachlan, *J. Embryol. Exp. Morphol. 55:*307(1980).
127. G. E. Wotring and A. R. Beaudoin, *Teratol. 21:*381(1980).
128. W. C. Ross, *J. Med. Chem. 10:*257(1967).
129. J. B. Clark, G. M. Ferris, and S. Pinder, *Biochim. Biophys. Acta 238:*82(1971).
130. S. Sholl, P. Goodwin, M. Halldersson, G. Khan, C. J. Skidmore, and C. Tsupanakis, *Biochem. Soc. Symp. 42:*103(1977).

PART II

NICOTINIC ACID AS A COFACTOR

The function of nicotinic acid at the metabolic level is pervasive. The very basis of our ability to utilize energy is through gradual oxidative processes. The nicotinic acid-derived coenzymes nicotinamide-adenine dinucleotide (NAD), and nicotinamide-adenine dinucleotide phosphate (NADP) are frequently the mediators in that first step of oxidation through their role as hydride ion acceptors. The recent discoveries of NAD participation in the generation of adenosine diphosphate ribose polymers, which have a role in DNA metabolism and function, broadened the impact of nicotinic acid on metabolism to all areas of cellular life.

It must therefore be obvious that any discussion of the subject of nicotinic acid as cofactor is arbitrarily limited to keep the task within boundaries. The limits are set by seeing the problem from the point of view of a nicotinic acid user: the physician who uses it as a drug, the nutritionist who must understand the rational use of nicotinic acid in the diet, and the biochemist who might want to utilize the intake of nicotinic acid for research purposes, whether used in pharmacological dose or in restricted amounts. The metabolic consequence of administration of nicotinic acid analogues have been alluded to in a previous chapter. The structural analogues of nicotinic acid coenzymes that are so derived have been extremely useful in elaborating the relationship between structure and function of the nicotinic acid coenzymes. Not only is a nicotinamide a component of the coenzymes, but it is the functionally active site of those molecules.

There are few areas of biochemistry in which pharmacology, nutritional biochemistry, and molecular biology are interwoven in such a coherent whole. Yet a stream of new discoveries is continually generating new understanding. Just as there must be an arbitrary limitation in scope, there must be a limitation in time. This review covers the literature through June, 1981.

8

THE HISTORY OF NICOTINIC ACID AS A COENZYME

I. THE DISCOVERY OF COENZYMES I AND II

The discovery of nicotinic acid coenzymes can be dated as far back as 1904 when Harden and Young noted that fermentation by yeast juice requires both a nondialyzable, heat-labile and a dialyzable, heat-stable fraction (1). The history has been reviewed in detail by Warburg and Christian (2). From that classical article one can follow the discoveries through the laboratories of the pioneers of classical biochemistry. First it was found that the heat-stable fraction contains two "coenzymes." Eventually Meyerhof and Lohman identified one coenzyme as adenosine triphosphate, the phosphate-transferring coenzyme. Von Euler studied the other factor, at that time usually designated as "cozymase." It was recognized as an adenine-containing substance (3), but the structure remained unclear until Warburg discovered a "hydrogen-transferring coenzyme" in red blood cells. Nicotinamide, adenine, and three moles of phosphate per mole of coenzyme were identified. The identification of nicotinamide as a component was suggested by Dr. Walter Schoeller who recognized that the melting point of the base isolated from the coenzymes was the same as that of nicotinamide, a substance already known for 50 years. The identity was proven by comparing melting points of the unknown and its derivatives with those of known nicotinamide and derivatives. The results were published in a very terse note (4). Thus, nicotinamide-adenine dinucleotide phosphate was identified (5). Von Euler and colleagues shortly afterwards identified nicotinamide as a component of cozymase (6,7). Following that discovery, the similarity between Warburg's hydrogen-transferring enzyme and cozymase was established both by von Euler and Schlenk (8) and Warburg (2). The only difference was one phosphate per mole, and von Euler and Adler discovered quite early that a fermenting yeast system could convert Warburg's coenzyme to cozymase (9).

Warburg had the genius to elucidate many of the chemical properties of the coenzymes, and the insight not only to relate these properties to the function of the coenzyme in metabolism, but to utilize the properties as tools for investigation of the coenzyme (5). Warburg and associates describe in that paper the catalytic function of the coenzyme as a participant in oxidation–reduction reactions. They discovered the absorption band that is seen only in the reduced coenzymes (Fig. 1), the fluorescence of the reduced coenzymes, and also showed that the enzymatically reduced and chemically reduced coenzymes were identical in properties. Using a model compound, N^1-methylnicotinamide, they were able to deduce that the nicotinamide was the likely site of reduction of the coenzyme. This was long before N^1-methylnicotinamide was discovered to be a metabolite of nicotinamide. Finally, in the same peper, they described the marked acid-lability of the reduced coenzymes. The metabolic function and the different roles for the two coenzymes (then called by Warburg "Diphospho-pyridinnucleotid" and "Triphosphopyridinnucleotid") were also sketched.

The discovery of the oxidative function of the nicotinamide coenzymes, together with the data on oxygen utilization of tissues, pro-

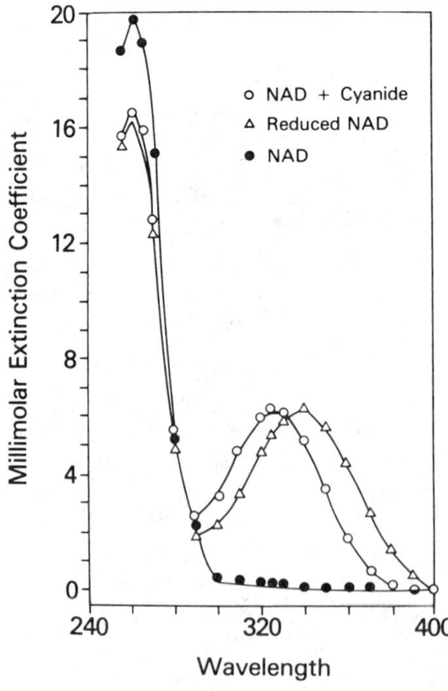

Figure 1 The spectrum of NAD, reduced NAD and the NAD-Cyanide complex.

vided the earliest evidence of the stepwise, sequential nature of biological oxidation. Up to that time some biochemists held the view that enzymes activated oxygen to allow it to react with the substrates, while others believed conversely that the enzyme acted on substrates to allow reaction with oxygen. Warburg reported that the reaction of glucose 6-phosphate with oxygen needed at least two enzymes in his system: the first, catalyzing the reduction of the pyridine nucleotide coenzymes by glucose 6-phosphate, was called the "Zwischenferment," the "enzyme in between." The reduced coenzyme so formed was then reoxidized by oxygen in a reaction catalyzed by a second enzyme, the "yellow enzyme." We now know that the yellow enzyme contains flavin nucleotides. For a period of time various similar flavin enzymes were discovered, and the term the "old yellow enzyme" was used. Ever since that series of researches, glucose 6-phosphate dehydrogenase has served as the major enzyme for the study of the triphosphopyridine nucleotide function.

The role of cozymase was studied primarily in yeast fermentation, where it was first discovered. Later, biochemical studies on muscle contraction, and alcohol and lactic acid fermentation suggested a similar role for cozymase.

The reduction of cozymase was accomplished by "carbohydrate." The fraction responsible for that reduction could be separated from a fraction that reoxidized the coenzyme with acetaldehyde or pyruvic acid. It was again in Warburg's laboratory that the sequence was resolved through the discovery of the reaction catalyzed by glyceraldehyde 3-phosphate dehydrogenase (10,11):

3-Phosphoglyceraldehyde + phosphate + cozymase
$\downarrow\downarrow$
1,3-Diphosphoglyceric acid + reduced cozymase.

The simultaneous incorporation of inorganic phosphate into a labile, "high-energy" phosphate bond demonstrated for the first time a mechanism for the trapping of oxidative energy into phosphate bonds.

II. OTHER NATURALLY OCCURRING NICOTINAMIDE NUCLEOTIDES

There was a short period during which the existence of a third nicotinamide coenzyme was postulated—coenzyme III (12,13). This coenzyme was thought to participate in sulfur oxidation. The proposed structure was nicotinamide ribose 5-pyrophosphate (13). Its function proved to be in error; nevertheless, the compound is enzymatically formed from nicotinamide-adenine dinucleotide in *Aspergillus niger* (14).

There are other naturally occurring nicotinamide nucleotides. Some of these are intermediates of nicotinamide-adenine dinucleotide biosynthesis: nicotinic acid mononucleotide and nicotinic acid-adenine dinucleotide. This will be discussed later.

More intriguing are the coenzyme analogues which contain the nicotinamide–riboside linkage in the α-configuration although the adenine–ribose linkage is still in the β-configuration. The existence of the so-called α-nicotinamide-adenine dinucleotide was discovered by Kaplan et al. as a component of the commercially available β-nicotinamide-adenine dinucleotide (15). The coenzyme was later found to occur naturally (16). In addition, α-nicotinic acid-adenine dinucleotide and α-nicotinic acid mononucleotide could be isolated from *Azotobacter vinelandii* cells. It is likely that these substances are intermediates in the known pathway of nicotinamide-adenine dinucleotide biosynthesis which results in the α-isomer. The α-isomer of the diphosphopyridine nucleotide is a substrate for the kinase enzyme that converts the di- to the triphosphopyridine nucleotide. The α-isomer of the triphosphopyridine nucleotide can also be identified in *A. vinelandii* cells, and appears to be widespread in nature (17). These coenzyme analogues are enzymatically quite limited in function. They can be chemically reduced but cannot replace the β-isomer in enzymatic dehydrogenase reactions. They act as substrates for NAD-degrading or -synthesizing enzymes *except* the nicotinamide glycohydrolases.

Another nicotinamide nucleotide found in nature is nicotinamide-adenine dinucleotide linked to adenosine diphosphate-ribose in a ribosidic linkage. This nucleotide is most likely a substance related to the formation of polyadenosine diphosphate (18).

Chemically the reduced coenzymes are more labile than the oxidized forms, especially in acid solutions. In the reduced form there is rapid anomerization of the β-coenzyme configuration to an equilibrium of the α- and β-coenzyme forms (19–22). Again, only the β-configuration is enzymatically active on the dehydrogenase. There is no certainty that reduced α-coenzymes exist naturally, though they can be reoxidized by cytochrome systems.

A coenzyme form, generally referred to as DPNH–X or NADH–X, was first discovered as a catalytic product of yeast glyceraldehyde 3-phosphate dehydrogenase when acting on reduced diphosphopyridine nucleotide at a pH of 5.0 (23). The rabbit muscle enzyme can execute the same reaction but at a slower rate. The turnover of the yeast enzyme is 1.4×10^{-4} times the rate of the turnover for the normal physiological reaction. Various additives accelerate or promote the reaction. When the reaction is executed in the presence of D_2O, deuterium is incorporated. When the substance is hydrogenated, the uptake suggests a single double bond in the nicotinamide moiety (24), in line with the absorption spectrum found for the nucleotide with a maximum at 290 nm (25). The structure is assumed to be a 2- or 6-hydroxy-tetrahydronicotinamide derivative. The action of acid on the reduced nicotinamide moiety yields a very similar product. However, the naturally found NADH–X is reconvertible to the reduced coenzymes by a yeast enzyme that requires ATP and magnesium (26). The acid reaction product is not a substrate in this reaction.

III. STRUCTURAL DETERMINATION OF NICOTINAMIDE COENZYMES

Once the pyridine nucleotide coenzymes were discovered, it was possible to purify them initially using the system through which they were identified as an assay system. As already mentioned, von Euler and colleagues published a series of papers in which a component of cozymase was identified as adenylic acid (3,27–31). The complete isolation of the components of coenzyme II by Warburg and co-workers (5) showed the composition to be" 1 mole of adenine, 1 mole of nicotinamide, 2 moles of pentose, and 3 moles of phosphoric acid. The resolution of the actual chemical structure followed soon. Adenylic acid was already known, and that structure was found to be a component of the coenzyme. Originally it was thought that nicotinamide-riboside and adenosine were linked through a triphosphate bridge. Schlenk proposed the correct formula in which the third phosphate is linked to the adenylic acid part of the molecule as in yeast adenylic acid (32). This linkage was later found to be specifically at the 2'-position of the adenosine component (33). Thereby the full structure of coenzyme II was finally elucidated. It is of specific interest that the 3'-phosphate isomer is not active in replacing coenzyme II (34).

Once the structure of coenzyme II was beginning to be understood, von Euler and co-workers identified coenzyme I, cozymase, as a "dinucleotide" composed of 1 mole of adenine, 1 mole of nicotinamide, 2 moles of phosphoric acid, and 2 moles of D-ribose per mole; the actual structure was proposed (6,36).

The pyrophosphate bond can be formed by means of specific dehydrating reactions utilizing nicotinamide and adenine mononucleotides. One of the first agents to be used in this way was trifluoroacetic anhydride (37).

IV. THE NOMENCLATURE OF NICOTINAMIDE COENZYMES

There have been many names applied to the coenzymes. None have been satisfactory. Cozymase became Coenzyme I or 1, while Warburg's hydrogen-transferring coenzyme became Coenzyme II or 2. After Warburg determined the elements of the structure, he named them diphospho-pyridinnucleotid and triphospho-pyridinnucleotid, which were then transliterated into diphosphopyridine nucleotide, abbreviated DPN, and triphosphopyridine nucleotide, or TPN. This was a misnomer, and the Commission on Enzymes of the International Union of Biochemistry proposed the names nicotinamide-adenine dinucleotide, or NAD, and nicotinamide-adenine dinucleotide phosphate, or NADP. These are now the accepted terms.

The reduced forms of the coenzymes are named DPNH or NADH, and TPNH or NADPH, respectively. Reduced NAD is generally written as NADH even though it is a two-electron process overall because the equation is generally balanced as follows:

$$RH_2 + NAD^+ \rightleftarrows R + NADH + H^+$$

V. THE STRUCTURE OF REDUCED NICOTINAMIDE COENZYMES

Warburg and Christian had determined the reducibility of the coenzymes by enzymatic reaction, and demonstrated that the same reduction could be accomplished with dithionite (5). By analogy to quaternary nicotinamide salts, Karrer et al. concluded that the reduction product was an *o*-dihydronicotinamide (38,39), though it was uncertain whether it was 1,2-dihydro- or 1,6-dihydronicotinamide. Warburg and Christian had demonstrated the characteristic absorption spectrum of reduced nicotinamide-adenine dinucleotide (5). Wallenfels demonstrated the separate absorption spectra of quaternary 1-, 4-, or 6-dihydronicotinamide derivatives (40). The spectrum of 1- or 6-dihydronicotinamide derivatives are very similar, with a peak at 350 nm; the 6-dihydro compound also has a peak at 260 nm, but that distinction cannot be made in the coenzyme because of the confounding 260-nm absorption of the adenine moiety. Therefore, the structure was not definitively established for a long time, even though that spectroscopic distinction was known at the time.

The use of deuterium isotopes, introduced in the study of the pyridine nucleotides by Westheimer and Vennesland, allowed a definitive determination of the structure of the reduced coenzyme. The work of Colowick along with students and colleagues established the structure definitively. First it was necessary to synthesize the 2- and 6-pyridines of N^1-methylnicotinamide as reference compounds by a single reproducible procedure (41). After that the coenzyme was reduced and reoxidized in such a way that deuterium was incorporated in the molecule.

Westheimer et al., using deuterium as a tracer, had demonstrated that the hydrogen was directly transferred from substrate to coenzyme without exchange with the medium (42). The coenzyme was chemically reduced using deuterium-containing reaction mixtures. The reduced coenzyme was then reoxidized with acetaldehyde and alcohol dehydrogenase leaving oxidized coenzyme with deuterium on the carbon that was reduced previously. The coenzyme was then hydrolyzed with the enzyme nicotinamide-adenine dinucleotide hydrolase (NADase) to yield free nicotinamide. This was converted with methyliodide to N^1-methylnicotinamide, which was itself converted to a mixture of 2- and 6-pyridones. Finally these were analyzed for deuterium. The reactions are summarized in Figure 2. If only the 2-pyridone contained deuterium, then the 6-position would be the site of reduction. If only the 6-pyridone contained deuterium, the 2-position would be the site of reduction, and if both the 2- *and* 6-pyridones contained deuterium, the reduction site would be the 4-position. Various cycles of chemical and enzymatic reduction were used, including the control reactions with enzymatic reduction and enzymatic reoxidation. Except in the latter case, in which no deuterium was left, all sequences yielded deuterium-labeled pyridones. This led to the inescapable conclusion that the reduction site is the *para* or 4-position (43). To prove the point further, it was also shown that direct chemical reduction of N^1-methylnicotinamide occurs in the 4-position (44). The reduction site was confirmed by a direct proof. 2-, 4-, or 6-deutero-labeled nicotinamides were synthesized, and these nitocinamides were incorporated into the coenzyme by the NADase-catalyzed exchange reaction. The deuterium-labeled coenzymes so obtained were chemically reduced with hydrosulfite and enzymatically reoxidized. The reduced substrates were analyzed for deuterium. Only in the coenzyme labeled in the 4-position of the nicotinamide could deuterium transfer be demonstrated (45). Reoxidation of similarly deuter-labeled N^1-benzylnicotinamide by malachite green confirmed the reduction at the 4-position (46).

Further direct evidence for the 4-position reduction comes from the fact that the model compounds 4-dihydro-N^1-substituted nicotinamides have a single absorption band at around 350 nm, while the 6-dihydronicotinamide derivatives have two absorption bonds at 260 and 360 nm

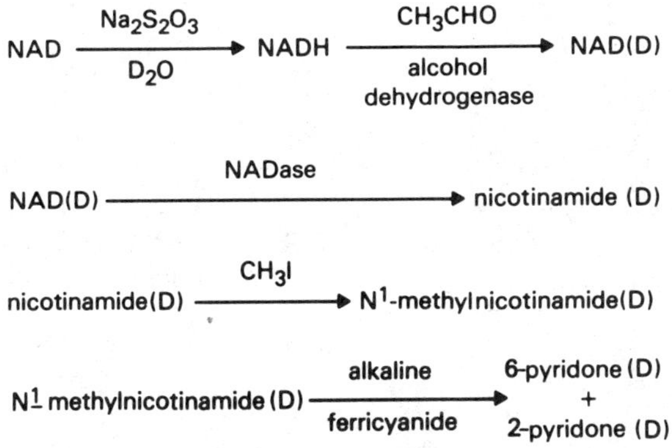

$$\text{NAD} \xrightarrow[\text{D}_2\text{O}]{\text{Na}_2\text{S}_2\text{O}_3} \text{NADH} \xrightarrow[\text{alcohol dehydrogenase}]{\text{CH}_3\text{CHO}} \text{NAD(D)}$$

$$\text{NAD(D)} \xrightarrow{\text{NADase}} \text{nicotinamide (D)}$$

$$\text{nicotinamide(D)} \xrightarrow{\text{CH}_3\text{I}} \text{N}^1\text{-methylnicotinamide(D)}$$

$$\text{N}^1 \text{ methylnicotinamide (D)} \xrightarrow[\text{ferricyanide}]{\text{alkaline}} \text{6-pyridone (D)} + \text{2-pyridone (D)}$$

Figure 2 The reaction sequence in the structure determination of reduced NAD.

(40). The final confirmation came from the determination of the nuclear magnetic resonance spectrum (47,48) and infrared spectroscopy (49) of deutero-labeled reduced quaternary nicotinamide derivatives.

While the 4-position hydrogens are equivalent in model compounds such as N^1-methylnicotinamide, this is not true in the coenzyme molecule. The coenzyme is in a folded configuration. This is clear from the observation that the fluorescence of reduced coenzyme can be excited both at 340 nm (the absorption of the dihydronicotinamide moiety) and at 260 nm (where adenine absorbs primarily) (21,50). When the configuration of the riboside linkage in the nicotinamide is in the α- in contrast to the β-configuration, this excitation at 260 nm is far less pronounced.

The reduction of the coenzyme is stereospecific. Even when the reduction is done chemically, utilizing deuterium or tritium, unequal proportions of the possible isomers are formed. The carbon at the 4-position of nicotinamide is a "meso" carbon, that is, a carbon of the configuration:

$$\begin{array}{c} R_2 \\ | \\ R_2 - C - R_3 \\ | \\ R_2 \end{array}$$

Although internally it is symmetrical, it is often handled asymmetrically in enzymatic systems. With hydrosulfite as reducing agent, the preference for one side (called side A) is nearly 2:1 (Fig. 3) (43,53).

The converse is also true. Reduced coenzyme is oxidized by ferricyanide preferentially on side A by a margin of 5:1 (54). Enzymatic reduction shows absolute specificity, as will be discussed later.

VI. THE TERTIARY STRUCTURE OF NICOTINAMIDE COENZYMES

The chemical nonequivalence of the hydrogens on the 4- carbon of the reduced nicotinamide moiety, and the spectral evidence suggested the folded structure of NAD and NADP. Even the configuration around the glycoside bonds becomes fixed. In solution there is no strong preference for the conformation around the glycosidic bond between nicotinamide and C_1 of the ribose (54). Biellman and Samama synthesized NAD analogues with 2-methyl or 6-methyl substitution on the nicotinamide. Because these substitutions fix the glycosidic bond configurations, they could make it likely that the bond in NAD bound on alcohol and lactic dehydrogenase would be in the *anti* configuration (56). Lappi et al. followed the same experimental design utilizing adenine-modified NAD analogues to study the confirmation about the adenine—riboside linkage. Those experiments suggested that lactate dehydrogenases preferred the *syn* configuration around that bond (57).

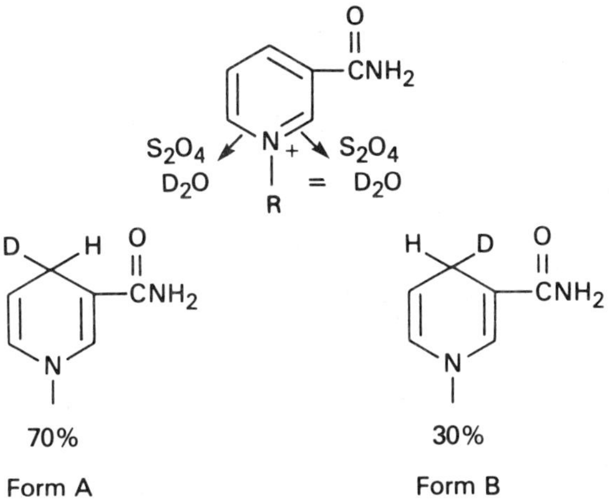

S_2O_4 D_2O S_2O_4 D_2O

70% Form A

30% Form B

Figure 3 The stereospecificity of the chemical reduction of NAD.

Now the three-dimensional structure of the coenzyme is known, and, in many cases, that of the enzymes to which they are bound is also known. There is clear evidence that the structure of enzyme-bound coenzyme many be different from that free in solution but there is great similarity in its bound structure with different enzymes. However, the confirmation is different when the enzyme has a specificity for side A reduction (e.g., lactic dehydrogenase) or side B reduction (e.g., glyceraldehyde 3-phosphate dehydrogenase) (54). The change in conformation of the coenzyme on binding to the enzymes is further demonstrated by the frequent changes in fluorescence or shifts in absorption maxima on binding to the enzymes.

REFERENCES

1. A. Harden and W. J. Young, *J. Physiol. 32:*Proc., Nov. 1904, 1905.
2. O. Warburg and W. Christian, *Biochem. Z. 287:*291(1936).
3. H. Von Euler and V. Myrbäck, *K. Physiol. Chem. 117:*28(1921).
4. O. Warburg and W. Christian, *Biochem. Z. 275:*464(1935).
5. O. Warburg, W. Christian, and A. Griese, *Biochem. Z., 282:*157 (1935).
6. H. von Euler, H. Albers, and F. Schlenk. *Z. Physiol. Chem. 237:* 180(1935).
7. H. von Euler, H. Albers, and F. Schlenk, *Z. Physiol. Chem., 240:* 113(1936).
8. F. Schlenk and H. von Euler, *Naturwissenschaften 24:*794(1936).

9. H. von Euler and E. Adler, *Z. Physiol. Chem. 252:*41(1938).
10. P. Negelein and W. Bromel, *Biochem. Z. 303:*132(1939).
11. O. Warburg and W. Christian, *Biochem. Z. 303:*40(1939).
12. J. Boldingh, *Rec. Trav. Chim. 69:*247(1950).
13. T. P. Singer and E. B. Kearney, *Biochim. Biophys. Act. 11:*290 (1953).
14. M. Kuwahara, in *Methiods in Enzymology Vol. 66* (D. B. McCormick, and L. D. Wright, eds.), Academic Press, New York, 1980, p. 123.
15. N. O. Kaplan, M. M. Ciotti, F. E. Stolzenbach, and N. R. Bachur, *J. Am. Chem. Soc. 77:*815(1955).
16. S. Suzuki, K. Suzuki, T. Imai, N. Suzuki, and S. Okuda, *J. Biol. Chem. 240:*PC554(1965).
17. K. Suzuki, H. Nakamo, and S. Suzuki, *J. Biol. Chem. 242:*3319 (1967).
18. T. Imai, S. Okuda, and S. Suzuki, *J. Biol. Chem. 244:*4547(1969).
19. C. Woenckhaus and P. Zumpe, *Biochem. Z. 343:*326(1965).
20. D. W. Miles, D. W. Chry, and H. Eyring, *Biochemistry 7:*2333 (1968).
21. E. L. Jacobson, M. K. Jacobson, and C. Bernofsky, *J. Biol. Chem. 248:*7891(1973).
22. N. J. Oppenheimer and N. O. Kaplan, *Arch. Biochem. Biophys. 166:*526(1975).
23. G. W. Rafter, S. Chaykin, and E. G. Krebs, *J. Biol. Chem. 208:* 799(1954).
24. S. Chaykin, J. O. Meinhart, and E. G. Krebs, *J. Biol. Chem., 220:*881(1956).
25. J. O. Meinhart and M. C. Hines *Fed. Proc. 16:*425(1956).
26. J. O. Meinhart, S. Chaykin, and E. G. Krebs, *J. Biol. Chem., 220:*821(1956).
27. H. von Euler and K. Myrbäck, *Z. Physiol. Chem. 136:*107(1924).
28. K. Myrbäck, *Ergeb. Enzymforsch. 2:*139(1933).
29. K. Myrbäck and H. von Euler, *Z. Physiol. Chem. 203:*236(1931).
30. K. Myrbäck and H. von Euler, *Z. Physiol. Chem. 203:*143(1931).
31. K. Myrbäck, H. von Euler, and Hellström, H., *Z. Physiol. Chem. 212:*7(1932).
32. F. Schlenk, *Symposium on Respiratory Enzymes*. Univ. Wisconsin Press, Madison 1942.
33. A. Kornberg and W. E. Price Jr., *J. Biol. Chem. 186:*557(1950).
34. T. P. Wang, L. Shuster, and N. O. Kaplan, *J. Biol. Chem. 206:* 299(1954).
35. L. Shuster and N. O. Kaplan, *J. Biol. Chem. 215:*183(1955).
36. H. von Euler and F. Schlenk, *Z. Physiol. Chem. 246:*64(1937).
37. L. Shuster, N. O. Kaplan, and F. E. Stolzenbach, *J. Biol. Chem. 215:*209(1955).
38. P. Karrer, B. H. Ringier, J. Büchii, H. Fritzsche, and U. Solmssen, *Helv. Chim. Act. 20:*55(1937).

39. P. Karrer, G. Schwarzenbach, F. Benz, and U. Sohnsse, *Helv. Chim. Act. 19:*811(1936).
40. K. Wallenfels, in *Steric Course of Microbiological Reactions* (G. E. W. Wolstenholme, and C. M. O'Connor, eds.), Churchill, London, 1959, p. 10.
41. M. E. Pullman and S. P. Colowick, *J. Biol. Chem. 206:*121(1954).
42. F. H. Westheimer, H. F. Fisher, E. E. Conn, and B. Vennesland, *J. Am. Chem. Soc. 73:*2403(1951).
43. M. E. Pullman, A. San Pietro, and S. P. Colowick, *J. Biol. Chem., 206:*129(1954).
44. G. W. Rafter and S. P. Colowick, *J. Biol. Chem., 209:*773(1954).
45. F. A. Loewus, B. Vennesland, and D. C. Harris, *J. Am. Chem. Soc. 77:*3391(1955).
46. D. Mauzerall and F. H. Westheimer, *J. Am. Chem. Soc. 77:*2261 (1955).
47. H. E. Dubb, M. Saunders, and J. H. Wang, *J. Am. Chem. Soc. 80:*1767(1958).
48. R. F. Hutton and F. H. Westheimer, *Tetrahydron 3:*73(1958).
49. M. C. Brown and H. C. Masher, *J. Biol. Chem. 235:*2145(1960).
50. G. Weber, *Nature 180:*1409(1958).
51. S. Shifrin and N. O. Kaplan, in *Light and Life* (W. D. McElroy and H. B. Glass, eds.), Johns Hopkins Press, Baltimore, 1961, p. 144.
52. S. Shifrin and N. O. Kaplan, *Nature 183:*1529(1959).
53. H. F. Fisher, E. E. Conn, B. Vennesland, and F. H. Westheimer, *J. Biol. Chem. 202:*687(1953).
54. A. San Pietro, N. O. Kaplan, and S. P. Colowick, *J. Biol. Chem. 212:*941(1955).
55. W. Egan, S. Forsen, and J. Jacobus, *Chem. Commun. 2:*42(1973).
56. J. F. Biellman and J. P. Samama, *FEBS Lett. 38:*175(1974).
57. D. A. Lappi, F. E. Evans, and N. O. Kaplan, *Biochemistry 19:* 3841(1980).
58. M. G. Rossman, A. Liljas, G. I. Bränden, and L. J. Banaszak, in *The Enzymes* (P. D. Boyer, ed.), 3rd Ed., Vol. XI, Part A, Academic Press, New York, 1978, p. 61.

9

THE BIOORGANIC CHEMISTRY OF NICOTINAMIDE COENZYMES

I. THE MECHANISM OF CHEMICAL REDUCTION

The nicotinamide coenzymes can be reduced chemically and physically in a variety of ways. The most commonly used method is through sodium hydrosulfite (dithionite) in a reaction which can be written as:

$$S_2O_4^{2-} + 2\ H_2O + NAD^+ \rightarrow 2\ SO_3^{2-} + NADH + 3H^+$$

The product of this reaction is enzymatically active (1). Other methods of reduction are less successful: sodium borohydride reduction yields a product that is only 50% active enzymatically (2,3). Reduction with molecular hydrogen, with a platinum catalyst, results in a hexahydronicotinamide compound (1). Physical methods include electrolytic reduction (4,5) or irradiation in the presence of ethanol (6). Neither the electrolytic reduction produced, nor the product derived from irradiation (7) was enzymatically active. However, both types of systems have been extensively investigated to find a chemical model of the enzymatic direct hydride transfer.

The electrochemical reduction of NAD and its analogues has been studied in many media. Many such compounds have been shown to have an initial reversible one-electron reduction. Such reduction has been studied with a variety of pyridinium ions. The radicals can often yield the dimer of the pyridine compound. For instance, N^1-propylnicotinamide, in water, yields the 4,4'-dimer (8). N^1-methylnicotinamide, in water, yields the 6,6'-dimer between pH 4 and 7, and the 1,4-dihydropyridine at pH values >7.0 (9). Nicotinamide-adenine dinucleotide yields the 4,4'-dimer (10). It appears that 4,4'-dimer or the 6,6'-dimer formation depends on the steric effects of the N^1-substituents (11). The median in which the reduction is carried out determines whether the dimer formed from the intermediate radical is further reduced or whether the anion as a product of further reduction is protonated.

The reverse of the process, electrochemical oxidation of NAD analogues, just forms the protonated pyridinyl radical which can disproportionate to yield the pyridinic salt and uncharacterized products (21).

When the reduction is executed chemically with sodium hydrosulfite, a transient yellow intermediate is seen. Under strongly alkaline conditions, this intermediate can be stabilized, but on neutralization it is converted to the dihydro derivative. Yarmolinski and Colowick (13) showed that this intermediate is an adduct of NAD and the sulfoxylate ion which form a sulfinate derivative of NADH (Fig. 1). When the sulfinyl derivative was hydrolyzed in deuterium oxide, deuterium was incorporated into the reduced coenzyme. But when the adduct was formed in deuterium oxide, but the hydrolysis done in H_2O, the reduced coenzyme was devoid of deuterium (13).

The yellow intermediate had previously been considered a free radical, but the work of Yarmolinski and Colowick ruled out that possibility. N^1-Pyridinium derivatives have a propensity to form charge transfer complexes, as was demonstrated first by Kosower et al. for alkylpyridium iodides (14,15). Kosower therefore proposed a charge transfer complex structure for the intermediate (16). However, from nuclear magnetic resonance spectral data on the analogous intermediate obtained from the reduction of N^1-benzylnicotinamide, the 4-sulfinic acid derivative, and not the charge transfer complex, is the true structure (17).

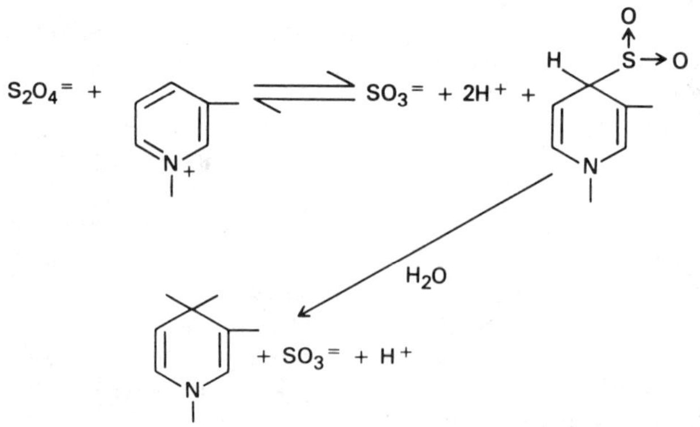

$S_2O_4^{=}$ + [pyridinium] ⇌ $SO_3^{=}$ + $2H^+$ + [4-sulfinate adduct]

$\xrightarrow{H_2O}$ [1,4-dihydropyridine] + $SO_3^{=}$ + H^+

Sum: $S_2O_4^{=} + NAD^+ \longrightarrow 2SO_3^{=} + 3H^+ + NADH$

Figure 1 The chemistry of the chemical reduction of NAD.

II. THE CHEMICAL PROPERTIES OF NAD(P) AND NAD(P)H

The properties of the quaternary pyridinium glycoside of the oxidized NAD and the tertiary glycoside of the reduced NADH are very different.

The oxidized NAD is alkali labile and the split occurs at the nicotinamide–ribose linkage (18). NADH is stable under these conditions. When the alkali is very concentrated (5 *N*) a split does not occur, but a stable fluorescent substance is formed (19). The same reactions are seen with the mononucleotide and N^1-methylnicotinamide. When the reaction is executed in deuterium oxide, the exchange occurs in the 2-position of the nicotinamide residue (20). The action of alkali on NADP yields adenosine-diphosphate-ribose-2'-phosphate (21) which further degrades to adenosine-2'-phosphate-5'-diphosphate (22).

Many nucleophilic substances will form adducts with NAD^+, and these adducts are likely to be in the 4-position. This could be proven for the NAD^+-cyanide complex which resembles NADH in absorption (23, 24). By studying the base-catalyzed deuterium exchange of the adduct, San Pietro proved the 4-position to be the site of reaction (Fig. 2) (25). These reactions are akin to the reactions observed with carbonyl reagents, such as the incubation of dihydroxy acetone with NAD in alkaline media (26–29). This adduct is again in the 4-position. Acetophenone interacts with N^1-benzoylpyridium salts at carbon atom 4 of the pyridinium ring (30). The adducts can be isolated by paper chromatography (31,32) and prepared by alcohol–ether precipitation (32). The substances can be converted to 4-substituted nicotinamide derivatives with ferricyanide oxidation (32).

These reactions occur with a wide variety of nucleophilic agents (33). Their importance is (a) as a means of determination of NAD (this was discussed in Chapter 5, and (b) as models for enzyme–coenzyme–substrate complexes, to be discussed later.

NADH has a very different set of stabilities. The nucleotide is extremely labile under acid conditions (1). First of all there is rapid anomerization around the nicotinamide-glycosidic linkage in NADH (34–37). NADPH is also very unstable under those conditions.

In acid, the 240-nm band shifts to form a band around 295 nm, which is itself further changed with complete loss of the absorption due to nicotinamide (38). In sulfurous acid the intermediate is stable and is called the primary acid product (38). The absorption spectrum is very similar to that previously discussed for NADH–X but not identical to it. In the model compound 1-benzyl-3-acetyl-1,4-dihydropyridine, the primary acid product is 1-benzyl-2-hydroxy-5-acetyl-1,2,3,4-tetrahydropyridine (39). This result agreed with the previous suggestion that it is an olefin addition product (40,41). The reactions in acid can be summarized, as shown in Figure 3.

Chemical oxidation of NADH can be accomplished in alkaline media utilizing hydrogen peroxide. In the presence of metal ions at moderate

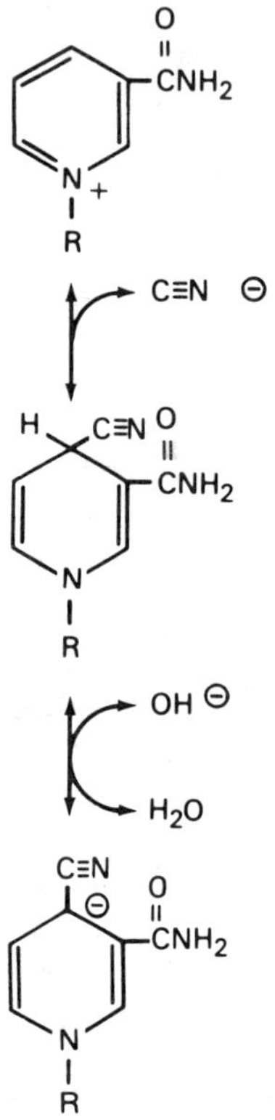

Figure 2 The chemical reaction between quarternary nicotinamide derivatives and cyanide.

alkaline pH values, the oxidation proceeds via NAD to further degradative products (42).

III. THE CHEMISTRY OF REDUCED N^1-SUBSTITUTED NICOTINAMIDES AND ANALOGUES

In the enzymatic reduction of NAD and NADP, a stereospecific direct hydrogen transfer occurs from the substrate. The evidence for this will be discussed later. To understand the organic chemical mechanism by which this occurs, and thereby to be able to formulate hypotheses for enzyme catalysis, a variety of model compounds have been studied. Usually substances have been selected to change the reactivity in the redox process so that reactions analogous to enzyme-catalyzed reactions can occur uncatalyzed. The models are stable, and their solubility in organic solvents facilitates certain reactions.

Two types of models are used: N^1-substituted nicotinamide derivatives or pyridine ring-modified substances, such as the so-called

Hantzsch esters: 1-substituted-2,6-dimethyl-3,5-dicarboxyldihydropyridines or their dimethyl esters.

For nicotinamide derivatives the nature of the N^1-substitution can drastically change the redox potential. For example N^1-*o*-carboxybenzyl-1,4-dihydronicotinamide reduces acridinium ions 100 times faster than the N^1-benzyl derivative (43).

These substances can readily reduce model compounds but they cannot reduce carbonyl groups without catalysis. To study such reactions, polar solvents such as dimethyl sulfoxide (DMSO) have been used. Moreover, the electrophilicity of the carbonyl carbon can be increased by using halogenated ketones. Both dihydronicotinamide and Hantzsch esters will reduce hexachloroacetone but not acetone (44,45). Another model substance is trifluoroacetophenone; it can be reduced while acetophenone cannot (46,47). Such reactions are, however, not

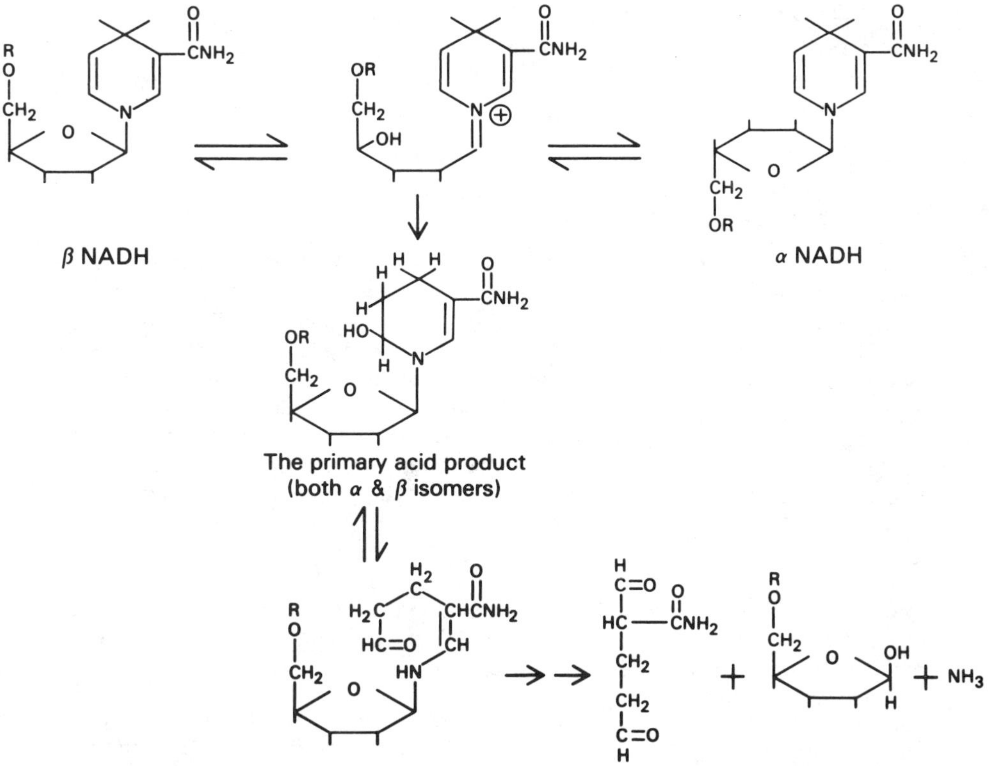

β NADH

α NADH

The primary acid product
(both α & β isomers)

Figure 3 The acid degradation of reduced NAD.

clearcut. Solvent variations or photolysis can lead to dehalogenation instead of reduction.

It is possible to generate more realistic model systems for enzymatic reduction using catalysts. One model utilizes metal ions. Since metal ions, especially zinc, are implicated in dehydrogenase actions, this is an attractive system. The reduction of trifluoroacetophenone with N^1-benzyldihydronicotinamide is accelerated by adding magnesium (22). Zinc is active when used in chelation, such as with 1,10-phenanthroline-2-carboxyaldehyde (48,49).

NAD-dependent dehydrogenases are involved in a wide variety of oxidoreductive processes, such as reduction of olefins, quinones, imines, flavins, disulfides, and simple transhydrogenase reactions wherein a nicotinamide coenzyme is reduced by a dihydronicotinamide coenzyme. Model reactions have been studied for all such reaction types. The importance of such reactions has been the study of hydride ion transfer and asymmetrical reductions. No direct clue to the actual mechanisms of dehydrogenases can be deduced because of the widely varying reactions in the model systems, depending on substrates, media, and physical conditions. Even the seemingly promising reaction in which NADH reduces acetaldehyde by ultraviolet irradiation probably occurs via the activation of NADH by irradiation, since NADH can be oxidized to NAD^+ without substrate by irradiation (50).

IV. ONE- OR TWO-ELECTRON REDUCTION OF PYRIDINE NUCLEOTIDES

In the two-electron reduction of pyridinium salts to form dihydropyridines, the mechanism could proceed via a free radical intermediate; alternatively, a hydride ion transfer could take place. It is quite feasible to generate pyridinyl radicals in model compounds, either by partial reduction of pyridines or partial oxidation of dihydropyridine derivatives. Kosower has summarized all possible reactions in an exhaustive review (51). The relevance of such reactions to biology is unclear, but the weight of evidence favors a hydride ion transfer with two electrons in a single step. The formation of dimers in a one-electron oxidation process makes the free radicals poor candidates for intermediates, even in the model system. Wallenfels observed the formation of leucocyanomalachite green with the NAD–cyanide adduct was oxidized with malachite green (52). In the model of thiobenzophenones direct hydrogen transfer is the most likely explanation of both isotope effects and the influences of substitution on the rate of reaction (53, 54).

On enzymes, however, the transfer of hydrogen is known to be direct (55,56), involving no exchange with the solvent (58). While complex kinetic isotope analyses of model reactions may suggest chemical intermediates, the relevance to enzyme reactions is uncertain. Kinetically there are multiple recognizable steps in the enzyme-catalyzed reactions (59).

V. STEREOSPECIFICITY OF THE COENZYMATIC REDUCTION

The enzymatic reduction and reoxidation of the pyridine nucleotide coenzymes shows absolute stereospecificity. This was first noted by Westheimer et al. utilizing 1,1-dideuteroethanol (56,57,60). When dideuteroethanol was equilibrated with NADH in a reaction catalyzed by yeast alcohol dehydrogenase, only one deuterium was incorporated in the reduced coenzyme. When that reduced, deuterium-containing coenzyme was oxidized with acetaldehyde, utilizing alcohol dehydrogenase, the reoxidized coenzyme was deuterium free. As discussed before, the two monodeutero NADH isomers are called form A and Form B. Form A is deuterated, reduced NAD formed with alcohol dehydrogenase catalysis. While with chemical reduction generation of form A is clearly preferred, there are as many enzymes that generate form B. There is likewise absolute stereospecificity for NADP-dependent enzymes.

The absolute configurations of the stereospecificity of the A and B forms have been determined by Canforth et al. (61).

> When an enzyme of Class A transfers hydrogen from a substrate to a pyridine nucleotide, the hydrogen is added to that side of the nicotinamide ring on which the ring atoms 1 to 6 appear to be in anticlockwise order.

The stereospecific reduction of the substrate by an NAD molecule can be imitated using model compounds in which the substituents of the nicotinamide are asymmetrical, e.g., N^1-benzyldihydronicotinamide, with optically active 1-phenylethane substituting on the carboxamide nitrogen. This substance reduced ethylbenzoylferrate to ethylmandelate with 38% optical purity (62), providing metal ions were present. Other model substances such as the equivalent N^1-propyl derivative reduce trifluoromethylphenylketone symmetrically; however, in the presence of magnesium, asymmetrical reduction occurs. Substitution on the Hantzsch ester with N^1-(–)-methyl groups resulted in a pronounced asymmetrical reduction of pyruvate and benzoylformate esters in the presence of zinc (63).

The stereospecific effects can also be observed with the rate of chemical reoxidation of α- and β-NADH. Whether the oxidant is *N*-methylacridinium ion, diamide, or 10-methyl-5'-deazaisoalloxazine, the oxidation of α-NADH is favored 10-fold over that of β-NADH, even though the overall rate of oxidation spans a 2000-fold difference (64). The same differences are seen with α- and β-nicotinamide-adenine dinucleotides (64).

The model systems are very sensitive to neighboring effects around the ring nitrogen. Therefore, some of the asymmetry effects may be explained by configurations favoring catalysis as opposed to configurations favoring asymmetrical intermediates.

To date the model reactions have not proven useful in organic synthetic reactions. They do provide insight into the chemical properties of nicotinamide on the coenzyme molecules, however.

REFERENCES

1. O. Warburg, W. Christian, and A. Griese, *Biochem. Z. 282:*157 (1935).
2. M. B. Matthews, *J. Biol. Chem. 176:*229(1948).
3. M. B. Matthews and E. E. Conn, *J. Am. Chem. Soc. 75:*5428(1953).
4. B. Ke, *Arch. Biochem. Biophys. 60:*505(1956).
5. B. Ke, *Biochim. Biophys. Acta 20:*547(1956).
6. A. J. Swallow, *Biochem. J. 54:*253(1953).
7. G. Stein and A. J. Swallow, *Nature 173:*937(1954).
8. Y. Paiss and G. Stein, *J. Chem. Soc. (Lond.):*2905(1958).
9. J. N. Burnett and A. L. Underwood, *J. Org. Chem. 30:*1154(1965).
10. R. W. Burnett and A. L. Underwood, *Biochemistry 7:*3328(1968).
11. C. O. Schmakel, K. S. V. Santhanam, and P. J. Elving, *J. Am. Chem. Soc. 95:*5482(1973).
12. W. J. Blaedel and R. G. Haas, *Anal. Chem. 42:*918(1970).
13. M. B. Yarmolinski and S. P. Colowick, *Biochim. Biophys. Acta 20:* 177(1956).
14. E. M. Kosower and P. E. Klinedinst, *J. Am. Chem. Soc. 78:*3493 (1956).
15. E. M. Kosower, *J. Am. Chem. Soc. 78:*3497(1956).
16. E. M. Kosower, in *The Enzymes* (P. O. Boyer, H. Lardy, and K. Myrbäback, eds.), Vol. 3, Academic Press, New York, 1960, p. 171.
17. W. S. Caughey and K. A. Schellenberg, *Fed. Proc. 23:*479(1964).
18. F. Schlenk, H. von Euler, H. Heiwinkel, W. Glein, and H. Nyshan, *Z. Physiol. Chem. 247:*23(1937).
19. N. O. Kaplan, S. P. Colowick, and C. C. Barnes, *J. Biol. Chem. 191:*461(1951).
20. A. San Pietro, *J. Biol. Chem. 217:*589(1955).
21. G. Ben-Hayyim, A. Hochman, and A. Avron, *J. Biol. Chem. 342:* 2837(1967).
22. C. Bernofsky, *Arch. Biochem. Biophys. 166:*645(1975).
23. O. Meyerhof, P. Ohlmeyer, and W. Mohle, *Biochem. Z. 297:*113 (1938).
24. S. P. Colowick, N. O. Kaplan, and M. M. Ciotti, *J. Biol. Chem. 191:*447(1951).
25. A. San Pietro, *J. Biol. Chem. 217:*579(1955).
26. D. M. Needham, L. Siminovitch, and S. M. Rapkine, *Biochem. J. 49:*113(1951).

27. O. Warburg, H. Klotzsch, and Z. Giawehn, *Naturforsch 96:*391 (1954).
28. R. M. Burton and N. V. Kaplan, *J. Am. Chem. Soc. 75:*1005(1953).
29. R. M. Burton and N. V. Kaplan, *J. Biol. Chem. 206:*283(1954).
30. W. E. von Doering and W. E. McEwen, *J. Am. Chem. Soc. 73:*2104 (1951).
31. R. M. Burton and A. San Pietro, *Arch. Biochm. Biophys. 48:*184 (1954).
32. R. M. Burton, A. San Pietro, and N. O. Kaplan, *Arch. Biochem. Biophys. 70:*87(1957).
33. S. P. Colowick, J. van Eys, and J. H. Park, in *Comprehensive Biochemistry,* (M. Florkin, and E. H. Stotz, eds.), *Vol. 14, Biological Oxidations,* Elsevier, Amsterdam, 1966, p. 1.
34. C. Woenckhaus and P. Zumpe, *Biochem. Z. 343:*326(1965).
35. D. W. Miles, D. W. Urry, and M. Eyring, *Biochemistry 7:*2333 (1968).
36. E. L. Jacobson, M. K. Jacobson, and C. Bernofsky, *J. Biol. Chem. 248:*7891(1973).
37. N. J. Oppenheimer and N. O. Kaplan, *Arch. Biochem. Biophys. 166:*526(1975).
38. E. Haas, *Biochem. Z. 288:*123(1936).
39. A. G. Anderson and G. Berkehammer, *J. Am. Chem. Soc. 80:*992 (1958).
40. P. Karrer, F. W. Kahnt, R. Epstein, W. Jaffe, and T. Ishii, *Helv. Chim. Act. 21:*223(1938).
41. P. Karrer and F. J. Stare, *Helv. Chim. Act. 20:*418(1937).
42. R. M. Burton and M. Lamborg, *Arch. Biochem. Biopshy. 62:*369 (1956).
43. J. Hadju and D. S. Sigman, *J. Am. Chem. Soc. 97:*3524(1975).
44. D. C. Dittmer and R. A. Fouty, *J. Am. Chem. Soc. 86:*91(1964).
45. D. C. Dittmer, L. J. Steffa, J. R. Poloski, and R. A. Fouty, *Tetrahedron Lett.* 827(1961).
46. J. J. Steffens and D. M. Chipman, *J. Am. Chem. Soc. 93:*6694 (1971).
47. Y. Ohnishi, T. Numakunai, and F. Ohno, *Tetrahedron Lett. 6:* 3813(1975).
48. D. J. Creighton and D. S. Sigman, *J. Am. Chem. Soc. 93:*6314 (1971).
49. D. J. Creighton, J. Hajdu, and D. S. Sigman, *J. Am. Chem. Soc. 98:*4614(1976).
50. D. Abelson, E. Panthé, K. W. Lee, and A. Boyle, *Biochem. J. 96:*840(1965).
51. E. M. Kosower in *Free Radicals in Biology* (W. A. Pryor, ed.), Academic Press, New York, 1976, p. 1.
52. K. Wallenfels, in *Steric Course of Microbiological Reactions,* (G. E. W. Wolstenholme and C. M. O'Connor, eds.), Churchill, London, 1959, p. 10.

53. R. H. Abeles, R. F. Hutton, and F. H. Westheimer, *J. Am. Chem. Soc. 79:*712(1957).
54. F. H. Westheimer, in *The Enzymes*, Vol. 1 (D. D. Boyer, H. Lardy, and K. Myrbäck, eds.), Academic Press, New York, 1959, p. 259.
55. F. H. Westheimer, in *The Mechanism of Enzyme Action*, (W. D. McElroy and B. Glass, eds.), Johns Hopkins Press, Baltimore, 1954, p. 321.
56. F. H. Westheimer, H. F. Fisher, E. E. Conn, and B. Vennesland, *J. Am. Chem. Soc. 73:*2403(1951).
57. F. H. Westheimer, H. F. Fisher, E. E. Conn, and B. Vennesland, *J. Biol. Chem. 202:*687(1953).
58. W. S. Allison, H. B. White, and M. J. Conners, *Biochemistry 10:* 2290(1970).
59. K. Dalziel, in *The Enzymes* (P. D. Boyer, ed.), 3rd Edition, Vol. 11, Academic Press, New York, 1975, p. 10.
60. B. Vennesland and F. M. Westheimer, in *The Mechanisms of Enzyme Action* (W. D. McElroy and H. B. Glass, eds.), Johns Hopkins Press, Baltimore, 1954, p. 357.
61. J. W. Conforth, R. Ryback, G. Popjack, G. Donninger, and G. Schroepfer, *Biochem. Biophys. Res. Commun. 9:*371(1962).
62. Y. Ohnishi, M. Kagami, and A. Ohro, *J. Am. Chem. Soc. 97:*4766 (1976).
63. K. Nishigama, N. Babu, J. Oda, and Y. Inouge, *J. Chem. Soc., Chem. Comm.:*101(1976).
64. J. Hajdu and D. S. Sigma, *Biochemistry 16:*2841(1977).

10

THE BIOSYNTHESIS OF NICOTINAMIDE COENZYMES

I. BIOSYNTHETIC PATHWAYS OF PYRIDINE NUCLEOTIDES

The biosynthesis of nicotinic acid and nicotinamide in the mammalian organism has been described in a previous chapter. The role of nicotinamide so formed is in the synthesis of nicotinamide coenzymes and in the regulation of their biosynthetic and degradative rate.

Nicotinamide coenzymes are derived from one of three sources: endogenously, from tryptophan via quinolinic acid, or from nicotinic acid or nicotinamide. Their pivotal role in energy metabolism, biosynthetic oxidoreductive pathways, and DNA repair makes the regulation of the coenzyme levels of vital importance. Such regulation must involve both the rate of utilization and the quantity available of the pyridine precursors.

Preiss and Handler (1) were first to propose a pathway for the biosynthesis of NAD. They suggested the following steps:

Nicotinic acid + phosphoribosyl pyrophosphate $\rightleftarrows$
Nicotinic acid mononucleotide + pyrophosphate

Nicotinic acid mononucleotide + ATP $\rightleftarrows$
Nicotinic acid-adenine dinucleotide + Pyrophosphate

Nicotinic acid-adenine dinucleotide + ATP + glutamine $\rightarrow$
Nicotinamide-adenine dinucleotide + AMP + pyrophosphate +
glutamic acid

While this pathway accounts for the incorporation of endogenously generated and diet-derived nicotinic acid, it does not readily show why nicotinamide can completely replace nicotinic acid in the function. Nicotinamide is the predominant form of circulating free vitamin (2,3). In bacteria, nicotinamide is freely deamidated. There are enzymes in mammalian cells that can execute the same reactions, but very few cells can do it readily (4), except the liver (5,6), and even that enzyme is not necessarily a specific nicotinamide deamidase. The enzyme hydrolyzes a variety of amides and esters. Suggestions have been made that mammals rely on bacterial deamidation in the intestines (7). That pathway does not need to be invoked anymore. There is an enzyme that converts nicotinamide with phosphoribosyl pyrophosphate (PRPP) into nicotinamide mononucleotide (8), which then can react with ATP to form NAD (1,9), or be deamidated to enter the Priess–Handler pathway (10). The latter reaction occurs, however, only in bacteria as far as is known (10–12).

Nevertheless, data seemed to suggest that nicotinamide is utilized for NAD biosynthesis obligatorily via nicotinic acid. When nicotinamide is administered to mice, rats, and cats, nicotinic acid derivatives are found (13–17). Those data were derived from experiments using pharmacological quantities of nicotinamide. Grunicke et al. (18) demonstrated that nicotinamide is a direct precursor under physiological conditions, and showed that unlabeled nicotinic acid does not dilute the incorporation of nicotinamide into NAD. Furthermore, azaserine, which is a potent inhibotor of the conversion of nicotinic acid-adenine dinucleotide to NAD (1), did not inhibit the incorporation of nicotinamide. Finally, after [^{14}C] nicotinamide administration, labeled nicotinic acid nucleotides did not accumulate. The latter findings were, as mentioned, in direct contradiction to previous data which suggested that nicotinic acid derivatives were intermediates (13,14,16,17,19). These studies all used amounts of nicotinamide that raised the concentrations in tissues above the K_m for nicotinamide deaminidase, which is very high (5,6,20). Physiological levels of nicotinamide are well below the concentration at which deamidation occurs (21,22). In Ehrlich ascites cells that level is 10–100 μM. At hyperphysiological levels the portal recirculation of nicotinamide excreted into the gut may in fact occur.

Figure 1 summarizes the pathways of NAD biosynthesis. The reaction that hydrolyzes NAD to nicotinamide and adenosine diphosphate-ribose (ADPR) will be discussed later in detail. In addition to the reactions described, nonspecific enzymes are found which dephosphorylate the mononucleotides and hydrolyze or phosphorylate the corresponding ribosides. Such enzymes generate free nicotinamide which is equivalent to other sources of nicotinamide.

The nicotinamide-riboside phosphorylases and hydrolases are nonspecific enzymes (23), and act also on adenosine. There is a structural similarity between adenosine and nicotinamide riboside that makes this a reasonable dual function of the enzymes (24). It appears that the rat liver nucleoside phosphorylase splits both nicotinic acid- and nicotinamide-riboside (25). The relative activities for nicotinic acid-riboside, nicotinamide-riboside, and adenosine are roughly equal. However,

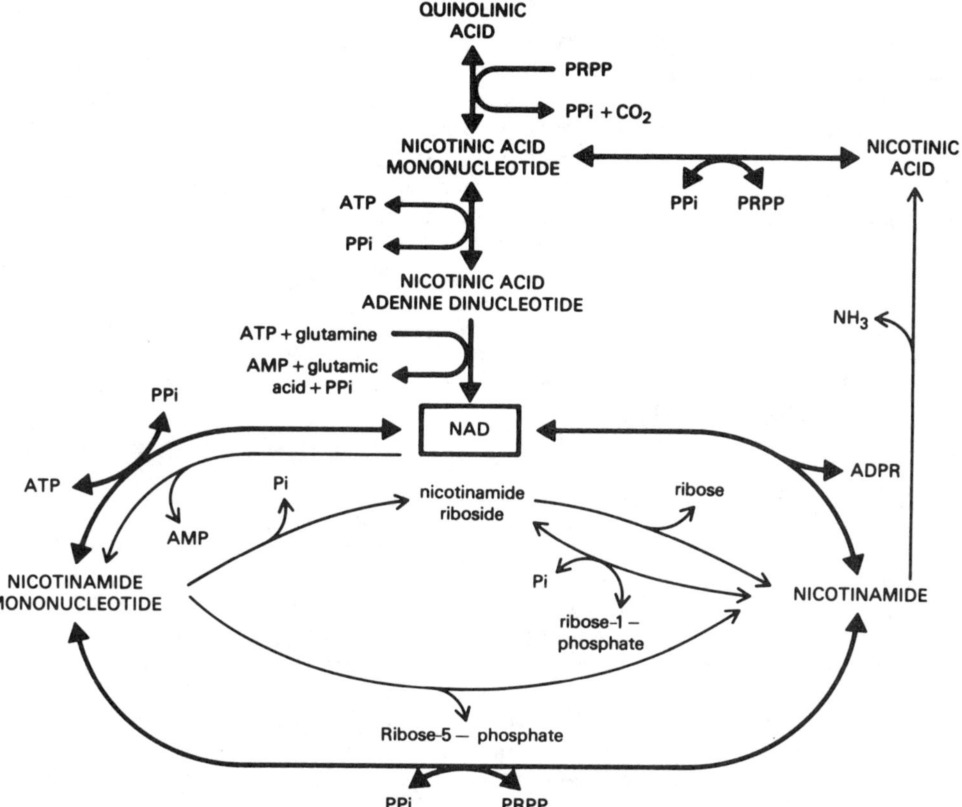

Figure 1 NAD metabolism.

the phosphorolysis of the pyridine ribosides is complete, while for purine nucleosides, the reverse, synthetic reaction is actually favored. The pyridine–riboside linkage is a high-energy bond (26).

The enzyme from human erythrocytes splits inosine- and nicotinamide-riboside, but acetylpyridine-riboside inhibits the latter and not the former activity, which suggests that two different enzymes are involved (27). The enzyme did not readily exchange nicotinamide with the riboside until ergothioneine was added (28). Interestingly, this component was also said to make the usually purely hydrolytic *Neurospora* NADase capable of exchanging nicotinamide. However, this is either an artifact or a nonphysiological reaction.

II. THE BIOSYNTHESIS OF NADP

The conversion of NAD to NADP is direct. The reaction is executed by a specific kinase, which generates a 2'-phosphate bond on the adenylic acid moiety of NAD. The enzyme is found in yeast (29,30), livers from pigeon (31,32) rabbit (33), guinea pig (34,35), rat (36), calf (37), rat brain (38), rat mammary gland (39), sea urchin (40), bacteria (41–43), and yeast (44). There is a kinase in yeast for NADH as well (42). The enzyme from *Azotobacteria vinelandii* will also phosphorylate the α-isomer of NAD to form α-NADP (45,46).

The NADP in turn can be converted to NAD by a phosphatase. The existence of this enzyme was already noted by von Euler (47). It is likely to be specific. In *E. coli* the conversion of NAD and NADP is a steady state equilibrium (41) which regulates nicotinic acid uptake. The 5'- and 3'-nucleotides do not attack NADP (45); however, nonspecific alkaline phosphatase does.

III. NAD DEGRADATION

The NAD cycle depicted has two degradative routes: via pyrophosphatase, to form AMP and nicotinamide mononucleotide, and via a hydrolysis, to yield nicotinamide and adenosine diphosphate-ribose (ADPR). Both enzymes have greater importance than merely degradative activity to balance synthesis and breakdown.

Pyrophosphatases that degrade NAD have been described by Kornberg (48–50). For preparative purposes the enzyme from potatoes is usually used; however, the enzymes are widespread in nature, have broad specificity, and are generally called nucleotide pyrophosphatases. There is also a pyrophosphatase that splits only NADH and NAD in pigeon liver (51,52). In bacteria the pyrophosphatase is membrane associated, and is considered integral to the mechanism of utilization of exogenous NAD in *E. coli* and *Salmonella* (52). A nicotinamide mononucleotide glycohydrolase, which yields nicotinamide and ribose-5'-phosphate, is part of the transport system which yields free intracellular nicotinamide (54). In animal cells there is compartmentation, and similar phenomena may occur.

The NAD glycohydrolases, which convert NAD to nicotinamide and adenosine diphosphate-ribose (ADPR), are a group of enzymes that have assumed major importance in molecular biology, having already been used broadly in the in vitro biosynthesis of NAD analogues of a wide variety.

The enzyme was first discovered in animal tissues by Handler and Klein (55). The first detailed studies of this enzyme activity were based on preparations from pig brain by McIllwain and Rodnight (56, 57). The third detailed investigation of this enzyme was undertaken using the NAD glycohydrolase from *Neurospora* (58), which was found to be especially high in activity after the mold was grown in a zinc-deficient medium (59). All enzymes utilized both NAD^+ and $NADP^+$ as substrate.

There were differences between the three enzymes. First, the *Neurospora* enzyme was uninhibited by nicotinamide while the pig-brain enzyme was (48). As already mentioned, it has been claimed that the addition of ergothioneine altered the *Neurospora* enzyme from nicotinamide insensitive to sensitive (28). The lack of exchange is not limited to mold enzymes. The enzyme from rat liver chromatin is uninhibited by nicotinamide, also (60).

The prevention of NAD degradation by nicotinamide was noted early: Von Euler (61) and Lennerstrand (62) had already made mention of breakdown of NAD which Mann and Quastel had found to be inhibited by nicotinamide (63). McIlwain studied the effect of nicotinamide and analogues of nicotinamide on the pig-brain enzyme (57,64). Spleen was found to be a richer source of this activity (65) and was therefore studied in detail for this effect. The landmark paper by Zatman et al. (65) showed the mechanism for this inhibition to be an exchange reaction, competing with hydrolysis:

$$\begin{array}{l} N^+ADPR + \text{enzyme} \rightleftarrows ADPR\text{-enzyme} + N + H^+ \\ \qquad\qquad\qquad\qquad\qquad\quad \downarrow H_2O \\ \qquad\qquad\qquad\qquad\qquad\quad ADPR + \text{enzyme} \end{array}$$

By executing the reaction in the presence of radioactive nicotinamide, labeled NAD could be isolated.

Some pyridine bases were more powerful inhibitors of the splitting of NAD by the beef spleen enzyme than was nicotinamide, but other enzymes, such as the pig-brain enzyme, were relatively insensitive to inhibition by such compounds. The model first studied was isonicotinic acid hydrazide (INH), and the terminology "INH-sensitive" and "INH-insensitive" as applied to enzymes was coined. The paradigms were the beef-spleen and the pig-brain enzyme, respectively (67). The pig-brain (68) and beef-spleen (69) enzymes formed an analogue of NAD wherein INH replaced the nicotinamide, but the yield of product was the beef-spleen enzyme was low because the analogue inhibited further breakdown of NAD. The importance of the analogues so formed will be discussed later. This enzyme activity as the basis of the action of pyridine antagonists to nicotinamide has already been discussed in Chapter 7.

The activity of NAD glycohydrolase took on new importance when Sugimura et al. discovered a reaction in eukaryotes which is best depicted as:

$$X + n\ N^+ADPR \rightarrow (ADPR)_n\text{-}X + nN + nH^+$$

where X is an acceptor. The reaction in prokaryotes is a mono-ADP ribosylation which occurs as:

$$X + N^{+}ADPR \rightarrow ADPR\text{-}X + N + H^{+}$$

In contrast to the exchange reaction, in which X is a pyridine base analogue and ADPR-X represents the corresponding NAD analogue, in these reactions X is a protein or other macromolecular acceptor. The reaction in bacteria has been described in many systems. The acceptor is a regulatory protein. Examples include the ribosylation of Elongation Factor 2 in eukaryotic protin-synthesizing systems by diphtheria toxin (70), and of membrane-bound adenylate cyclase by cholera toxin (71, 72). On the other hand, the acceptor for poly-ADPR is a histone, and the reaction is involved with DNA metabolism. The reaction is DNA dependent. The polymerization of ADPR and the initial cleavage of NAD have not been clearly separated. The enzyme activity has been found in nuclei of many eukaryotic species, and the enzyme has been purified to a homogeneous state from Ehrlich ascites cells (73) and from calf thymus (74–76). It has also been purified from rat liver (77), and is found in plants (78) and slime molds (79), as well as in animal cells.

The roles of these enzymes will be discussed later. Suffice it to say that these enzymes have an absolute requirement for DNA and histone, and therefore their activity is distinct from pure glycohydrolase activity. Nicotinamide is a potent inhibitor of the reaction (76).

IV. CELLULAR LOCALIZATION OF NAD BIOSYNTHESIS AND DEGRADATION

Compartmentation in the cells is exemplified by NAD. Mitochondria are largely impermeable to the pyridine nucleotides, and hydrogen transfer from cytoplasm to mitochondria occurs via substrates that can permeate the mitochondrial membranes or via transhydrogenation between pyridine nucleotides.

It is therefore reasonable to expect nicotinamide biosynthesis to occur in mitochondria. The bulk of the synthesis from nicotinic acid and quinolinic acid occurs in the cytoplasm. Nicotinamide incorporation into nicotinamide mononucleotide likewise occurs via a cytoplasmic enzyme.

The pyrophosphorylase enzyme, which catalyzes the reaction

$$\text{Nicotinamide mononucleotide} + \text{ATP} \rightleftarrows \text{NAD} + \text{pyrophosphate}$$

is found primarily in nuclei (80). This enzyme is the same as the corresponding nicotinic acid-mononucleotide-utilizing enzyme which forms nicotinic acid-adenine dinucleotide (1,9). The enzyme has been brought to near crystalline state of purification (9,81). Because the enzyme does not occur in the cytoplasm, erythrocytes from mammals are the only tissue that cannot utilize nicotinamide for NAD biosynthesis (1). The conversion of the nicotinic acid analogue to NAD (38) and the conversion of NAD to NADP (83) again occur in the cytoplasm.

Pyrophosphatases are found in the cytoplasm and in microsomes (32), and their relative activities toward NADH and NAD are different.

NADases are distributed over many fractions, but the activity is highest in nuclei and microsomes. It is likely that the measured activity in nuclei is a combination of NADase per se and poly-ADPR synthesis, since most assays rely on the measurement of NAD disappearance.

Grunicki (18) has presented evidence for separate mitochontrial NAD synthesis, through incorporation of labeled nicotinamide, but nicotinic acid did not serve as precursor of mitochondrial NAD.

V. REGULATION OF NAD BIOSYNTHESIS

The complex nature of NAD functions necessitates that NAD levels be regulated. All tissues can incorporate nicotinamide into NAD (8,83), and actually prefer nicotinamide to nicotinic acid.

The role of the liver in the biosynthesis of nicotinamide from nicotinic acid and 5-phospho-α-D-ribose 1-pyrophosphate (PRPP) could be the consequence of the NAD cycle:

Nicotinic acid + PRPP → nicotinic acid mononucleotide
Nicotinic acid mononucleotide + ATP → nicotinic acid-adenine dinucleotide + PP_i
Nicotinic acid-adenine dinucleotide + ATP + glutamine → NAD + AMP + PP_i + glutamic acid
NAD → nicotinamide + adenosine diphosphate-ribose

Sum: Nicotinic acid + PRPP + 2 ATP + glutamine → nicotinamide + AMP + ADPR + 2 PP_i + glutamic acid

Nicotinamide then is a product of liver metabolism. It is the primary circulating form of the vitamin (2,3).

At the cellular level the tissue levels are under feedback control. For instance, nicotinamide phosphoribosyl transferase is under strong feedback inhibition by nicotinamide coenzymes in oxidized and reduced forms (85). Bacteria also have elaborate regulatory mechanisms at the enzyme and at the genetic level. The details of such mechanisms are too far afield in this context. For review, the paper by Foster and Moat (86) can be consulted.

The homeostasis in whole animals is of greater importance. This problem has intrigued biochemists and nutritionists for a long period of time. A classical early review details much of the older data (87). The primary regulatory substance is nicotinamide itself. Nicotinamide levels in blood are buffered by the liver by being converted into a storage form of NAD. In liver, total NAD is greater than functional NAD (88). Excess NAD can be hydrolyzed by the NAD glycohydrolases to free nicotinamide. Excess nicotinamide can be converted to any of the numerous products that have been described in Chapter 4. The level of hepatic NAD is strongly affected by the level of nicotinamide administered. The administration of nicotinamide to mice results in increases of up to 10-fold in NAD content (89). The magnitude of increase in synthesis after nicotinic acid or tryptophan administration is much lower.

In tissue culture studies an increase in NAD content of individual cells after nicotinamide exposure can also be demonstrated. There is a rapid turnover of the nicotinamide moiety of NAD (82,90). It is likely that this turnover is proportional to the formation of poly-ADPR (91) and protein ADP ribosylations (92). Nicotinamide inhibits that process. In fact, nicotinamide is toxic to cells at concentrations that increase NAD above normal (93).

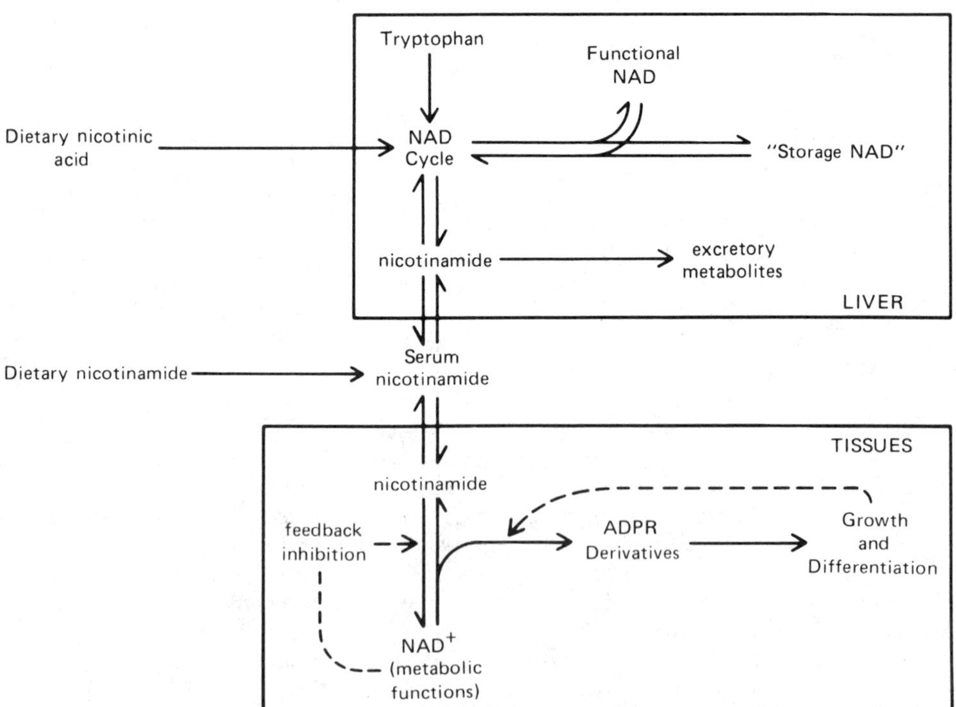

Figure 2 Metabolic utilization of dietary nicotinamide and nicotinic acid.

Nicotinamide is also toxic when fed at levels of 5.3–6.8 m*M*/kg, and at levels >10 m*M*/kg, rats actually lose weight (94). The LD_{50} of nicotinamide for young rats is 13.8 μ*M*/kg (95). It was thought that such amounts inhibit RNA biosynthesis (96), but it is quite possible that these amounts create a competition for ATP and phosphoribosylpyrophosphate as proposed by Ferris and Clark (97). However, inhibition of ADPR synthesis may be a factor.

The suggestion that the NAD content of liver is in part a storage form of nicotinamide is intriguing. The increase of NAD^+ following nicotinamide administration is small in the liver (86) and what little change occurs is seen in the level of NADH. The maximum is reached in 8–12 hr and then declines over 24–32 hr back to normal levels (86, 98,99). The lack of equilibration argues for a separate compartment for the NAD since the NAD^+/NADH ratio responds rapidly to the redox state of the cell. The ratio of NAD^+/NADH does not change if there were an absolute separation of newly formed NAD from previously existing, functional NAD^+ (100).

These studies are complicated by the enzyme-inducing effects of excess nicotinamide. The studies on hypophysectomized animals by Gveengard (100,101) are also complicated by the conversion of tryptophan to niacin (102).

In spite of all these provisos the schema advanced by Bernofsky are a model for the regulation of NAD levels in whole animals (88). This regulation is then paralleled at the cellular level by feedback inhibition of biosynthesis and regulation of the rate of NAD^+ utilization because of changes in DNA activity during cell growth and differentiation (Fig. 2).

REFERENCES

1. J. Preiss and P. Handler, *J. Biol. Chem. 233:*488(1958).
2. S. Chaykin, M. Dagani, L. Johnson, and M. Samli, *J. Biol. Chem. 240:*932(1965).
3. S. Chaykin, M. Dagani, L. Johnson, M. Samli, and Bataile, J., *J. Biochm. Biophys. Act. 100*:351(1965).
4. J. Kirchner, J. G. Watson, and S. Chaykin, *J. Biol. Chem. 241:* 953(1966).
5. R. K. Cholson, I. Ueda, M. Ogasawara, and L. M. Henderson, *J. Biol. Chem. 239:*1208(1964).
6. B. Peshack, P. Greengard, A. Craston and H. J. Kalinsky, *Biochem. Biophys. Res. Comm. 13:*478(1963).
7. H. Ijichi, A. Ichiyama, and O. Hayaishi, *J. Biol. Chem. 241:*3701 (1966).
8. L. S. Dietrich, L. Fuller, I. L. Yero, and L. Martinez, *J. Biol. Chem. 241:*188(1966).
9. M. R. Atkinson, J. Jackson, and R. K. Morton, *Nature 192:*946 (1961).

10. H. C. Friedman, in *Methods in Enzymology Vol. 18*, (D. B. McCormick and L. D. Wright, eds.) Academic Press, New York, 1971, p. 192.
11. S. A. Fyfe and H. C. Friedman, *J. Biol. Chem. 244:*1659(1969).
12. T. Imai, *J. Biochem.* (Tokyo) *73:*139(1973).
13. N. O. Kaplan, A. Goldin, S. R. Humphreys, M. M. Ciotti, and F. E. Stolzenbach, *J. Biol. Chem. 219:*287(1956).
14. T. A. Langan, N. O. Kaplan, and L. Shuster, *J. Biol. Chem. 234:*2161(1959).
15. H. Ijichi, A. Ichiyama, and O. Hayaishi, *J. Biol. Chem. 241:*3701 (1966).
16. C. Ricci and V. Pallini, *Biochem. Biophys. Res. Comm. 112:*282 (1965).
17. V. Pallini and C. Ricci, *Arch. Biochem. Biophys. 112:*282(1965).
18. H. Grunicke, H. J. Keller, M. Liersch, and A. Benaguid, *Adv. Enz. Reg. 12:*397(1974).
19. S. A. Narrod, V. Bonavita, E. R. Ehrenfeld, and N. O. Kaplan, *J. Biol. Chem. 236:*931(1961).
20. B. Peshach, P. Greengard, A. Craston, and F. Sheppy, *J. Biol. Chem. 240:*1725(1965).
21. R. Fussgänger, *Bestimmung und Stoffwechselverhalten von Nikotinsäure und Nikotinsäureamid in Ehrlich Ascites-Tumorzellen.* Dissertation, Med. Fakultät, Universität Freiburg (1965), quoted in 18.
22. F. Märki and P. Greengard, *Biochim. Biophys. Act. 113:*587(1966).
23. J. W. Rowen and A. Kornberg, *J. Biol. Chem. 193:*497(1951).
24. J. van Eys, *J. Bacteriol. 80:*386(1960).
25. Y. Nishizuka, in *Methods in Enzymology, Vol. 18* (D. B. McCormick and L. D. Wright, eds.) Academic Press, New York, 1971, p. 204.
26. L. J. Zatman, N. O. Kaplan, and S. P. Colowick, *J. Biol. Chem. 200:*197(1953).
27. L. Grossman and N. O. Kaplan, *J. Biol. Chem. 231:*717(1958).
28. L. Grossman and N. O. Kaplan, *J. Biol. Chem. 231:*727(1958).
29. H. von Euler and F. Adler, *Z. Physiol. Chem. 252:*41(1938).
30. A. Kornberg, *J. Biol. Chem. 182:*805(1950).
31. A. H. Mehler, A. Kornberg, S. Grisolia, and S. Ochoa, *J. Biol. Chem. 174:*961(1974).
32. B. Katchman, J. J. Betheil, A. I. Shepartz, and D. R. Sanadi, *Arch. Biochem. Biophys. 34:*437(1951).
33. T. P. Wang and N. O. Kaplan, *J. Biol. Chem. 206:*311(1954).
34. V. L. Nemchinskaya, V. P. Kishner, Y. M. Bozhkov, E. I. Turcheako, and S. E. Tukachinskii, *Biochimiya 31:*306(1966).
35. D. K. Apps, *Biochem. J. 104:*358(1967).
36. A. L. Greenbaum, J. B. Clark, and P. McLean, *Biochem. J. 95:* 167(1965).

37. L. S. Dietrich and I. L. Yers, in *Methods in Enzymology, Vol. 18* (D. B. McCormick and L. D. Wright, eds.), Academic Press, New York, 1971, p. 156.
38. S. Pinder, J. B. Clark, and A. L. Greenbaum, in *Methods in Enzymology, Vol. 18,* (D. B. McCormick, and L. D. Wright, eds.), Academic Press, New York, 1971, p. 20.
39. A. L. Greenbaum and S. Pinder, *Biochem. J. 107:*55(1968).
40. C. H. Blomquist, *J. Biol. Chem. 248:*7044(1973).
41. R. Lundquist and B. M. Olivera, *J. Biol. Chem. 246:*1107(1971).
42. J. Imsande and A. B. Pardee, *J. Biol. Chem. 237:*1305(1962).
43. A. E. Chun, in *Methods in Enzymology, Vol. 18,* (D. B. McCormick and L. D. Wright, eds.), Academic Press, New York, 1971, p. 149.
44. D. K. Apps, *Eur. J. Biochem. 13:*223 (1970).
45. K. Suzuki, H. Nakano, and S. Suzuhi, *J. Biol. Chem. 242:*3319 (1967).
46. C. Bernofsky and W. J. Gallagher, *Anal. Biochem. 67:*611(1975).
47. L. Shuster and N. O. Kaplan, *J. Biol. Chem. 215:*183(1955).
48. A. Kornberg and W. E. Pricer, *J. Biol. Chem. 182:*763(1950).
49. A. Kornberg and D. Lindberg, *J. Biol. Chem. 176:*665(1948).
50. G. W. E. Plaut and K. A. Plaut, *Arch. Biochem. Biophys. 48:*189(1954).
51. K. B. Jacobson and N. O. Kaplan, *J. Biol. Chem. 226:*427(1957).
52. K. B. Jacobson and N. O. Kaplan, *J. Biophys. Biochem. Cytol. 1:*31(1957).
53. J. W. Foster, D. M. Kinney, and A. G. Moat, *J. Bacteriol. 137:*1165(1979).
54. A. J. Andreoli, T. W. Okita, R. Bloom, and T. A. Grover, *Biochem. Biophys. Res. Comm. 12:*92(1963).
55. P. Handler and J. R. Klein, *J. Biol. Chem. 143:*49(1942).
56. H. McIllwain and R. Rodnight, *Biochem. J. 44:*470(1949).
57. H. McIllwain and R. Rodnight, *Biochem. J. 45:*337(1949).
58. N. O. Kaplan, S. P. Colowick, and A. Nason, *J. Biol. Chem. 191:*473(1952).
59. A. Nason, N. O. Kaplan, and S. P. Colowick, *J. Biol. Chem. 188:*397(1951).
60. V. Meda, M. Fukushima, M. Okayama, and O. Hayashi, *J. Biol. Chem. 250:*7541(1975).
61. H. von Euler, H. Heiwinkel, and F. Schlenk, .. *Physiol. Chem. 247:*iv(1937).
62. A. Lennerstand, *Biochem. Z. 172:*287(1936).
63. P. J. G. Mann and J. H. Quastel, *Biochem. J. 35:*502(1941).
64. H. McIlwain, *Biochem. J. 46:*612 (1950).
65. J. H. Quastel and L. Zatman, *J. Biochim. Biophys. Acta 000:*000(0000).
66. L. J. Zatman, N. O. Kaplan, and S. P. Colowick, *J. Biol. Chem. 200:*197(1953).

67. L. J. Zatman, N. O. Kaplan, S. P. Colowick, and M. M. Ciotti, *J. Biol. Chem. 209:*453(1954).
68. L. J. Zatman, N. O. Kaplan, S. P. Colowick, and M. M. Ciotti, *J. Biol. Chem. 209:*467(1954).
69. D. S. Goldman, *J. Am. Chem. Soc. 76:*2841(1954).
70. T. Honjo, Y. Nishizuka, J. Kato, and O. Hayaishi, *J. Biol. Chem. 246:*4251(1971).
71. D. Cassel and T. Pfeuffer, *Proc. Nat. Acad. Sci. 78:*2669(1978).
72. N. H. Iglewski, J. Sadoft, M. L. Bjorn, and E. S. Maxwell, *Proc. Nat. Acad. Sci. 78:*2669(1978).
73. T. Kristensen and J. Holtlund, *Eur. J. Biochem. 70:*441(1976).
74. P. Mandel, H. Okazaki, and C. Niedergang, *FEBS Lett. 84:*331 (1977).
75. K. Yoshihara, T. Hashida, Y. Tanaka, H. Ohgushi, H. Yoshihara, and T. Kamiya, *J. Biol. Chem. 253:*6459(1978).
76. S. Ito, Y. Shizuta, and O. Hayaishi, *J. Biol. Chem. 254:*3647 (1979).
77. H. Okayama, C. Edson, M. Fukushima, R. Ueda, and O. Hayaishi, *J. Biol. Chem. 252:*7000(1977).
78. J. F. Payne and A. K. Bal, *Exp. Cell. Res. 99:*428(1976).
79. M. Brightwell and S. Shall, *Biochem. J. 125:*768(1971).
80. G. H. Hogeboom and W. C. Schneider, *J. Biol. Chem. 197:*611 (1952).
81. M. R. Atkinson and R. K. Morton, *Nature 188:*58(1961).
82. V. Stollar and N. O. Kaplan, *J. Biol. Chem. 236:*1836(1961).
83. C. Streffer and J. Benes, *Eur. J. Biochem. 21:*357(1971).
84. P. B. Collins and S. Chaykin, .. *Biol. Chem. 247:*778(1972).
85. L. S. Dietrich, O. Muniz, and M. Powando, *J. Vitaminol. 14:* (Suppl.):123(1968).
86. J. W. Foster and A. G. Moat, *Microbiol. Rev. 44:*83(1980).
87. S. Chaykin, *Ann. Rev. Biochem. 36:*149(1967).
88. C. Bernofsky, *Mol. Cell. Biochem. 33:*135(1980).
89. N. O. Kaplan, A. Goldin, S. R. Humphreys, M. M. Ciotti, and F. E. Stolzenbach, *J. Biol. Chem. 219:*287(1956).
90. L. Shuster, T. A. Langan, N. O. Kaplan, and A. Goldin, *Nature 182:*512(1958).
91. T. Sugimura, *Prog. Nucl. Acid. Res. Mol. Biol. 13:*127(1973).
92. T. Honjo and O. Hayaishi, *Curr. Top. Cell. Reg. 7:*87(1973).
93. D. A. Gardner, G. H. Sato, and N. O. Kaplan, *Develop. Biol. 28:*84(1972).
94. P. Handler and W. J. Dann, *J. Biol. Chem. 146:*357(1943).
95. F. G. Brozda and R. A. Coulson, *Proc. Soc. Exp. Biol. Med. 62:* 19(1946).
96. M. Revel and P. Mandel, *Cancer Res. 22:*486(1962).
97. G. M. Ferris and J. R. Clark, *Biochm. J. 121:*655(1971).
98. A. Bonsignore and C. Ricci, *Scientia Med. Ital. 6:*650(1958), as quoted in ref. 88.

99. P. Feigelson, J. N. Williams, and C. A. Elvehjem, *Proc. Soc. Exp. Biol. Med. 78:*34(1951).
100. P. Greengard, G. P. Quinn, and M. P. Reid, *J. Biol. Chem. 239:* 1887(1964).
101. P. Greengard, B. Peshach, and H. Kalinsky, *J. Biol. Chem. 242:* 152(1967).
102. P. Greengard, H. Kalinsky, T. J. Manning, and S. B. Zak, *J. Biol. Chem. 243:*4216(1968).

11

NICOTINAMIDE COENZYME ANALOGUES

I. SYNTHESIS OF COENZYME ANALOGUES

The reactions of NAD glycohydrolase enzymes occur with an adenosine diphosphate-ribose (ADPR)−enzyme complex intermediate, allowing a variety of bases to react with the intermediate to form coenzyme analogues:

$$\text{ADPR–N}^{+} + \text{base} \rightarrow \text{ADPR–base} + \text{N}$$

The original finding of an exchange reaction between nicotinamide and NAD (1) allowed the formation of ^{14}C-labeled NAD. The first analogue prepared was the isonicotinic acid hydrazide-adenine dinucleotide utilizing the pig-brain enzyme (2). This reaction served as a model for the preparation of all analogues. The NAD glycohydrolases that were primarily inhibited by isonicotinic acid hydrazide (INH) form the analogue also (3). The analogue thus formed can be freed from residual NAD by hydrolyzing that nucleotide with *Neurospora* NAD glycohydrolase which has a near absolute specificity for NAD and NADP. This has been discussed in some greater detail in Chap. 7.

The exchange reaction occurs with a wide variety of bases. Not only pyridine derivatives react, but also pyrimidines, pyrazines, pyridazines, imidazoles, pyrazoles, thiazoles, and thiadiazoles. The analogues would have remained curiosities if it were not for two observations. First, many substances that exchanged for nicotinamide were nicotinamide antagonists. This reaction was thought to be the fundamental mechanism of action of these compounds, an observation discussed in detail in Chap. 7. Second, analogues of the coenzymes proved very valuable as tools in the study of the mechanism of action of dehydrogenases. A very wide number of such NAD analogues have been made through the exchange reactions (4). The first to be extensively studied was the acetylpyridine-adenine dinucleotide (5).

The exchange reaction is only one mode of replacing or modifying the nicotinamide moiety. Already mentioned is the oxidation of the 4-adduct with dihydroxyacetone to yield 4-substituted nicotinamide coenzymes (6). A pyridine-purine dinucleotide can be modified. The original NAD analogue, nicotinamide-hypoxanthine dinucleotide, was prepared by nitrous acid deamination of NAD (7). Later, it was found that enzymatic deamination was also possible (8), but the chemical preparation is simpler (9). Some mononucleotides react with ATP, deoxyATP, or other polyphosphates in the nucleoside phosphorylase reaction (10).

Chemical reactions with the purine and pyridine moieties have yielded a variety of analogues, as has the synthesis of the dinucleotides via the anhydride formation of mononucleotides. It would serve little purpose to tabulate all the analogues prepared to date. In 1966, 64 were reported. Now there is no limit to their preparation. They are made as needed for the specific enzymological experiment.

II. FUNCTIONAL CAPACITY OF COENZYME ANALOGUES

Many pyridine derivatives can be reduced when incorporated into analogues. The most widely used ones have been the acetylpyridine, pyridine 3-aldehyde, and nicotinamide derivatives. The acetylpyridine-adenine dinucleotide is the paradigm.

The analogue can be chemically reduced, and it can react with cyanide. The spectrum is similar, but shifted towards longer wavelength when compared to NADH or the NAD–cyanide complex (5). Most interestingly, the analogue can substitute for NAD in a variety of enzymatic reactions, first noted by Kaplan et al. (12). On occasion, the reaction occurs more efficiently than with NAD. More importantly, the redox potential of the analogue may be found to be more positive. The initial assignment of the potential for the system of acetylpyridine-adenine dinucleotide and its reduced form is −0.248 volts as compared to an E_0' of −0.320 volts for NAD/NADPH (12). This means that the equilibrium

$$\text{Ethanol} + \text{coenzyme} \rightleftarrows \text{acetaldehyde} + \text{reduced coenzyme}$$

favors ethanol oxidation when the coenzyme is the acetylpyridine analogue.

Another type of analogue that can be prepared includes derivatives which can be made to react chemically with the active site of the enzyme. Two examples are the 4-methyl-5-β-hydroxyethylthiazole analogue (13), and the 3-aminopyridine analogue of NAD, which can be made through chemical modification of NAD (14) or through exchange with NAD glycohydrolase (15). The thiazole analogue can be bound to horse liver alcohol dehydrogenases through ferricyanide oxidation (13), while the 3-amino analogue can be diazotized and thereby linked to the proteins (16,17). There are other examples. This will be discussed in more detail later.

An understanding of the ready availability of such analogues is necessary to appreciate their use in the evaluation of the mode of interaction of NAD or NADP with their enzymes which will be discussed in the next chapter.

REFERENCES

1. L. J. Zatman, N. O. Kaplan, and S. P. Colowick, *J. Biol. Chem. 200:*97(1953).
2. L. J. Zatman, N. O. Kaplan, S. P. Colowick, and M. M. Ciotti, *J. Biol. Chem. 209:*467(1954).
3. D. S. Goldman, *J. Am. Chem. Soc. 76:*2841(1954).
4. S. P. Colowick, J. van Eys, and J. H. Park, in *Comprehensive Biochemistry, Vol. 14, Biological Oxidations* (M. Florkin, and E. M. Stotz, eds.), Elsevier, Amsterdam 1966, p. 1.
5. N. O. Kaplan and M. M. Ciotti, *J. Biol. Chem. 221:*823(1956).
6. R. M. Burton, A. San Pietro, and N. O. Kaplan, *Arch. Biochem. Biophys. 70:*87(1957).
7. F. Schlenk, H. Hellström, and H. von Euler, *Berichte 71:*1471 (1938).
8. N. O. Kaplan, S. P. Colowick, and M. M. Ciotti, *J. Biol. Chem. 194:*579(1952).
9. M. E. Pullman, S. P. Colowick, and N. O. Kaplan, *J. Biol. Chem. 194:*593(1952).
10. C. P. Fawcett and N. O. Kaplan, *J. Biol. Chem. 237:*1709(1962).
11. H. Klenow and B. Anderson, *Biochim. Biophys. Act. 23:*92(1957).
12. N. O. Kaplan, M. M. Ciotti, and F. E. Stotzenbach, *J. Biol. Chem. 221:*833(1956).
13. J. van Eys, R. Kretszehmar, N. S. Tseng, and L. W. Cunningham, *Biochem. Biophys. Res. Comm. 8:*243(1962).
14. T. L. Fisher, S. V. Vercellotti, and B. M. Anderson, *J. Biol. Chem. 248:*4293(1973).
15. B. M. Anderson, J. H. Yuan, and S. V. Vercellotti, *Mol. Cell. Biochem. 8:*89(1975).
16. J. K. Chan and B. M. Anderson, *J. Biol. Chem. 250:*67(1975).
17. N. K. Amy, R. H. Garrett, and B. M. Anderson, *Biochim. Biophys. Acta 480:*83(1977).

12

OXIDATION–REDUCTION FUNCTIONS OF NICOTINAMIDE COENZYMES

There are two basic roles for nicotinamide-adenine nucleotides: (1) as acceptor or donor for hydride ions or electrons in oxidation–reduction reactions, and (2) as a high-energy bond compound, acting as donor for the adenosine diphosphate-ribose (ADPR) moiety. Of these two fundamentally different functions, the coenzyme role in redox reactions is the one originally discovered, and the one most widely studied.

NAD and NADP can be viewed as coenzymes of reaction sequences and substrates for enzymes that catalyze hydride or electron donor or acceptor reactions. These tasks are the same at the molecular level, but they are different when viewed at the physiological level. The first understood function of NAD, discovered as cozymase, was as an intermediate in yeast fermentation, and later, in glycolysis. The net reaction of fermentation and glycolysis is without oxygen uptake, so that the function of NAD is a cyclic oxidation and reduction in a reaction sequence through participation as a substrate on two enzymes: glyceraldehyde 3-phosphate dehydrogenase and alcohol dehydrogenase for fermentation, or lactate dehydrogenase for glycolysis.

The effect of nicotinamide deprivation, to the degree that functional coenzyme levels are decreased, is a perturbation of the integrated enzyme systems and not of the individual apoenzyme function. The influence of nicotinamide coenzymes on metabolism is all-pervasive and pivotal in many reaction sequences. Therefore, the role of the coenzymes is complex.

The actual ratio of coenzyme to enzyme on a molar basis is low, compared to the conditions under which enzyme reactions are studied in vitro when the emphasis is on the catalytic function of the protein. When coenzyme–enzyme interactions are studied, more realistic protein–coenzyme ratios are used. There are so many nicotinamide coenzyme-utilizing reactions in cells that the bulk of coenzyme can be bound to proteins. In fact, there are unoccupied binding sites in many areas of the cell. The distribution of nicotinamide coenzymes over the various available binding sites of proteins is dependent on a variety of intracellular conditions. First, the binding constant is usually different for the oxidized and reduced coenzyme. Therefore, the NAD/NADH ratio is a factor determining the coenzyme distribution. Second, the degree of binding is often quantitatively different in the presence of substrate. Therefore, the availability of metabolic substrates and the metabolic flux through given areas of metabolism determine the occupied binding sites. Third, the amount of apoenzyme is not necessarily fixed, but can fluctuate with hormonal levels or nutrition state. Fourth, the subcellular particles provide compartmentation for nicotinamide coenzymes, so that the cytoplasmic metabolic state may be different from the mitochondrial environment. Lastly, the ratio between NAD and NADP can vary. Most dehydrogenases have absolute specificity for either NAD or NADP. Therefore, the functions of the two coenzymes in metabolic processes differ. These topics will be briefly reviewed in Chap. 14.

The metabolic function of the coenzymes can be more readily understood if one appreciates the enzymology of individual reactions. The volume of knowledge about individual enzymes that utilize nicotinamide coenzymes is overwhelming. In many cases, the full three-dimensional structure of the enzyme is known, the enzyme—coenzyme substrate complexes have been defined chemically and physically, and the mechanism of catalysis has also been elaborated. Therefore, only the highlights of the dehydrogenase functions which can contribute to an appreciation of the metabolic impact of nicotinamide will be discussed in this chapter. One enzyme type has been selected to be discussed in some detail to illustrate the principles involved.

I. TYPES OF OXIDOREDUCTION FUNCTIONS OF NICOTINAMIDE COENZYMES

There are a large number of reactions in which pyridine nucleotides participate as electron (hydride) acceptor or donor. To a degree, the enzymes that have similar functions have similar properties. It is therefore useful to group these enzymes into 8 sets according to function.

A. Dehydrogenases with Direct Hydrogen Transfer to the Coenzyme

In this group of enzymes there is a direct hydrogen transfer from substrate to coenzyme, for example:

$XH_2 + NAD^+ \rightleftarrows X + NADH + H^+$

In this group are all enzymes that catalyze reactions such as:

Alcohols ⇄ aldehydes or ketones
α-Hydroxyacids ⇄ α-ketoacids
Amines ⇄ imines ⇄ ketones or aldehydes + ammonia.

Where the substrate is an aldehyde being converted to an acid, i.e.,

$-CHO \rightleftarrows -CO_2H$

the substrate is the hydrated aldehyde. (Other types of aldehyde dehydrogenases have flavin prosthetic groups.) A variant reaction is the oxidation of the aldehyde to a thioester, for example:

Formaldehyde + glutathione ⇄ *S*-formylglutathione.

There may be secondary catalytic reactions on the enzyme resulting in more complex overall reactions. For instance, there are two types of malate dehydrogenases:

Malate ⇄ oxalacetate

and

Malate ⇄ pyruvate + CO_2

In contrast to the former, which uses NAD, the latter type tends to use NADP.

In some instances the reaction proceeds in two steps on one enzyme:

Uridine diphosphate-glucose → uridine diphosphate-glucuronic acid

All these reactions are simple hydrogen transfers. However, while enzymes have specificity, the specificity is not absolute, and therefore, classifying enzymes exclusively on the basis of the reaction catalyzed can be misleading. Lactate dehydrogenase is an example. Ordinarily this enzyme catalyzes the reaction

(1) Pyruvate + NADH + H^+ ⇄ L-lactate + NAD^+
(2) Glyoxalate + NAD^+ → oxalate + NADH + H^+

For lactate dehydrogenases from pig heart and pig muscle, the turnover numbers for these substrates are equivalent to those of lactate/pyruvate interconversions (1,2).

B. Acyl-Forming Enzymes

A second group of dehydrogenases are more complex, in that acyl formation occurs during the reaction sequence. The paradigm is glyceraldehyde 3-phosphate dehydrogenase, which catalyzes the following reaction:

$$\text{3-Phosphoglyceraldehyde} + NAD^+ + H^+ + P_i \rightleftarrows \text{2,3-Diphosphoglyceric acid} + NADH$$

Such complex enzymes have many secondary or experimentally artificial catalytic functions.

C. Flavin-Containing Nicotinamide Coenzyme-Dependent Dehydrogenases

A third group of dehydrogenases do not catalyze a direct hydride transfer from substrate to nicotinamide, but utilize a carrier intermediate, usually a flavin. The classical reaction catalyzed by this type of dehydrogenase is the reduction of double bonds:

$$-CH{=}CH- \rightleftarrows -CH_2-CH_2-$$

There is a wide variety of enzymes that use this system of dehydrogenation. Glutathione reduction also involves this class of enzymes. These enzymes have often unexpected substrate preferences. For instance, aldehyde dehydrogenase from liver is responsible not only for aldehyde oxidation, but also for the conversion of a number of aromatic nitrogen compounds to hydroxylated derivatives; for example, N^1-methylnicotinamide is converted to the pyridones. Many such enzymes have metal cofactors as well, such as iron or molybelenum.

D. Complex Dehydrogenases

These enzymes catalyze a very complex interconversion in which nicotinamide coenzymes are involved. The reaction is an oxidative decarboxylation of an α-ketoacid to form an acyl-coenzyme A derivative. The paradigm is the pyruvate decarboxylase reaction:

$$\text{Pyruvate} + NAD^+ + \text{coenzyme A} \rightarrow \text{acetyl-coenzyme A-}CO_2 + NADH^+ + H^+$$

Several other prosthetic groups participate, namely, thiamine pyrophosphate, flavin, and lipoic acid.

E. Electrotransport Sequences

The reduced nicotinamide may be the first donor in an electron transfer sequence. Usually flavin coenzymes are intermediates in this first step, followed by quinone or cytochrome acceptors. However, in bacteria, metals and inorganic acceptors can often serve. Nitrate reductases exemplify this pattern. As will be discussed later, such electron transport sequences are frequently highly organized, bound to membranes, and compartmentalized in organelles, as, for example, the primary electrotransport sequence in mitochondria through which oxidative phosphorylation takes place.

F. Mixed Function Oxidases

A series of reactions oxidize substrates with molecular oxygen, utilizing the driving force of NADPH oxidation. The overall reaction can be written as:

$$RH + NADPH + H^+ + O_2 \rightarrow ROH + NADP^+ + H_2O$$

Such reactions are extremely important in the metabolism of a variety of natural metabolic intermediates. However, in medicine this reaction type is particularly important for the metabolism of many drugs. Mixed function oxidases situated in the microsomal fraction of cells are under intense study by a wide number of pharmacologists, biochemists, and toxicologists.

G. Transhydrogenases

Reduced pyridine nucleotides equilibrate through hydride ion transfer:

$$NAD^+ : NADPH \rightleftarrows NADH + NADP^+$$

The divergent functions in metabolism for NAD^+ and $NADP^+$, as well as their different roles in generating high-energy phosphate, make this type of reaction of pivotal importance in metabolic regulation.

H. Photosynthesis and Related Reactions

Photo- and chemoautotrophic plants and bacteria can use physical or chemical reactions to drive the generation of reductive power. Illuminated grana generate reduced diphosphopyridine nucleotides in reactions driven via metal-containing proteins such as ferridoxin.

II. NOMENCLATURE OF PYRIDINE NUCLEOTIDE-DEPENDENT DEHYDROGENASES

The sets of nicotinamide coenzyme-requiring enzymes described in Section II are not necessarily together in this manner in the organized nomenclature style that has been accepted by the International Union of Biochemistry (3). In that nomenclature system, enzymes are grouped in classes, subclasses, sub-subclasses, and then as specific enzymes. Nicotinamide coenzyme-utilizing enzymes belong to the class oxidoreductases, which is an abbreviation for donor:acceptor oxidoreductases. The term *dehydrogenase* is used whenever possible, but in cases of obvious irreversibility the term *reductase* can be used. Oxidase is used only when oxygen is the acceptor. The enzymes are then grouped according to the substrate which can undergo oxidation. In reactions utilizing a nicotinamide coenzyme, the latter is always regarded as the acceptor, even if the direction of the reaction toward coenzyme reduction is not readily demonstrated. The obvious exception is the subclass in which NAD(P)H is the donor and some other redox catalyst the acceptor.

There is a numbering system, in which the first number is the class, the second the subclass, and the third, the sub-subclass. The enzymes under consideration here are grouped in the class oxidoreductases; the subclass stipulates the donor and the sub-subclass the acceptor. The class of oxidoreductases is always 1. and the sub-subclass with NAD^+ or $NADP^+$ as acceptor is always 1.–.1. Therefore, the classes of enzymes utilizing NAD^+ or $NADP^+$ are as follows:

- 1.1.1 Acting on the —CHOH— group of donors
- 1.2.1 Acting on the aldehyde or oxo groups of donors
- 1.3.1 Acting on the —CH=CH— group of donors
- 1.4.1 Acting on the $-CH-NH_2$ group of donors
- 1.5.1 Acting on the —CH—NH— group of donors
- 1.6.– Acting on NADH or NADPH as donors
  - 1.6.1 With NAD^+ or $NADP^+$ as acceptors
  - 1.6.2 With a cytochrome as acceptor
  - 1.6.4 With a disulfide as acceptor
  - 1.6.5 With a quinone or related compound as acceptor
  - 1.6.6 With a nitrogenous compound as acceptor
  - 1.6.7 With an iron sulphur protein as acceptor
  - 1.6.99 With other acceptors
- 1.8.1 Acting on a sulphur group of donors
- 1.10.1 Acting on diphenols and related substances as donors
- 1.12.1 Acting on hydrogen as donor
- 1.14.– Acting on paired donors with incorporation of molecular oxygen
  - 1.14.12 With NADH or NADPH as one donor and incorporating two atoms of oxygen into one donor

1.14.13 With NADH or NADPH as one donor and incorporating one atom of oxygen into one donor
1.17.1 Acting on $-CH_2-$ groups.

Following these numbers the specific number assigned to the enzyme is given. For instance, transhydrogenase is classified 1.6.1.1, and is thereby specified to catalyze

$$NADH + NADP^+ \rightleftarrows NAD^+ + NADPH$$

Another example is 1.1.1.27, which is a hydroxy-NAD(P) oxidoreductase, specifically, lactic dehydrogenase. Since the reactions of all lactic dehydrogenases are similar, one number is given. Verbally the source of the enzyme must be specified, since, as will be discussed later, subtle difference in properties adapt the function of the enzyme to the special needs of that organ in that species. Therefore, one should say, for instance, pig-heart lactic dehydrogenase (1.1.1.27).

Neither the grouping nor the classification is entirely satisfactory in organizing the enormous bulk of knowledge on dehydrogenases that exists. As an example, the group of oxidoreductases utilizing a $-CHOH-$ donor and $NAD(P)^+$ as acceptor has 176 entry numbers, of which six previously recognized were recently deleted or transferred, identifying 170 separate enzyme activities in that sub-subclass alone. That group, the largest of NAD-related oxidoreductases, illustrates the complexities of the enzymes that utilize pyridine nucleotide coenzymes.

III. PROPERTIES OF DEHYDROGENASES

The functions of the reduced pyridine nucleotides in electron transport and in mixed-function oxidases are primarily metabolic and compartmentalized. The functions of the dehydrogenases that catalyze individual catabolic or anabolic steps in intermediate metabolism are best understood if some of the enzymological properties are appreciated in detail. The path of electron and hydrogen transfer in metabolism proceeds via nicotinamide coenzymes. The specificity of the enzymes with regard to substrates provides some understanding of the consequences of competing substrates. The mechanism of binding indicates the affinity for NAD or NADP, and the abortive ternary complexes with substrate analogues illustrate potential metabolic probes and inhibitors.

A. Mechanism of Hydrogen Transfer

This has already been discussed in some detail. With most simple dehydrogenases it can be demonstrated that there is a direct hydrogen transfer from substrate to coenzyme. The first experiments utilized

1,1-dideuteroethanol and NAD as substrates for alcohol dehydrogenase in normal water. The resulting NADH incorporated one atom of deuterium, and could yield the deuterium to acetaldehyde (4,5). Many simple dehydrogenases carry out such a direct hydride ion transfer. There is no evidence that enzymes such as aocohol or lactate dehydrogenases have a reduced enzyme-protein intermediate.

B. Stereospecificity of Reduction

It has also been discussed previously that the reduction of nicotinamide coenzymes is stereospecific with regard to the hydrogen on the carbon-4 of the pyridine ring. Some enzymes have stereospecificity for the A side, others for the B side. A tabulation has been provided by You et al. (6). As already discussed, the absolute stereospecificity for one side or the other is known. When a reaction sequence utilizes NAD with similar stereospecificity for the hydrogen, the hydrogen on the enzymes involved, as deuterium or tritium, can be traced through the reaction sequence. However, in glycolysis, glyceraldehyde 3-phosphate dehydrogenase has a B-side preference (6–8), and lactate dehydrogenase has an A-side preference (6,9). Therefore, the hydrogen does not get transferred from glyceraldehyde 3-phosphate to lactate until after the NAD has cycled several times.

Vennesland (9,10) has made some generalizations regarding this stereospecificity: (a) When an enzyme utilizes a variety of substrates or coenzyme analogues, the stereospecificity of the coenzymes does not vary. (b) The dehydrogenase catalyzed oxidation of an alcohol group on a carbon atom adjacent to a carboxyl group occurs with transfer of hydrogen to the 4-pro, R-position (A side); however, dehydrogenases that act on phosphorylated carbohydrate derivatives are 4-pro, S-specific (B side); (c) When the natural substrate of a dehydrogenase is a nonphosphorylated organic compound having three or fewer carbon atoms, the reaction is 4-pro, R-specific with respect to the coenzyme. Sequences of readings involving NAD or NADP often have the same specificity. There are exceptions to these rules, which are entirely empirical. For instance, there is the obvious exception to the side-specificity of phosphorylated carbohydrate derivatives in that phosphoglycerate dehydrogenase is A-side specific (11). Yet such generalizations are useful as a basis for generating hypotheses for the mechanism of the full catalytic event.

C. Substrate Specificity

Simple dehyrogenases do have substrate specificity, but the specificity is often very broad. Furthermore, as a consequence of this broad substrate specificity a wide range of reactions can be catalyzed, but all seem to be characterized by oxidation and reduction. The dehydrogenases from different tissues that seemingly catalyze the same reactions can have different substrate specificities, resulting in different catalytic properties of the enzymes.

The best studied example of this divergence in properties of enzymes with seemingly identical physiological functions is the alcohol dehydrogenase from mammalian liver and from yeast. Both catalyze the reaction:

$$\text{Ethanol} + \text{NAD}^+ \rightleftarrows \text{acetaldehyde} + \text{NADH} + \text{H}^+$$

Since the equilibrium of the reaction is thermodynamically determined, both show a preference for acetaldehyde reduction at equilibrium. At pH 7.2 and 20°C and ionic strength of 0.1, the equilibrium constant is:

$$K_{eq} = \frac{[\text{NADH}][\text{CH}_3\text{CHO}][\text{H}^+]}{[\text{NAD}^+][\text{CH}_3\text{CH}_2\text{OH}]} = 8.01 \pm 0.14 \times 10^{-12} \quad (12)$$

For the yeast enzyme acetaldehyde reduction is the physiological function during fermentation. Mammalian liver contains a large amount of alcohol dehydrogenase, but the specificity is broader and its actual physiological function and role in ethanol metabolism remain a question of great interest.

Even among mammalian liver alcohol dehydrogenases, there is not uniformity of properties. The most studied enzymes are those from horse, rat, and human liver. The horse enzyme is far more an ethanol dehydrogenase than are all the others. Horse-liver alcohol dehydrogenase shows a preference for long-chain aliphatic alcohols. In addition, the enzyme has two subunits and two genetically determined types of subunits. Since these freely associate, there are three isozymes known: EE, ES, and SS. These isozymes have been named by Theorell ethanol-active and steroid-active subunits (13,14). However, all isozymes of human liver have activity toward hydroxysteroids (15), but it is low compared to the activity toward ethanol.

For human alcohol dehydrogenase, which is also a dimer, the suggested model is three polypeptide chains, called α, β, and γ (16,17). In addition, the γ-chains have genetic polymorphism through two common alleles called γ_1 and γ_2. The development of three alcohol dehydrogenases is ordered, with the early fetus showing an α,α-form, progressing with development to α,β- and β,β-forms. γ-Chains develop after birth (17–19). Mixed protein species can exist, resulting in a complex isozyme pattern (20). In addition, atypical variants are present in various degrees among population groups (20–23). Finally, there are enzyme species that do not fit the current genetic models (24).

Such isozymes and genetic variants differ in reactivity, substrate specificity, and sensitivity to inhibitors (25–27). These differences can be enhanced by using unnatural coenzymes and substrates. For horse-liver alcohol dehydrogenases, using 2-methyl-NAD as coenzyme and cinnamul alcohol as substrate results in a 10-fold difference between the maximal velocities of isozymes (28).

Because of the wide substrate specificity, liver alcohol dehydrogenase can oxidize substrates like vitamin A (29). The affinity of the mammalian enzyme for higher aliphatic alcohols is actually greater than for ethanol (30,31), and cyclic alcohols are also substrates (31). Coupled oxidations and reductions of alcohols and aldehydes can occur under suitable conditions, for example (32):

$$\text{Ethanol + pyridine 3-aldehyde} \rightleftarrows \text{acetaldehyde + pyridine 3-carbinol}$$

Similarly, the dismutation of aldehydes to esters (33):

$$\text{2 Formaldehyde} \xrightarrow{NAD^+} \text{methylformate}$$

More obscure reactions catalyzed by liver alcohol dehydrogenase which may have a physiological function are the oxidation of histidinol to histidine (34) or coproporphyrinogen to protoporphyrinogen in which propionyl side chains are oxidized to vinyl groups (34). In both cases pyridoxal phosphate is said to be involved.

The yeast enzyme, having a different specificity, can catalyze the oxidation of a variety of alcohols, but prefers short-chain alcohols except methanol (35,36)(Fig. 1), and primary over secondary alcohols (37,38). If the alkyl groups on the secondary alcohol are both larger than methanol, the enzyme is inactive (39,40); however, the enzyme will utilize lactate, albeit slowly (35). The specificity of the enzyme

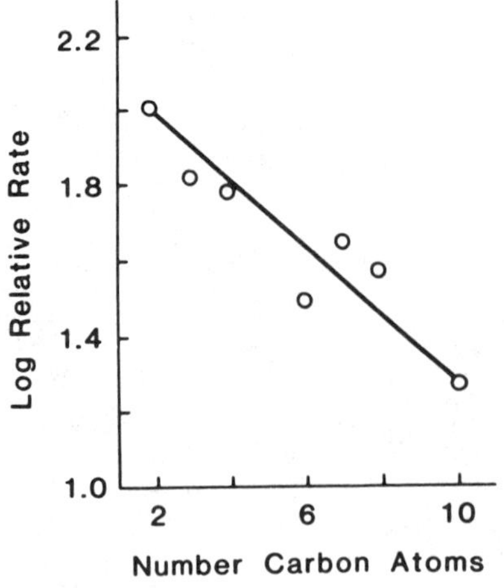

Figure 1 The effect of increasing chain length on the rate of oxidation: alcohol concentration 2.5×10^{-3} *M*, Tris buffer 0.05 *M*, pH 9.3, NAD 3.5×10^{-4}*M*. The value for amyl alcohol is not given, since the alcohol contained an inhibitor, which even on repeated fractional distillation could not be removed completely. (From Ref. 35.)

Table 1 The Effect of Coenzyme on Substrate Specificity of Yeast Alcohol Dehydrogenase[a]

Coenzyme	Relative rate of coenzyme reduction		
	Ethanol	Methanol	Ratio of M/E (× 100)
NAD	100	0.8	0.80
Acetylpyridine AD	22.3	0.017	0.08
Pyridine 3-aldehyde AD	7.3	0.86	11.77

[a]Conditions in these experiments were: coenzyme $1.5 \times 10^{-4} M$; substrate $5.0 \times 10^{-3} M$, Tris buffer 0.05 *M*, pH 9.3. Rate with NAD and ethanol set arbitrarily at 100. AD = alcohol dehydrogenase.
Source: Ref. 35.

can vary to a degree with the coenzyme analogue used (Table 1). The enzyme shows absolute stereospecificity toward its substrates. For the yeast enzyme, this can be deduced from the observation that yeast cells reduce 2-octanone to (+)octan-2-ol (41). This can be confirmed using yeast alcohol dehydrogenase by testing the specific oxidation of resolved octane-2-ol (35). Yeast alcohol dehydrogenase shows absolute specificity for (−1)(1*S*)-1-deuteroethanol (36). All these reactions are compatible with only one configuration of substrate.

D. Coenzyme Specificity

As is the case for substrates, the alcohol dehydrogenases show a wide tolerance for variation in coenzyme binding and activity. Again, yeast and liver enzymes differ (Table 2) (42). Coenzyme analogues that bind but do not function are by and large competitive inhibitors. When systematic changes were made in the NAD molecule, the coenzymes so obtained have been evaluated for activity on horse-liver alcohol dehydrogenase. For instance, the hypoxanthine analogue of NAD, deoxy-analogues of NAD, or coenzymes with substitution of the adenine or of the pyridine ring or with alkylation of the adenine nitrogens all show binding and activity that agree with the data surmised for the NAD-binding site deduced from x-ray crystallography. Two interesting analogues include 3-iodopyridine-adenine dinucleotide, which can replace NAD for the liver enzyme (43) and can also serve as a heavy atom label in x-ray crystallography (44), and the N^6-ethynyl analogue of NAD^+ which is fluorescent and active (45). Some coenzyme analogues have reactive groups that allow affinity labeling of the active site. These include the 4-methyl-5-β-hydroxyethylthiazole analogue of NAD (46,47),

Table 2 Coenzyme Specificity for Horse Liver and Yeast Alcohol Dehydrogenase[a]

Coenzyme	Horse-liver alcohol dehydrogenase	Yeast alcohol dehydrogenase
NAD	100	100
3-Acetylpyridine AD	205	10
Pyridine 3-aldehyde AD	68	2
Thionicotinamide AD	348	16

[a]Rates are relative to NAD as 100. The relative rates vary with coenzyme concentrations, pH, and especially alcohol concentrations. The relative rates stated apply for a coenzyme concentration of $1 \times 10^{-4} M$, ethanol at 0.01–0.05 *M*, pH 9.5 in Tris buffer, 0.1 *M*.
Source: Ref. 48.

the 3-aminopyridine analogue (48), the 3-chloroacetylpyridine analogue (49,50), and a variety of bromoacetyl derivatives in which alkyl chains are interspersed between the ribose and ring nitrogen (51). Analogues have been prepared which bind the adenine moiety of NAD to the enzyme (52). Such an analogue can be shown to be catalytically active by using cycling substrates in a coupled reaction (52,53).

Attempts at photoaffinity labeling using the 3-diazoacetoxymethylpyridine analogue (54) or the 3-azido analogue of NAD (55) have failed, but the 3-diazirino analogue of NAD can be covalently bound, at least with lactate dehydrogenase (56).

Finally, it is possible to use immobilized coenzymes to generate affinity chromatography and solid state catalysts (51).

The coenzyme analogues that function change the *thermodynamic properties* of the reaction drastically. For instance, the equilibrium constant of ethanol and acetaldehyde interconversion, which was cited previously, is much higher for the 3-acetylpyridine analogue of NAD (58). However the *chemical* course of the reaction is not influenced. The reaction remains a hydride ion transfer and the stereospecificity of the pyridine ring remains the A-side (59).

There are many compounds that act as competitive inhibitors of the coenzymes. Some of these are zinc chelators, but x-ray crystallography and kinetic evidence suggests that they act by binding at the NAD binding site. The coenzymes bind at a specific site, one per subunit. The binding induces a configurational change in the coenzyme which may be distinguished from the configuration of the free coenzyme in solution. The absorption spectrum of the reduced coenzyme changes on binding to liver alcohol dehydrogenases, a finding first reported with the horse-liver enzyme (60,61). The yeast enzyme does not show this phenomenon, but the absorption spectrum of adducts, which are favored by the presence of the enzyme, shifts for certain complexes on both the liver (62) and the yeast (63) enzymes. From such primitive studies, multiple additional reactions were discovered and utilized in physical and enzyme kinetic studies to determine the geography of the enzyme active site. An important finding which stems from this specific binding of the coenzyme is that the equilibrium constant for the reaction with NAD/NADH bound to the enzyme is different from K_{eq} measured with free coenzyme (60).

E. Substrate Analogues

Just as there are inactive coenzymes that nevertheless inhibit by binding at the active site for the natural coenzyme, there are also substrate analogues that have the analogous effect. Some of such reactions seem to represent primarily a facilitated addition reaction. The liver alcohol dehydrogenase–NAD–hydroxylamine complex has already been alluded to (62,64). That reaction is more than the formation of an analogue of NADH, since hydroxylamine is a competitive inhibitor of ethanol on both the liver and yeast enzyme (65). The NAD–cyanide complex can also be formed, but cyanide does not act as a competitive inhibitor (66). Utilizing mercaptans, the specificity for substrate analogues could be demonstrated for a number of simple dehydrogenases (67).

This phenomenon can be exploited physiologically. Complexes can be formed on lactate dehydrogenases between NADH, enzyme, and carboxylic acids. One specific inhibitor, oxamate, has been exploited as a cancer chemotherapeutic agent in patients (68). Tumor cells are more readily inhibited by oxamate because they have a high reliance on aerobic glycolysis (69).

There are a number of substances that form ternary complexes with the enzyme–coenzyme complex, but that are not necessarily direct substrate analogues, though they can function as competitive inhibitors. Complexes between liver alcohol dehydrogenase–NADH and amides are readily formed. Furthermore, complexes between fatty acids or pyrazoles and alcohol dehydrogenase–NAD are seen; moreover, these pyrazole complexes seem specific for alcohol dehydrogenases (70,71). Analogous to oxamate in cancer therapy, the pyrazoles have been used experimentally to combat the effects of alcohol abuse (72,73).

F. Metal Content

Many dehydrogenases, especially the alcohol dehydrogenases, contain zinc. Liver alcohol dehydrogenase contains two zinc atoms per subunit (74), which are not equilvalent (75). One zinc atom is necessary for catalysis, but not for coenzyme binding, subunit association, or ternary complex formation. The zinc binds water and participates in the final catalytic event (76,77). The liver enzyme can be manipulated to contain cobalt at its active site, but this protein is catalytically inactive; however, it binds coenzyme analogues and substrates (78). The yeast enzyme, which contains 4 moles of zinc per mole of enzyme can be modified to contain cobalt (79) or manganese (80) and retain catalytic activity. Nevertheless, zinc deficiency affects dehydrogenase activity in molds.

G. Kinetic Analysis

The dehydrogenases have been extensively studied by detailed kinetic analyses, and Dalziel has summarized many such studies (81). The data show an ordered addition of substrate and coenzyme, and tighter binding of reduced than oxidized coenzyme. Different enzymes catalyzing similar reactions have kinetic parameters which are in line with the expected functions of the enzymes. This phenomenon has been studied especially for the lactate dehydrogenases. The degree of pyruvate inhibition through formation of an abortive lactic dehydrogenase–pyruvate–NAD complex is greater for the heart than for the muscle enzyme (82,83); thus the balance between glycolysis and respiration in the aerobic heart is different from that in the aerobic muscle. The physical binding constants correspond well with the kinetically derived values. Thus studies of alcohol dehydrogenases are interwoven into a coherent whole. The evidence for overall catalytic function of the enzyme is derived from enzymatic studies, from the kinetics, physical binding, ternary complex formation of the enzyme; the effects of metal substitution on binding, primary structure determination, and teriary structure analysis determined from enzyme, enzyme–substrate, enzyme–coenzyme, or enzyme–coenzyme–substrate analogue complexes. methods have included spectroscopy, x-ray crystallography, optical rotatory dispersion, nuclear magnetic resonance, steady state kinetics, low temperature studies, studies of isotope effects, chemical or physical modification of the protein, and covalent linkage of coenzyme to enzyme (84). Since those reviews, many further investigations have been published. Equal detail is known for the lactate dehydrogenases, the malic dehydrogenases and many others.

IV. GLYCERALDEHYDE 3-PHOSPHATE DEHYDROGENASE

One dehydrogenase occupies a special place in the nicotinamide physiology: glyceraldehyde 3-phosphate dehydrogenase. The enzyme was crystallized early in the history of enzymology, and was investigated in great detail by Warburg (85). The enzyme was found to catalyze the reaction

$$\begin{array}{c}\text{Glyceraldehyde 3-phosphate} + \text{NAD}^+ + \text{P}_i \\ \uparrow\downarrow \\ \text{1,3-Diphosphoglyceric acid} + \text{NADH} + \text{H}^+\end{array}$$

The enzyme is the pivotal step in glycolysis as the "oxydierenden Gärungsferments," the oxidizing glycolytic enzyme. It is the first enzyme to incorporate high-energy phosphate into metabolic intermediates, and therefore, it has remained a conceptual model for oxidative phosphorylation. It is also one of the first enzymes for which an enzyme–substrate covalent bond was proposed, and is therefore significantly different from simple dehydrogenases. Warburg and Christian had proposed a simple mechanism in which a phosphohemiacetal was formed which acted as substrate (87). Racker showed that an acetyl-enzyme is formed which acted as substrate (87). Racker showed that an acetyl-enzyme is formed (88) which could actually be isolated (89) (Fig. 2). In the presence of arsenate the product is 3-phosphoglyceric acid. The enzyme will catalyze the arsenolysis of 1,3-diphosphoglycerate to 3-phosphoglyceric acid. The enzyme has complex catalytic behavior. It can be made to promote a variety of reactions ranging from the usual catalysis with NADH formation, already previously discussed, to hydrolysis of acetylphosphate and *p*-nitrophenylacetate. It reduces dyes in the presence of NADH. These reactions were summarized in an early review (90) and are enumerated in Table 3.

In contrast to its association in simple dehydrogenases, NAD is tightly bound to muscle glyceraldehyde 3-phosphate dehydrogenase and is found in the crystalline preparation. While this is untrue for the yeast enzyme, even here the binding of NAD is tighter than the binding of NADH (90). This allows the sequence

Glyceraldehyde 3-phosphate $\rightarrow$ lactate

to smoothly favor the direction in which glycolysis proceeds.

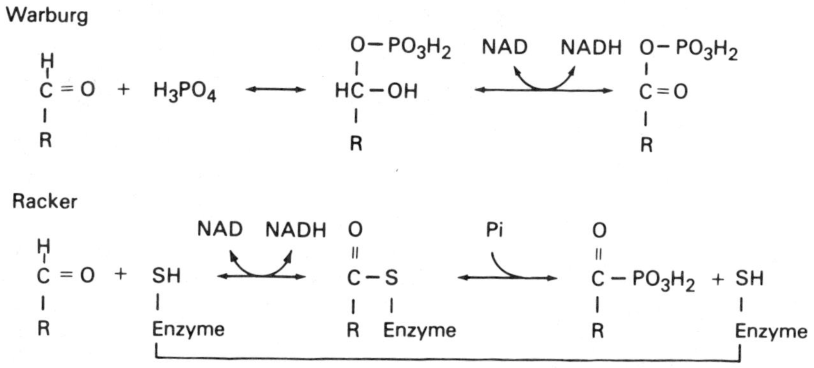

Warburg

$$\text{H–C(R)=O} + H_3PO_4 \longleftrightarrow \text{HC(R)(OH)–O–}PO_3H_2 \xleftrightarrow{\text{NAD} \rightarrow \text{NADH}} \text{R–C(=O)–O–}PO_3H_2$$

Racker

$$\text{H–C(R)=O} + \text{HS–Enzyme} \xleftrightarrow{\text{NAD} \rightarrow \text{NADH}} \text{R–C(=O)–S–Enzyme} \xleftrightarrow{\text{Pi}} \text{R–C(=O)–}PO_3H_2 + \text{HS–Enzyme}$$

Figure 2 Historical proposals for the mechanism of action of glyceraldehyde-3-phosphate dehydrogenase.

Table 3 Summary of Activities of 3-Phosphoglyceraldehyde Dehydrogenase[a]

Activity	NAD or NADH involved in the reaction	SH group on enzyme required
Oxidation and phosphorylation	+	+
Transferase, ^{32}Pexchange, or arsenolysis	+	+
Hydrolysis of acetylphosphate	+	0
Hydrolysis of *p*-nitrophenylacetate	0	+
NADH formation	+	+
Diaphorase action	+	+
Transphosphorylase	+	+

[a]The last two reactions are artificial constructs requiring enzyme as participant.
Source: Ref. 81.

Yet there are great similarities in the structures of glyceraldehyde 3-phosphate dehydrogenase and simple dehydrogenases. There is a specificity for the B side of nicotinamide on reduction, as already mentioned. This is not expressed by major changes in NAD conformation when bound to the enzyme but seems to be expressed simply by a rotation of 180° about the glycosidic bond of the pyridine ring (91).

The enzyme is a measurable percentage of all cytoplasmic proteins and therefore binds a significant amount of all available NAD. In yeast the enzyme is 5% of the dry weight (92) and binds up to 20% of all NAD in the cell (93). In muscle the enzyme approaches 70 μmol concentration of active site (94). In the past it was argued that glyceraldehyde 3-phosphate dehydrogenase-bound NADH could react directly with other dehydrogenases (95,96), but that probably represents primarily the differential affinity for NAD and NADH between glyceraldehyde 3-phosphate and lactate or alcohol dehydrogenases. Nevertheless, the NAD reservoir bound to this enzyme is a major fraction of the coenzyme participating in glucose metabolism. The acyl-enzyme and catalytic intermediate is quantitatively a significant metabolic intermediate, as well. The enzyme is active in both gluconeogenesis and glycolysis; its concentration is nearly rate-limiting in some tissues, such as adipose tissue (97), but usually exceeds the rate-limiting activity by an order of magnitude. Nevertheless, because of its binding to this enzyme, NAD participation in metabolism depends on the availability of inorganic phosphate as well as on more refined allosteric modulation of glycolytic enzymes that are sensitive to ADP and ATP levels.

V. CONCLUDING REMARKS

To summarize this discussion of nicotinamide coenzyme physiology as it pertains to whole cell or animal metabolism, (1) nicotinamide coenzymes are involved in an extremely wide variety of catabolic or anabolic reactions; (2) the dehydrogenases bind NAD(P) and NAD(P)H specifically, and the number of binding sites exceeds the available NAD; (3) multiple substrate analogues can affect the reaction, and substrate inhibition or activation occurs, making the relative rates of dehydrogenases a complex interplay of factors; and (4) these functions of nicotinamide coenzymes are exploitable for modifying metabolism.

REFERENCES

1. R. J. S. Duncan and K. F. Tipton, *Eur. J. Biochem. 11:*58(1969).
2. W. A. Warren, *J. Biol. Chem. 145:*1675(1970).
3. International Union of Biochemistry Committee, *Enzyme Nomenclature,* 1978. Academic Press, New York, 1979.
4. H. F. Fisher, E. E. Conn, B. Vennesland, and F. H. Westheimer, *J. Biol. Chem. 202:*687(1953).
5. F. H. Westheimer, H. F. Fisher, E. E. Conn, and B. Vennesland, *J. Am. Chem. Soc. 73:*2403(1951).
6. U. You, L. T. Arnold Jr., W. S. Allison, and N. O. Kaplan, *Trends in Biochemical Sciences 3:*265(1978).
7. H. R. Levy and B. Vennesland, *J. Biol. Chem. 228:*85(1957).
8. H. R. Levy, P. Talalay, and B. Vennesland, in *Progress in Stereochemistry, Vol. III* (P. D. B. de La Mare and W. Kilyve, eds.), Butterworth, London 1962, p. 239.
9. B. Vennesland, *Proc. Intern. Cong. Biochem.* Symposium 5, preprint 145, 1961, p. 1.
10. B. Vennesland, *Topics in Current Chemistry 481:*39(1974).
11. I. V. Winico, *Biochim. Biophys. Acta 397:*288(1975).
12. K. J. Bäcklin, *Act. Chem. Scand. 12:*1279(1958).
13. R. Pietruszko and H. Theorell, *Arch. Biochem. Biophys. 131:*288 (1969).
14. H. Theorell, in *Pyridine Nucleotide-Dependent Dehydrogenases* (H. Sund, ed.) Springer Verlag, Berlin, 1970, p. 121.
15. R. Pietruszko, H. Theorell, and C. De Zalenski, *Arch. Biochem. Biophys. 153:*279(1972).
16. M. Smith, D. A. Hopkinson, and H. Harris, *Ann. Hum. Gen. 34:* 251(1971).
17. M. Smith, D. A. Hopkinson, and H. Harris, *Ann. Hum. Gen. 35:* 243(1972).
18. P. Pikkaroinen and N. C. R. Raikra, *Nature 222:*563(1969).
19. R. F. Murray and A. Motulsky, *Science 171:*71(1971).
20. M. Smith, D. A. Hopkinson, and H. Harris, *Ann. Hum. Gen. 36:*401 (1973).

21. T. -K. Li and L. J. Magnes, *Biochem. Biophys. Res. Comm. 63:* 202(1975).
22. J. P. von Wartburg and D. M. Schurch, *Ann. N. Y. Acad. Sci. 151:*936(1968).
23. G. Stamatoyannopoulos, S. -H. Chen, and M. Fukui, *Am. J. Hum. Gen. 27:*789(1975).
24. W. F. Bosron, L. G. Lange, W. Pafeldeher, T. -K. Li, and B. L. Vallee, *Biochem. Biophys. Res. Comm. 74:*85(1977).
25. T. M. Schenker, L. J. Teeple, and J. P. von Wartburg, *Eur. J. Biochem. 24:*271(1971).
26. J. P. von Wartburg, J. Pagenberg, and H. Pebri, *Can. J. Biochem. 43:*889(1968).
27. M. Smith, D. A. Hopkinson, and M. Harris, *Ann. Hum. Gen. 37:* 49(1973).
28. J. F. Biellmann and J. P. Samama, *FEBS Lett. 38:*175(1974).
29. A. F. Bliss, *Arch. Biochem. Biophys. 31:*197(1951).
30. A. D. Winer, *Act. Chem. Scand. 12:*1695(1958).
31. A. D. Merritt and G. M. Tomkins, *J. Eiol. Chem. 234:*778(1959).
32. N. O. Kaplan and M. M. Ciotti, in *Methods in Enzymology, Vol. III* (S. P. Colowick and N. O. Kaplan, eds.), Academic Press, New York, 1957, p. 253.
33. R. H. Abeles and H. A. Lee Jr., *J. Biol. Chem. 235:*1499(1960).
34. J. Jeffery, in *Dehydrogenases Requiring Nicotinamide Coenzymes, Experienta. Suppl., Vol. 36* (J. Jeffery, ed.), Birkhäuser Verlag, Basel, 1980, p. 85.
35. J. van Eys and N. O. Kaplan, *J. Am. Chem. Soc. 79:*2782(1957).
36. H. R. Levy, F. A. Loewus, and B. Vennesland, *J. Am. Chem. Soc. 79:*2949(1957).
37. F. M. Pickinson and G. B. Monger, *Biochem. J. 131:*261(1973).
38. W. Schöpp and H. Aurich, *Act. Biol. Med. Ger. 31:*19(1973).
39. F. M. Dickinson and K. Dalziel, *Nature 214:*31(1967).
40. F. M. Dickinson and K. Dalziel, *Biochem. J. 104:*165(1967).
41. C. Neuberg and F. F. Nord, *Berichte 52:*2237(1919).
42. J. van Eys, *Mol. Cell. Biochem. 8:*43(1975).
43. M. A. Abdallah, J. F. Biellmann, J. P. Samama, and A. D. Wrixon, *Eur. J. Biochem. 64:*351(1976).
44. J. P. Samama, E. Zeppezaner, J. F. Biellman, and C. J. Bränden, *Eur. J. Biochem. 81:*403(1977).
45. J. R. Burrio, J. A. Scerist, and N. J. Leonard, *Proc. Nat. Acad. Sci. 69:*2039(1972).
46. J. van Eys, R. Kretszchmer, Nan-Sen Tseng-Lui, and L. W. Cunningham Jr., *Biochem. Biophys. Res. Comm. 8:*243(1962).
47. Tseng-Jui Nan Sen, Ph.D. Thesis, Vanderbilt University, Nashville, 1965.
48. T. L. Fisher, S. V. Vercellotti, and B. M. Anderson, *J. Biol. Chem. 248:*4293(1973).

49. J. F. Biellmann, G. Branlant, B. Y. Foucaud, and M. J. Jung, *FEBS Lett. 40:*29(1974).
50. J. F. Biellmann, P. R. Goulas, J. C. Nicholas, B. Descomps, and A. Crastes de Paulet, *Eur. J. Biochem. 99:*81(1979).
51. R. J. Guillory and S. J. Jeng, in *Methods in Enzymology, Vol. 46* (W. B. Jaboby and M. Wilderech, eds.), Academic Press, New York, 1977, p. 259.
52. M. O. Mänsson, P. O. Larsson, and K. Mosbach, *Eur. J. Biochem. 86:*415(1978).
53. M. O. Mänsson, P. O. Larsson, and K. Mosbach, *FEBS Lett. 98:* 309(1979).
54. D. T. Browne, S. S. Hixon, and F. H. Westheimer, *J. Biol. Chem. 246:*4477(1971).
55. S. S. Hixon and S. H. Hixon, *Photoderm. Photobiol. 18:*135(1973).
56. D. N. Standring and J. R. Knowles, *Biochem. 19:*2811(1980).
57. K. Mosbach, in *Pyridine Nucleotide-Dependent Dehydrogenases* (H. Sund, ed.), Walter de Gruyter, Berlin, 1977, p. 263.
58. N. O. Kaplan, M. M. Ciotti, and F. E. Stolzenbach, *J. Biol. Chem. 221:*833(1956).
59. J. -F. Biellmann, C. G. Hirth, M. J. Jung, N. Rosenheimer, and A. D. Wrixon, *Eur. J. Biochem. 41:*517(1974).
60. H. Theorell and R. Bonnichsen, *Acta Chem. Scand. 5:*1105(1951).
61. P. D. Boyer and H. Theorell, *Acta Chem. Scand. 10:*447(1954).
62. R. M. Burton and N. O. Kaplan, *J. Biol. Chem. 211:*447(1954).
63. J. van Eys, M. M. Ciotti, and N. O. Kaplan, *Biochim. Biophys. Acta 23:*581(1957).
64. N. O. Kaplan and M. M. Ciotti, *J. Biol. Chem. 211:*431(1954).
65. N. O. Kaplan and M. M. Ciotti, *J. Biol. Chem. 201:*785(1953).
66. N. O. Kaplan, M. M. Ciotti, and F. E. Stolzenbach, *J. Biol. Chem. 211:*419(1954).
67. J. van Eys, F. E. Stolzenbach, L. Sherwood, and N. O. Kaplan, *Biochim. Biophys. Acta 27:*63(1958).
68. V. H. Reynolds, personal communication.
69. J. Papaconstantinou and S. P. Colowick, *J. Biol. Chem. 236:*285 (1961).
70. M. Reynier, *Acta Chem. Scand. 23:*1119(1969).
71. T. -K. Li and H. Theorell, *Acta Chem. Scand. 23:*892(1969).
72. R. D. Hawkins and H. Kalant, *Pharmacol. Rev. 24:*67(1972).
73. R. Blomstrand and H. Theorell, *Life Sci., Part II 9:*631(1970).
74. A. Akeson, *Biochem. Biophys. Res. Comm. 17:*211(1964).
75. D. D. Ulmer and B. L. Vallee, *Adv. Chem. Ser. 100:*187(1971).
76. S. A. Evans and J. D. Shore, *J. Biol. Chem. 255:*1509(1980).
77. P. P. Anderson, J. Krassman, A. Lindström, B. Olden, and B. Petterson, *Eur. J. Biochem. 108:*303(1980).
78. H. Dietrich, W. Marel, L. Wallen, and M. Zeppezauer, *Eur. J. Biochem. 100:*267(1979).

79. A. Cordel and Iwatsubo, *FEBS Lett. 1:*133(1968).
80. P. L. Coleman and M. Weiner, *Biochemistry 12:*3466(1973).
81. K. Dalziel, in *The Enzymes, Vol. XI,* (P. D. B. Boyder, ed.), 3rd Ed., Academic Press, New York, 1975, p. 1.
82. P. G. W. Plagemann, K. F. Gregory, and F. Wroblewski, *Biochem. Z. 334:*37(1961).
83. A. Pesce, T. Fondy, F. E. Stolzenbach, D. Castillo, and N. O. Kaplan, *J. Biol. Chem. 242:*2155(1967).
84. C. I. Bränden, H. Jörnvall, M. Ekulund, and B. Furugren, in *The Enzymes, Vol. XI,* (P. D. B. Boyer, ed.), 3rd Edition, Academic Press, New York, 1975, p. 103.
85. J. Jeffery, Dehydrogenases requiring nicotinamide coenzymes, *Experientia Suppl. 36* Birkhäuser Verlag, Basel, 1980, p. 85.
86. O. Warburg, *Weiterentwicklung der Zellphysiologischen Methoden,* G. Thieme Verlag, Stuttgart, 1962, p. 47.
87. O. Warburg and W. Christian, *Biochem. Z. 303:*40(1939).
88. E. Racker and J. Krimsky, *J. Biol. Chem. 198:*731(1952).
89. J. Krimsky and E. Racker, *Science 122:*319(1955).
90. S. P. Colowick, J. van Eys, and J. H. Park, in *Comprehensive Biochemistry,* Vol. 14, (N. Florking and E. M. Stotz, eds.), Elsevier, Amsterdam, 1966, p. 1.
91. M. G. Rossmann, A. Liljas, C. F. Bränden, and L. J. Banaszak, in *The Enzymes* Vol. XI, 3rd Ed., (P. D. B. Boyer, ed.), Academic Press, New York, 1975, p. 61.
92. E. G. Krebs, G. W. Raffer, and J. M. Tange, *J. Biol. Chem. 200:* 479(1953).
93. B. Chance, in *The Mechanism of Enzyme Action,* (W. D. McElroy and H. B. Glass, eds.), Johns Hopkins Press, Baltimore, 1954, p. 445.
94. C. S. Furfine and S. F. Velick, *J. Biol. Chem. 240:*844(1965).
95. A. P. Nygaard and W. Rutter, *Acta Chem. Scand. 9:*1300(1955).
96. L. Aschachan, S. P. Colowick, and N. O. Kaplan, *Biochim. Biophys. Acta 24:* 141(1957).
97. C. E. Shank and C. E. Boxer, *Cancer Res. 24:*709(1964).

13

NICOTINAMIDE–ADENIDE–DINUCLEOTIDES AS HIGH-ENERGY DONORS FOR TRANSRIBOSYLATION REACTIONS

As discussed previously, NAD is not only a carrier of hydrogen in oxidation–reduction sequences, but is also a reservoir of ribosyl bonds in high-energy linkage through the quaternary nitrogen—glycosidic bond. Therefore, the biosynthesis of NAD and the subsequent transglycosylation reaction are the important steps in overall nicotinamide metabolism. Quantitatively, these reactions are the most important determinants of NAD turnover. The product, adenosine diphosphate-ribosylated (ADPR) proteins, are involved in DNA metabolism. Alternatively, the ADP-ribosylation modulates the activity of regulatory or synthetic enzymes. Again, as for previous topics, a vast literature exists. Even though the discovery of these reactions dates back only to the mid-1960s, recent interest in the expression of DNA and in molecular biology has made this area of NAD biochemistry even more productive than the enzymology of dehydrogenases. It is therefore impossible to generate an exhaustive review. The emphasis will be placed on the involvement of NAD and the effect of nicotinamide on the reactions. Only indirectly will the functions of ADPR be reviewed.

In the previous chapter on NAD metabolism (Chap. 12), the fundamental reactions involved in this aspect of NAD function were reviewed. These reactions are part of a NAD glycohydrolase activity in which the reaction of the type

$$NAD^+ + X \rightarrow ADPR-X + N + H^+$$

is catalyzed. When X is an alternative pyridine base, NAD analogues are synthesized; when X is a macromolecule, mono-ADP-ribosylation occurs; in eukaryotes, poly-adenosine diphosphate-ribose is made (Fig. 1). The enzymes that catalyze mono-ADP-ribosylation in prokaryotes are effective in modulating protein activity. Many of the proteins that

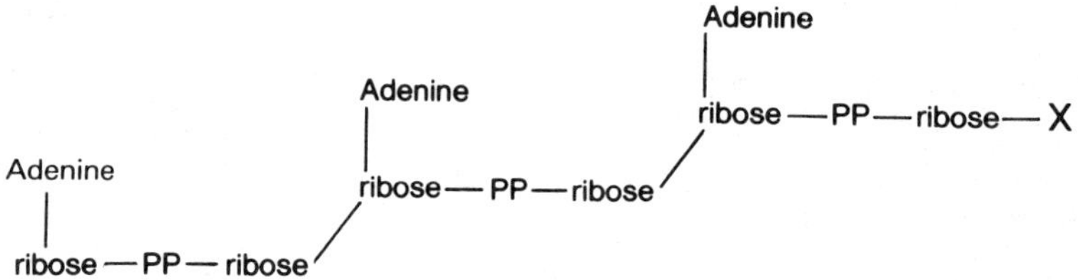

Figure 1 The structure of polyadenosine diphosphate ribose.

catalyze the ADP-ribosylation were discovered by their physiological function before their enzymatic functions were known. The best example that can be used to illustrate the ribosylation is the action of diphtheria toxin.

I. MONO-ADP-RIBOSYLATION AS EXEMPLIFIED BY DIPHTHERIA TOXIN

Corynebacterium diphtheriae produces an exotoxin that is extremely toxic to mammals. As little as 2–3 μg is fatal to humans. The toxin is liberated only when the *C. diphtheriae* is infected with a β-bacteriophage.

Diphtheria toxin inhibits exclusively protein synthesis without affecting nucleic acid synthesis, glycolysis, or respiration (1,2). This inhibition of protein synthesis requires NAD^+ (3). The mechanism of action is an inactivation of the translocating enzyme elongation factor-2 (EF-2), responsible for translocation of peptidyl-transfer RNA on ribosomes (4,5). The diphtheria toxin as elaborated by the bacterium has a molecular weight of 62,000 but is not active per se. Only after proteolytic and reductive cleavage which liberates a fragment, called fragment A, is the toxin capable of this inactivation (6) even though fragment A is by itself only 0.01% as toxic as the whole toxin molecule (7). The fragment A is responsible for all the activity of ADP-ribosylation that can be attributed to the toxin (8,9).

This fragment catalyzes the reaction:

$$NAD^+ + EF\text{-}2 \rightleftarrows ADPR{-}EF\text{-}2 + H^+ + \text{nicotinamide}$$

The reaction favors the ADP-ribosylation reaction strongly (10,11). However, with purified EF-2, the reaction can be made to proceed in the reverse direction (12,13). By adding excess nicotinamide and fresh fragment A, HeLa cells, inhibited from protein synthesis, can be reactivated (12,14). The same can be demonstrated for intoxicated guinea-pig heart (15).

The enzymatic activity of the toxin is specific for EF-2. It requires NAD^+ specifically. Few NAD analogues are active. Among the many analogues tested, only nicotinamide had any real activity, while acetyl-pyridine-adenine dinucleotide and nicotinamide-hypoxanthine dinucleotide (deamino-NAD) has slight activity (16).

The reaction is inhibited by nicotinamide, not only by displacing the equilibrium, but also by direct inhibition of the forward reaction (11, 17). Adenine is also a powerful inhibitor (17–20). Fragment A on the whole toxin is a slow NAD^+ glycohydrolase:

$$H_2O + NAD^+ \rightarrow ADPR + \text{nicotinamide} + H^+$$

At high pH, the toxin acts as acceptor and becomes inactivated by ADP-ribosylation (21). This set of reactions has the same pyridine nucleotide specificity as the original reaction (17).

Similar reactions occur with a *Pseudomonas* toxin (18). The fragment A of diphtheria toxin is resistant to boiling, as is the *Pseudomonas* toxin glycolydrolase activity. The isolation of a NAD glycolydrolase from a *Proteus* species by boiling, which inactivates the heat-labile inhibitor, was reported in 1956 (23), and has since been observed in *Mycobacterium* (24,25) and *Pseudomonas* (22,26). This is entirely analogous to the diphtheria toxin, which is not enzymatically active but contains the fully active heat stable fragment A and a heat-labile fragment B. Other mono-ADP-ribosylating enzymes such as those found in bacteriophages have other protein targets (27). In some systems the membrane-bound adenylate cyclase is the target. While such mono-ADP-ribosylation reactions occur primarily via prokaryotic enzymes, a turkey erythrocyte cytoplasmic protein can ADP-ribosylate cyclic-AMP synthetase (28). Cholera toxin is likewise capable of executing that reaction (29,30). The similarity in the behavior of interferons and cholera toxins prompted an evaluation of ADP-ribosylation as catalyzed by interferon. It was claimed that mouse fibroblast interferon catalyzes protein ADP-ribosylation, while rabbit kidney interferon does not (31). The impurity of interferon preparations makes such claims open to question.

Therefore, the majority of enzymes that catalyze mono-ADP-ribosylation are of prokaryotic origin but they are not limited to that group of organisms. Furthermore, their action can profoundly affect the mammalian cell, as exemplified by diphtheria and cholera toxins.

II. POLY-ADP-RIBOSE SYNTHESIS

The synthesis of poly-ADPR is of the type:

$$n\text{NAD}^+ + \text{acceptor} \rightarrow \text{acceptor-(ADPR)}_n + n\text{nicotinamide} + nH^+$$

Most poly-ADPR is attached covalently to an acceptor molecule (32–34), although free poly-ADPR can exist (35). The chain length is variable and ranges from 1 to 30 monomers (36–38). The structure is complex. The basic structure is a (1→2)-ribosylribose linkage (39). When hydrolyzed by pyrophosphatase, an adenosine diphosphate-ribose is formed that is called 1-iso-ADP-ribose (Fig. 2). The linkage between the ribose is α-(1→2) (40). There is branching, indicated by the discrepancy between molecular weight and the chain length calculated from hydrolysis products. If the poly-ADPR were linear, hydrolysis with pyrophosphatase would yield one 5'-AMP and iso-ADPR. The average chain length would conform to the equation

$$\text{Average chain length} = \frac{\text{5'-AMP + iso-ADPR}}{\text{5'-AMP}}$$

Discrepancies of as much as a factor of 3 have been found (41), however.

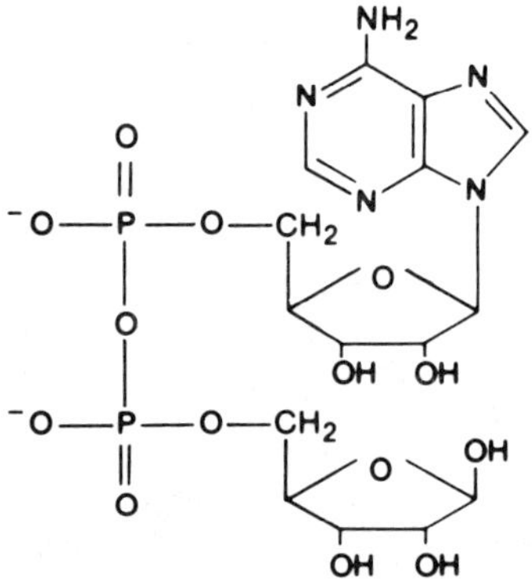

(a)

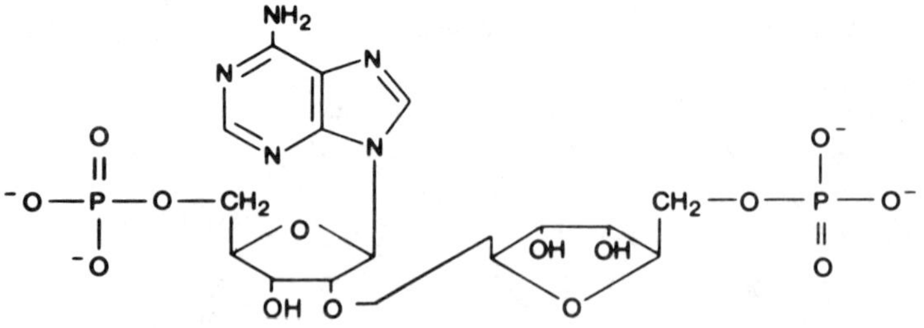

(b)

Figure 2 (a) Adenosine diphosphate-ribose (ADPR); (b) Iso-adenosine diphosphate ribose [2'-(5"-phosphoribosyl)-5'-adenylic acid (PR-AMP)].

The synthesis of poly-ADPR can be achieved using deoxy-NAD in which the adenylic-acid moiety has a 2'-deoxy sugar (42,43). The product is by definition a (1→3)-ribosyl-ribose structure, and branching is impossible. The targets of the two NAD analogues are different in the ribosylation: NAD^+ preferentially ADP-ribosylates histones, while 2-deoxy-NAD ADP-ribosylates nonhistone proteins (44). The average chain length with the latter as donor was 1.2, but with NAD^+ it was 5.0–7.2 for histone proteins and 1.7 for nonhistone proteins (44).

Specific enzymes degrade the poly-adenosine-ribose to ADPR (45–49). There is also an enzyme that splits adenosine diphosphate-ribose from the acceptor histones (50).

The combined action of these enzymes with the biosynthetic enzyme is identical to that of nicotinamide-adenine dinucleotide glysohydrolase:

$$\begin{array}{l} n\mathrm{NAD}^+ + \mathrm{X} \rightarrow n\mathrm{N}^+ + n\mathrm{H}^+ + \mathrm{X(ADPR)}_n \\ \mathrm{X(ADPR)}_n \rightarrow \mathrm{XADPR} + (n-1)\mathrm{ADPR} \\ \mathrm{XADPR} \rightarrow \mathrm{X} + \mathrm{ADPR} \\ \hline n\mathrm{NAD}^+ \xrightarrow{\mathrm{X}} n\mathrm{N}^+ + n\mathrm{H}^+ + n(\mathrm{ADPR}) \end{array}$$

When NADase activity is measured in cells, most of the activity is found in nuclei (51), and a rapid turnover of NAD in nuclei has been observed (52). The degradative enzyme that splits poly-ADPR can debranch as well.

Another set of enzymes, the phosphodiesterases, hydrolyze the pyrophosphate bonds. Some can liberate the iso-ADPR (snake vemon and tobacco phosphodiesterases) (53,54). An enzyme from rat liver liberates the terminal 5'-AMP (55).

The polymer is present in vivo as well as in vitro (56–61).

III. FUNCTION OF POLY-ADP-RIBOSE

The biosynthesis of poly-ADPR is strongly inhibited by nicotinamide (62). Therefore, the effect of nicotinamide on whole cells can be the result of inhibition of this enzyme activity. However, it must be kept in mind that nicotinamide can also be a methyl trap through the biosynthesis of N^1-methylnicotinamide, thereby affecting methylation of t-RNA. Other inhibitors of poly-ADPR synthesis include thymidine (62), 5'-methylnicotinamide (63), actinomycin D (64), and theophillin (65).

The primary target of the ADPR-ribosylation is histone protein (44, 65,66), and experiments have often verified this association both in vivo and in vitro. It had been suggested that cellular proliferation depends on the NAD^+ content because the NAD^+ content of rapidly dividing cells is low (68). The poly-ADPR is tightly bound to chromatin, as is the enzyme that synthesizes it. Without detailing the experiments that prove the assertion, poly-ADPR has been implicated in the function of the cell at many levels: DNA synthesis (69), gene expression (70–72), chromatin structural change (73,74), cross-linkage of histones, chromatin acidic protein modification (75), and modulation of RNA polymerase activity (76). Whatever the primary function, the effects are pervasive, and the consequent effects of nicotinamide on whole cells are equally profound.

IV. THE EFFECT OF NICOTINAMIDE ON CELL GROWTH AND DNA METABOLISM

Nicotinamide has a profound effect on cell growth, cell differentiation, and the modulation of cytotoxic agents, such as alkylating chemotherapeutic agents. It is tempting to speculate that all of these effects are mediated via the inhibition of adenosine diphosphate-ribosylation. However, as already mentioned, excess nicotinamide can also affect DNA/RNA metabolism by acting as a methyl trap through formation of N^1-methylnicotinamide. The actions can be differentiated in part by evaluating whether the effect can be imitated by other inhibitors of poly-ADP-ribose synthesis, such as thymidine (62) or N^1-methylnicotinamide (63).

The actions of alkylating agents and their interaction with nicotinamide serve as model for observing the effects of nicotinamide in vivo. The most widely studied are the nitrosoureas, especially streptozotocin. The data on this agent will therefore be summarized in some detail. BCNU [1,3-bis(2-chloroethyl)-1-nitrosourea] inhibits the growth of mouse Ehrlich ascites tumors, and the change in growth seems dependent on the NADase activity, which is increased, and the NAD content of the cells, which is decreased (77). In the light of current knowledge, the drop in NAD content of tissues is quite likely related to the increased utilization of NAD for poly-ADP-ribose synthesis. A variety of DNA-targeted antineoplastic agents have the effect of lowering tissue NAD in various tissues, for example: BCNU and cyclophosphamide change NAD glycolydrolase activity and lower NAD in normal mouse tissues and in leukemias (78). Methyl- and dialkylnitrosoureas do the same (79,80). Streptozotocin lowers NAD levels in mouse livers (80), while triaziquone [Trenimon, or 2,3,5-tris-1-aziridinyl)-*p*-benzoquinone] does so in lenses of calves (81) and Ehrlich's ascites tumor cells (82,83). Thio-TEPA (triethylenethiophosphoramide) also lowers NAD levels in rat livers (84), as does radiotherapy in Ehrlich ascites cells (82). Many other examples can be found in the literature.

These findings were reported before the role of nicotinamide in DNA repair was known. In those years, glycolytic rates were evaluated to test the hypothesis that such drugs were active through glycolytic inhibition. However, using Trenimon as an example, that did not seem to be the case (83). Furthermore, not all drugs had such NAD-lowering effects (85). In many instances, nicotinamide counteracted the

NAD-lowering effect. The complex relationship between a chemotherapeutic agent and nicotinamide can be illustrated by detailing the interaction between streptozotocin and nicotinamide.

Streptozotocin is an *N*-methylnitrosourea derivative of glucose which was isolated from culture bottles of *Streptomyces acromogenes* but has also been synthesized (86). The substance seems to act on DNA synthesis, primarily (87,88); however, as already mentioned, it lowers tissue NAD (80).

It is antineoplastic in a number of tumors, especially pancreatic islet cell carcinoma (89), and gastrointestinal cancer (90), but is also specifically diabetogenic in animals (91), and in humans as well (90). There is a general toxicity, though bone marrow toxicity is limited (89, 90). A transient renal toxicity, as well as hepatic toxicity, have been noted, with nausea and vomiting. Finally, like many antineoplastic drugs, the compound is oncogenic, and produces pancreatic tumors (92) and renal tumors (93).

Nicotinamide modifies the action of streptozotocin in many ways. First, it was discovered that nicotinamide at 500 μg/kg protected both mice and rats against the diabetogenic action of streptozotocin (94,95) without loss of antitumor activity (94). This effect was noted with nicotinic acid, N^1-methylnicotinamide, or nicotinuric acid (93). The mechanism of streptozotocin-induced diabetes is complex. It was initially postulated that the lowering of tissue NAD was responsible (80, 91); however, the phenomenon of isletitis which occurs in mice after streptozotocin administration is prevented with nicotinamide, but also with 3-*O*-methylglucose and antilymphocyte serum. The effects of antilymphocyte serum and nicotinamide were additive to a degree, but protection against diabetes in these Charles River CD-mice was not complete (96). Nicotinamide is routinely used in humans receiving streptozotocin to protect against diabetes (97,98). However, the effective protection against diabetes in humans that several French investigators have reported (98) has not been duplicated in American clinical settings (90,99). Such discrepancies between species and between investigators are frequent in research on cancer chemotherapeutic agents and are almost always the result of schedule of administration and the different therapeutic margins between species.

Streptozotocin is also oncogenic. The induction of tumors is markedly affected by nicotinamide. In rats, the incidence of pancreatic islet cell tumors was 83% in rats receiving both compounds, none when nicotinamide was administered alone, but only one tumor was seen in rats receiving streptozotocin alone (96). Later studies showed that streptozotocin by itself has tumorogenic activity on the pancreas but that nicotinamide enhances the effect (100,101), as does picolinamide (100). Pancreatic islet cell tumors are endocrinologically active (102), and are seen in species other than rats and mice, e.g., the dog (101). The combined administration of streptozotocin and nicotinamide is now the standard model for producing pancreatic islet cell tumors in experimental studies. These tumors are transplantable (104,105), and remain endocrinologically active (104).

Conversely, after streptozotocin administration renal tumors decrease in size with nicotinamide (100). To further illustrate the complexity, another example may be given. The compound heliotropin induces liver tumors. Pretreatment of rats with nicotinamide prevents liver necrosis and therefore promotes tumor formation by allowing the animals to survive (106). In mice both 3-*O*-methylglucose and nicotinamide protect against streptozotocin toxocity in mice. Both prevent the NAD decline in tissues. Therefore, it has been suggested that the protective effect is the result of maintaining NAD levels (107). Nicotinamide pretreatment markedly increases the retention of streptozotocin in pancreatic islets (108). The mechanism of this oncogenesis is not at all clear, and the effect of nicotinamide is equally uncertain. There are tumor models in which nicotinamide inhibits tumor formation, for example, in methane-induced pulmonary adenomas (109,110). Nicotinic acid does not protect against adenoma formation while nicotinamide does. Some of the effects of streptozotocin appear to be mediated through the cellular NAD level. Streptozotocin enhances the induction of L-tyrosine aminotransferase by dexamethasone in rat liver. *N*-methyl-N^1-nitrosoguanidine was 100 times as effective. NAD levels dropped, but NAD turnover was not increased even though nicotinamide inhibits the reaction. Nicotinamide does not affect the increase in hormonal activation of adenylate cyclase observed with treatment of cultured cells with nitrosoureas (112).

The antitumor action of anticancer agents is also modulated by nicotinamide. Nicotinamide potentiates nitrogen mustard activity, and this proliferation is coincidental with an increased cellular NAD content (113). The exposure of cells to agents such as methylnitrosourea increases NAD degradation, and this is affected by nicotinamide and other inhibitors of poly-ADP-ribose biosynthesis (114). Methylnitrosourea causes extensive single-strand breaks in HeLa cell DNA. Inhibition of poly-ADP-ribose synthesis with nicotinamide results in extensive degradation of the altered DNA, while in the absence of nicotinamide, DNA repair prevails (115,116). 3T3 lymphoma cells can be depleted of NAD in vitro by growing them in the absence of precursors. Such NAD-depleted cells are incapable of unscheduled DNA synthesis after *N*-methyl-N^1-nitrosoguanidine exposure (117). This defect in DNA synthesis in 3T3 cells depleted of NAD is noted primarily in DNA repair synthesis and not in replicative DNA synthesis (118). Detailed studies suggest strongly that the sequence of events leading up to the lowering of cytoplasmic NAD is the synthesis of poly-ADP-ribose in response to DNA repair provoked by nitrosourea exposure (119). This response is illustrated in Figure 3. In human lymphocytes nicotinamide stimulates repair of DNA damage from radiation, dimethylsulfate, and

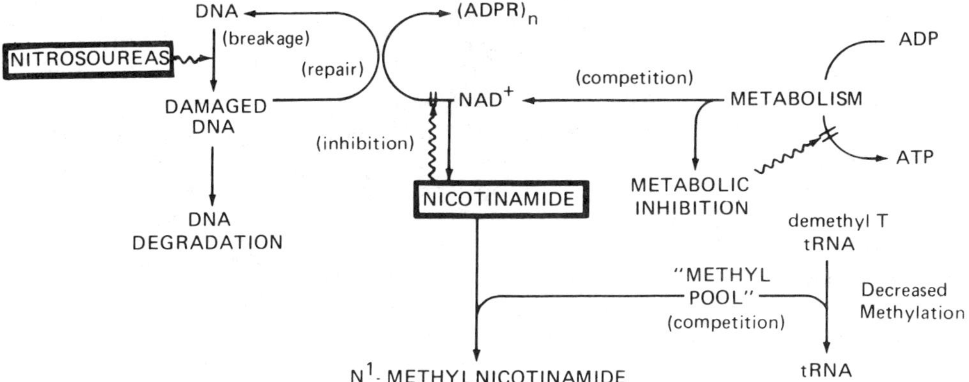

Figure 3 Effects of nitrosoureas and nicotinamide on DNA/RNA homeostasis and on metabolism.

nitrosoguanidine (120). In addition to its effect on DNA repair synthesis, nicotinamide inhibits phytohemagglutinin-stimulated DNA and RNA biosynthesis (121). NAD stimulates DNA synthesis in human bone marrow cells (122).

Nicotinamide can induce hemoglobin synthesis and differentiation in murine erythroleukemia cells (123,124). All compounds that inhibited poly-ADP-ribose synthesis induce differentiation, but N^1-methylnicotinamide, which is not an inhibitor, is also effective (124). Nicotinamide causes an increase in sister chromatid exchange in Chinese hamster lung fibroblasts (125). All such studies bring us tantalizingly close to proposing a unifying hypothesis for the mechanism of the effect of nicotinamide on DNA repair—it inhibits poly-ADP-ribose synthesis and thereby modulates DNA expression and integrity.

However, again there is some evidence that nicotinamide can act as a methyl trap and thereby modulate RNA synthesis: nicotinamide inhibits the transfer-RNA methylase in human KB-tumor tissue, rat liver, and W-256 tumor cells (126). Using excretion of pseudouridine as a measure, L-tryptophan and nicotinamide were found to decrease t-RNA turnover in bladder cancer patients (127,128). Herpes simplex virus type 1, when grown in primary rabbit kidney cells, is methylated. This phenomenon is inhibited by nicotinamide (129). The two effects of nicotinamide are interrelated. Ethionine, which is carcinogenic may inhibit nicotinamide methylation (by forming *S*-ethyladenosine), higher levels of nicotinamide; these higher levels then inhibit poly-ADP-ribose biosynthesis, blocking DNA repair (130). In *E. coli*, nicotinamide mononucleotide stimulation is specific for DNA polymerase I-mediated DNA repair synthesis. That system is active in ultraviolet irradiation damage (131). It may be remembered here that analogues of nicotinamide inhibit embryogenesis, as was discussed in a previous chapter. Nicotinamide raises NAD content in chick mesodermal cells in vitro and inhibits chondrogenic expression, while 3-acetylpyridine, which lowers NAD, potentiates chondrogenic expression (132).

The effects of nicotinamide on cellular homeostasis are just too complex to interpret in vivo, or in intact-cell experiments. Nicotinamide by itself is not mutagenic (135), but it can potentiate response in the Ames test, presumably by promoting biotransformation to active substances (136). In fact, NAD itself was called cytostatic in tumor-bearing mice (135). However, interest in the effects of nicotinamide remains. Recent experiments showed that nicotinamide protects against the carcinogenesis from Bracken Fern (136).

As a last complication, nicotinamide *N*-oxide is a radiation sensitizer (137,138). It is formed during radiation exposure experiments in which nicotinamide is administered. Thus, the effects of nicotinamide become even more difficult to interpret. It has been suggested that the action of antimetabolites be potentiated with nicotinamide, but even this extrapolation of the incomplete data may prove futile in one of the most stubborn of childhood cancers—neuroblastoma. Neuroblastoma secretes a nicotinamide deamidase (139,140).

REFERENCES

1. N. Strauss and E. D. Hendee, *J. Exp. Med. 109:*145(1959).
2. I. Kato and A. M. Pappenheimer Jr., *J. Exp. Med. 112:*329(1960).
3. R. J. Collier, and A. M. Pappenheimer Jr., *J. Exp. Med. 120:* 1019(1964).
4. R. J. Collier, *J. Mol. Biol. 25:*83(1967).
5. R. S. Goor and A. M. Pappenheimer Jr., *J. Exp. Med. 126:*899 (1967).
6. R. J. Collier and H. H. Cole, *Science 164:*1179(1969).
7. J. Zanen, G. Muyldermans, and N. Beugier, *FEBS Lett. 66:*261 (1976).
8. D. M. Gill and A. M. Pappenheimer, Jr., *J. Biol. Chem. 246:*1492 (1975).
9. R. E. Drazin, P. Kandel, and R. J. Collier, *J. Biol. Chem. 246:* 1504(1971).
10. T. Honjo, Y. Nishizuka, O. Hayaishi, and I. Kato, *J. Biol. Chem. 243:*3553(1968).
11. D. M. Gill, A. M. Pappenheimer Jr., and J. B. Baseman, *Cold Spring Harbor Symp. Quant. Biol. 34:*595(1969).
12. D. M. Gill, A. M. Pappenheimer Jr., R. Brown, and J. T. Kurnick, *J. Exp. Med. 129:*1(1969).
13. R. S. Goor, A. M. Pappenheimer Jr., and E. Ames, *J. Exp. Med. 126:*923(1967).
14. T. J. Moehring and J. M. Moehring, *J. Bacteriol. 96:*61(1968).

15. C. G. Bowman and P. F. Bonventre, *J. Exp. Med. 131:*659(1970).
16. R. S. Goor and A. M. Pappenheimer Jr., *J. Exp. Med. 126:*913 (1967).
17. J. Kandel, R. J. Collier, and D. W. Chung, *J. Biol. Chem. 249:* 2088(1974).
18. L. Montamaro and S. Sperti, *Biochem. J. 105:*635(1967).
19. S. Sperti and L. Montanaro, *Biochem. J. 107:*730(1968).
20. R. S. Goor and E. S. Maxwell, *J. Biol. Chem. 245:*616(1970).
21. D. M. Gill and D. Steinhaus, *J. Hyg. 18:*316(1974).
22. B. H. Igleioshi and D. Kahat, *Proc. Natl. Acad. Sci. USA: 72:* 2284(1975).
23. M. W. Swiartz, N. O. Kaplan, and M. E. Frech, *Science 123:*50 (1956).
24. M. Kern and R. A. Natale, *J. Biol. Chem. 231:*41(1958).
25. K. P. Gopinathan, M. Sirsi, and C. S. Valdyanatha, *Biochem. J. 91:*277(1964).
26. I. H. Mather and M. Knight, *Biochem. J. 129:*141(1972).
27. A. Pesce, C. Casoli, and G. C. Schilo, *Nature 262:*412(1976).
28. J. Moss and M. Vaughan, *Proc. Natl. Acad. Sci. USA 75:*3621 (1978).
29. D. Cassel and T. Pfeuffer, *Proc. Natl. Acad. Sci. USA 75:*2669 (1978).
30. D. M. Gill and R. Meren, *Proc. Natl. Acad. Sci. USA 75:*3050 (1978).
31. J. B. Trepel, L. M. Neckers, D. M. Chuang, and N. H. Neff, *Pharmacology 21:*213(1979).
32. P. Adamietz and H. Hilz, *Biochem. Soc. Trans. 3:*1118(1975).
33. P. Adamietz and H. Hilz, *Z. Physiol. Chem. 357:*527(1976).
34. K. Ueda, S. Narumiya, and O. Hayaishi, *Z. Physiol. Chem. 353:* 846(1972).
35. D. Richwood, M. J. McGillivray, and W. J. D. Whish, *Proc. 10th FEBS Meeting 216,* 1975.
36. P. R. Stone and H. Hilz, *FEBS Lett. 57:*204(1975).
37. P. R. Stone, R. Bredehorst, M. Kittler, H. Lengyel, and H. Hilz, *Z. Physiol. Chem. 357:*51(1976).
38. M. Tanaka, K. Hayashi, H. Sakura, M. Miwa, T. Natushima, and T. Sugimura, *Nucleic Acids Res. 5:*3183(1978).
39, J. Doly and F. Petek, *Campte Rendu Soc. Biol. (Paris) 263:*1341 (1966).
40. M. Miwa, H. Saito, H. Sakura, N. Saikawa, T. Watanabe, T. Matsushima, and T. Sugimura, *Nucleic Acids Res. 4:*3997(1977).
41. T. Sugimura, M. Miwa, H. Saito, Y. Kanai, M. Ikejima, M. Terada, M. Yamada, and T. Utakoji, *Adv. Enzyme Regul. 18:*195(1980).
42. M. B. Lennon, D. M. Lichtenwalner, and R. J. Suhaldonik, *Fed. Proc. 35:*1756(1976).

43. R. J. Suhaldonik, R. Baur, D. M. Lichtenwalner, T. Uematsu, J. H. Roberts, S. Sudhakar, and M. Smulson, *J. Biol. Chem. 252:* 4134(1977).
44. D. M. Lichtenwalner and R. J. Suhadolnik, *Biochem. 18:*3749 (1979).
45. M. Miwa and T. Sugimura, *J. Biol. Chem. 246:*6362(1971).
46. K. Ueda, J. Oka, S. Narumiya, N. Miyakawa, and O. Hayaishi, *Biochem. Biophys. Res. Comm. 46:*516(1972).
47. M. Miwa, M. Tanaka, T. Matushima, and T. Sugimura, *J. Biol. Chem. 249:*3475(1974).
48. M. Miwa, K. Nakatsugawa, K. Hara, T. Matushima, and T. Sugimura, *Arch. Biochem. Biophys. 167:*54(1975).
49. M. Tanaka, M. Miwa, T. Matshushima, T. Sugimura, and S. Shall, *Arch. Biochem. Biophys. 172:*224(1976).
50. H. Okayama, M. Honda, and O. Hayaishi, *Proc. Natl. Acad. Sci. USA: 75:*2254(1978).
51. M. Fukushima, H. Okayama, Y. Takahashi, and O. Hayaishi, *J. Biochem. 80:*167(1976).
52. M. Rechsteiner, D. Hillyard, and B. M. Olivera, *Nature 259:*695 (1976).
53. H. Matsubara, S. Hasegawa, S. Fujimura, T. Shina, T. Sugimura, and M. Futai, *J. Biol. Chem. 245:*3606(1970).
54. H. Shinshi, M. Miwa, K. Kato, M. Noguchi, T. Natsushima, and T. Sugimura, *Biochem. 15:*2185(1976).
55. H. Matsubara, S. Hasegawa, S. Fujimura, T. Shima, T. Sugimura, and M. Futai, *J. Biol. Chem. 245:*4317(1970).
56. Y. Nishizuka, K. Ueda, T. Honjo, and O. Hayaishi, *J. Biol. Chem. 243:*3765(1968).
57. H. Otake, M. Miwa, S. Fujimura, and T. Sugimura, *J. Biochem. (Japan) 65:*145(1969).
58. K. Ueda, A. Omachi, M. Kawaichi, and O. Hayaishi, *Proc. Natl. Acad. Sci. USA: 72:*205(1975).
59. J. H. Roberts, P. Stark, C. P. Giri, and M. Samulson, *Arch. Biochem. Biophys. 171:*305(1975).
60. S. Tanuma, T. Emoto, and Y. Yamada, *Biochem. Biophys. Res. Comm. 74:*599(1977).
61. M. G. Ord and L. A. Stocken, *Biochem. J. 161:*583(1977).
62. J. Preiss, R. Schlaeger, and H. Hilz, *FEBS Lett. 19:*244(1971).
63. J. B. Clark, G. M. Ferris, and S. Pinder, *Biochem. Biophys. Acta 238:*82(1971).
64. K. Yoshihara, *Biochem. Biophys. Ves. Comm. 47:*119(1971).
65. Y. Nishizuka, K. Ueda, T. Honjo, and O. Hayaishi, *J. Biol. Chem. 243:*3765(1968).
66. H. Otake, M. Miwa, S. Fujimura, and T. Sugimura, *J. Biochem. 65:*145(1969).
67. N. C. Wong, G. G. Poirier, and G. H. Dixon, *Eur. J. Biochem. 77:*11(1977).

68. R. K. Morton, *Nature 181:*540(1958).
69. L. Burzio and S. S. Koide, *Biochem. Biophys. Res. Comm. 53:* 572(1973).
70. A. F. Caplan and M. J. Rosenberg, *Proc. Natl. Acad. Sci. USA 72:*1852(1975).
71. W. E. G. Müller, A. Totsuka, I. Nussei, J. Obermeier, H. J. Rohde, and R. K. Zahn, *Nucleic Acids Res. 1:*1317(1974).
72. W. H. Proctor and J. E. Casida, *Science 190:*580(1975).
73. C. P. Giri, N. H. P. West, and M. Simulson, *Biochemistry 17:*3495 (1978).
74. C. P. Giri, M. H. P. West, and M. Simulson, *Biochemistry 17:*3501 (1978).
75. R. A. Colyer, K.E. Burdette, and W. R. Kidwell, *Biochem. Biophys. Res. Comm. 53:*960(1973).
76. W. E. G. Müller and R. K. Zahn, *Mol. Cell. Biochem. 12:*147(1976).
77. S. Green and A. Dobrojansky, *Cancer Res. 27:*226(1967).
78. S. Tsukagoshi, M. H. Kao, and H. Goldin, *Cancer Chemother. Rep. 52:*569(1968).
79. P. S. Schein, *Cancer Res. 29:*1226(1969).
80. P. S. Schein and S. Loftus, *Cancer Res. 28:*1501(1968).
81. W. Schmack, O. Hockwin, and K. K. Adusa-Amankwa, *Ber. Dtsch. Ophthalmol. Ges. 70:*376(1970).
82. E. Kohen, C. Kohen, and B. Thorell, *Acta Pharmacol. Toxicol. 26:*556(1968).
83. N. E. Fusenig and P. Z. Obrecht, *Krebsforsch. 69:*351(1967).
84. E. D. Bidenko, *Ukr. Biokhim. Zh. 40:*436(1968).
85. C. G. Schmidt, *Z. Krebsforsch 73:*223(1970).
86. E. Hessler and H. Jahnke, *J. Org. Chem. 35:*245(1970).
87. R. Reusser, *J. Bacteriol. 105:*580(1971).
88. H. Rosenkrantz and H. Carr, *Cancer Res. 30:*112(1970).
89. L. E. Broder and S. K. Carter, *Ann. Int. Med. 79:*108(1973).
90. C. G. Moertel, R. J. Reitemeir, A. J. Schutt, and R. G. Hahn, *Cancer Chem. Rep. 55:*303(1971).
91. N. Katsilambos, Y. A. Rahman, M. Hinz, R. Fussganger, K. E. Schroder, K. Shaub, and E. F. Pfeiffer, *Horm. Metab. Res. 2:* 268(1970).
92. N. Rakieten, B. S. Gordon, A. Beaty, D. A. Cooney, R. D. Davis, and P. I. Schein, *Proc. Soc. Exp. Biol. Med. 151:*356 (1976).
93. N. Rakieten, B. S. Gordon, A. Beaty, D. A. Cooney, and P. S. Schein, *Proc. Soc. Exp. Biol. Med. 151:*356(1976).
94. P. S. Schein, D. A. Cooney, and M. L. Vernon, *Cancer Res. 27:* 2324(1967).
95. P. S. Schein and R. W. Bates, *Diabetes 17:*760(1968).
96. A. A. Rossini, N. A. Like, W. L. Chick, M. C. Appel, and G. F. Cahill, *Proc. Natl. Acad. Sci. USA: 74:*2485(1977).

97. J. Warter, D. Phillippides, B. Gillet, A. Weryha, and B. Meknini, *Ann. Med. Interne 124:*729(1973).
98. G. Deby, *Ann. Gastroenterol. Hepatol. 13:*249(1977).
99. N. W. du Priest, W. H. Massey, and M. J. Fletcher, *Ann. Surg. 38:*514(1972).
100. T. Kazumi, G. Yoshino, S. Fujii, and S. Baba, *Cancer Res. 38:* 2144(1978).
101. T. Kazumi, G. Yoshino, Y. Yoshida, K. Doi, M. Yoshida, S. Kaneko, and S. Baba, *Endocrinol. 103:*1541(1978).
102. P. K. Dixit and G. E. Bauer, *Proc. Soc. Expl. Biol. Med. 152:* 232(1976).
103. D. J. Meyer, *Am. J. Vet. Res. 37:*1221(1976).
104. G. Yoshino, T. Kazumi, S. Fujii, S. Kaneko, M. Yoshida, and S. Baba, *Tohoku J. Exp. Med. 130:*297(1980).
105. W. L. Chick, M. C. Appel, G. C. Weir, A. A. Like, V. Lauris, J. G. Porter, and R. N. Clute, *Endocrinology 107:*989(1980).
106. R. Schoental, *Cancer Res. 35:*2020(1975).
107. M. M. Wick, A. Rossini and D. Glynn, *Cancer Res. 37:*3901(1977).
108. E. Johansoon-Brittebo and H. Tjolve, *Cancer Lett. 8:*169(1979).
109. F. A. French, *J. Int. Res. Comm. 1:*33(1973).
110. F. A. French, *Adv. Exp. Med. Biol. 91:*281(1977).
111. J. Hoshino, U. Kuhne, and H. Kroger, *Z. Physiol. Chem. 358:* 1222(1977).
112. G. S. Johnson and W. B. Anderson, *J. Natl. Cancer Inst. 63:* 1451(1979).
113. G. Calcutt, S. W. Ting, and A. W. Preece, *Brit. J. Cancer 24:* 380(1970).
114. C. J. Skidmore, M. I. Davies, P. M. Goodwin, H. Halldorsan, P. J. Lewis, S. Shall, and A. A. Ziaee, *Eur. J. Biochem. 101:* 135(1979).
115. M. Smulson, P. Schein, B. Shnider, and D. Mullins, *Proc. Am. Ass. Cancer Res. 117:*000(1976).
116. M. E. Smulson, P. Schein, D. W. Mullins, and S. Sudhaker, *Cancer Res. 37:*3006(1977).
117. E. L. Jacobson and G. Narasiurhan, *Fed. Proc. 38:*619(1979).
118. E. L. Jacobson, G. Narasimhan, and M. K. Jacobsen, *Tex. J. Sci. 31:*85(1979).
119. M. K. Jacobson, V. Levi, H. Juarez-Salinas, R. A. Barton, and E. L. Jacobson, *Cancer Res. 40:*1797(1980).
120. N. A. Berger and G. W. Sikorski, *Biochem. Biophys. Res. Comm. 95:*67(1980).
121. C. Rochette-Egly, M. E. Ittel, J. Biles, and P. Mandel, *FEBS Lett. 120:*7(1980).
122. L. P. Schacter and D. J. Burke, *Biochem. Biophys. Res. Comm. 80:*504(1978).
123. K. Morioka, T. Tanaka, T. Nokuo, M. Ishizawa, and T. Ono, *Gann 70:*37(1979).

124. M. Terada, H. Fujik, P. A. Marks, and T. Sugimura, *Proc. Natl. Acad. Sci. USA 76:*6411(1979).
125. T. Utakoji, K. Hosoda, U. Umezuwa, M. Sawamura, T. Matsushima, M. Miwa, and T. Sugimura, *Biochem. Biophys. Res. Comm. 90:* 1147(1979).
126. L. Buch, D. Streeter, R. M. Halpern, L. N. Simon, M. G. Stout, and R. A. Smith, *Biochemistry 11:*393(1971).
127. H. Wolf, H. R. Nieltsen and R. R. Brown, *Proc. Am. Ass. Cancer Res. 17:*6(1976).
128. H. R. Nielsen, H. Wolf and R. R. Brown, *J. Natl. Cancer Inst. 59:*855(1977).
129. S. Sherma and N. Biswal, *Virology 82:*265(1977).
130. R. Cox and C. M. Schmidt, *Proc. Am. Ass. Cancer Res. 21:*69 (1980).
131. J. W. Dorson and R. E. Moses, *J. Biol. Chem. 253:*665(1978).
132. M. J. Rosenberg and A. J. Caplan, *J. Embryol. Exp. Morph. 33:* 947(1975).
133. Litton Biometrics, Inc., *Mutagenicity Evaluation [Niacin (nicotinic acid)]*, 1977, 49 pp. Available through National Technical Information Service, Springfield, VA as PB-278 472/69A, FDA/BF-78/ 68, FDA 75-88.
134. R. Braun, J. Schoneich, and D. Ziebarth, *Exp. Pathol. 18:*314 (1980).
135. S. Nolde and H. Hilz, *Brit. J. Cancer 26:*299(1972).
136. A. M. Pamukcu, U. Milli, and G. T. Bryan, *Nutrition and Cancer 3:*86(1982).
137. D. Brown and A. Goldfeder, *Proc. Am. Ass. Cancer Res. 21:*268 (1980).
138. D. M. Brown, *Diss. Abst. Int. [B] 40:*5164(1980).
139. M. Wintzerith, A. Dierich, and P. Mandel, *J. Neurochem. 33:*677 (1979).
140. M. Wintzerith, A. Dierich, and P. Mandel, *Biochim. Biophys. Acta 613:*191(1980).

14

CELLULAR BIOCHEMISTRY OF NICOTINAMIDE COENZYMES

As explained in a previous chapter, the pyridine nucleotide coenzymes are coenzymes of systems. In their behavior with individual enzymes, they seem more like substrates. While pyridine nucleotide coenzymes freely dissociate from enzyme sites, most cellular NAD and NADP is protein bound.

The overall metabolism of a cell is an intricate intertwining network of sequentially catalyzed transformations of organic molecules, regulated by allosteric modulation, feedback inhibition, and hormonally induced alterations on enzyme levels. But there is a basic regulation that is predicated by substrate availability and compartmentation of reactions. The cell needs to liberate and utilize chemical energy and generate organic building blocks to make complex cellular components. The overall sequences of metabolic reactions can be divided into many distinct sets. Quantitatively, the extraction of energy from carbohydrate dominates metabolism, and therefore glycolysis and respiration will be used to illustrate the principles of cellular metabolism. In that metabolic sequence, the glycolytic sequence is cytoplasmic and the energy-yielding oxidative steps are mitochondrial. In the utilization of carbohydrates (as for all substrates) there is a separation of carbon flow and hydrogen flow (Fig. 1). In that sequence the reduced pyridine nucleotide coenzymes are the common intermediates for the metabolic flow of hydrogen. However, to transfer the hydrogen from the cytoplasm to the mitochondria, carbon and hydrogen flows need to be rejoined. This chapter will briefly restate the catalytic role of the pyridine nucleotides in this overall process, trace the movement of hydrogen between compartments, and discuss the role of pyridine nucleotides in the energy-generating process.

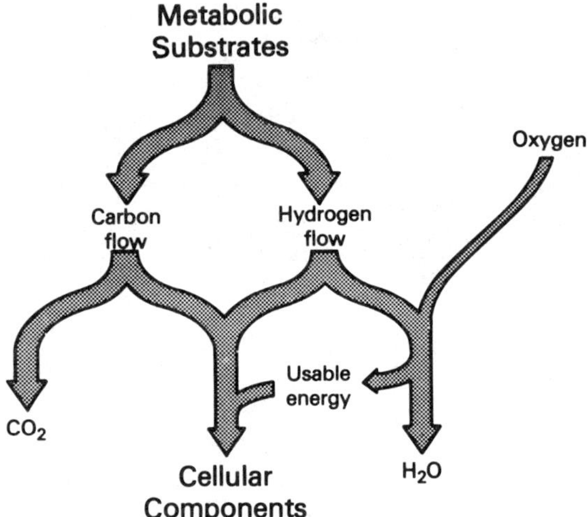

Figure 1 The distinction between carbon and hydrogen flow in metabolism.

I. THE DIVERSE FUNCTIONS OF NAD AND NADP

In the clycolytic sequence from glucose to lactata, NAD is alternatively reduced and reoxidized. There is no net oxidation. The carbohydrate, as glucose 6-phosphate, can also be oxidized by NADP through catalytic action of glucose 6-phosphate dehydrogenase to 6-phosphogluconic acid and NADPH. There is no internally closed loop for NADPH reoxidation as there is for NAD. There is evidence that some lactate dehydrogenases can utilize NADPH for pyruvate reduction (1), but NAD inhibits that reaction strongly and therefore only with very low levels of NAD would the reaction be able to proceed (2,3). Normally, this is not the situation. The NADPH that accumulates is used to drive biosynthetic reactions, especially lipid biosynthesis. When the pyruvate is oxidized in mitochondria and the NADH hydrogen is also transferred to intramitochondrial pyridine nucleotides for further oxidation, there is no alternative but to use NADPH for biosynthetic processes (Fig. 2). In general, therefore, NAD/NADH is used in catabolic processes and NADP/NADPH in anabolic processes for which hydrogen is required.

In a steady state there is an equilibrium between NADH and NAD as well as between NADPH and NADP. In tissues, the experimental conditions of oxygen availability, the rapidity of interruption of metabolic sequences, and the level of allosteric modifiers all affect the actual reduced/oxidized coenzyme ratios. However, their values as measured document the differential formation of NAD and NADP. In early experiments cytoplasmic NAD/NADH ratios were found to be 5–7/1 (4,5), which is reflected in a lactate/pyruvate ratio of about 20/1 (6). This is not a thermodynamic equilibrium, achieved between NAD and lactate as catalyzed by traces of lactic dehydrogenase, because most pyridine nucleotides are bound to dehydrogenases, thus giving a considerably different potential to the NAD/NADH complex.

In the cytoplasm the equilibrium between NAD and NADP is not as predicted from their redox potential. In the reaction

$$NADH + NADP^+ \rightleftarrows NAD^+ + NADPH$$

catalyzed by nonenergy-dependent transhydrogenase, the equilibrium constant is 0.79 (7). This is in agreement with the difference in equilibrium ratios of the two coenzyme pairs when glutamate is oxidized through the action of glutamate dehydrogenase, utilizing NAD or NADP, which function equally well on liver glutamate dehydrogenase (8).

However, when the ratios are deduced from substrate couples such as malate/pyruvate + CO_2, a NADP/NADPH ratio of 1/240 is obtained.

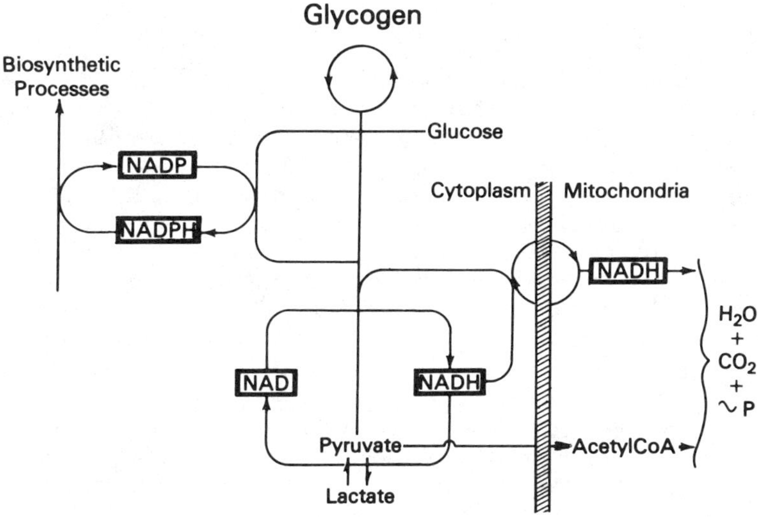

Figure 2 The distinctive roles of NAD and NADP.

There is no cytoplasmic transhydrogenase so this discrepancy can be understood (9). The use of such couples to calculate coenzyme levels is justified by the fact that NAD/NADH ratios calculated from various substrate couples in one compartment agree (9).

The NAD/NADH ratio in the soluble fraction and in mitochondria differs, however (4,5,9). More importantly, in mitochondria the ratio of NADPH/NADH markedly favors NADPH, even when the steady state reduction levels of the pyridine nucleotides are measured in intact mitochondria (10). Overall in liver the ratio of NADPH/NADP, as measured directly, ranges from 2.9/1 to more accurate values of about 7.7/1 (11). A pronounced alteration in the NAD/NADH ratio is seen after ethanol ingestion: an NADH excess is produced; α-glycerophosphate accumulates and thus there is abundant substrate for triglyceride biosynthesis; gluconeogenesis is inhibited. These events derange overall homeostasis and strongly suggest the importance of the normal redox potential of the NAD/NADH couple (15).

In the cytoplasm, the rate of glycolysis is dependent on the rate of oxidation since many of the enzymes in glycolysis are under allosteric control by the intermediates determining the energy charge of the cell (i.e., AMP, ADP, and ATP). The biosynthesis of fat and the catabolism of glucose are intimately linked. When the metabolism is overpowered by an abnormal substrate, the ratio of NAD/NADH may fall significantly. For instance, the pentahydroxy alcohol xylitol, when given in excessive doses, alters the liver cytosol NAD/NADH ratio toward a higher reducing state (12). There is simultaneously an accumulation of α-glycerophosphate, probably because of this changing ratio of NAD/NADH (13). The same phenomenon occurs in humans after xylitol and ethanol ingestion (14).

The two coenzymes NAD and NADP are in constant equilibrium, via a series of reactions. Of these, the biosynthetic interconversions have been covered previously. From a metabolic point of view the transhydrogenase reaction would seem to be the most important (Fig. 3). However, in mammalian systems that reaction does not occur in the cytoplasm and therefore cannot serve to equilibrate NAE and NADPH levels. Thus the biosynthetic drive of NADPH is different from the energy production function of NADH. The NADPH level is regulated, as can be shown in the NADPH-generating pentose phosphate pathway, by allosteric inhibitions by NADPH at the initial step catalyzed by glucose 6-phosphate dehydrogenase (16). However, in general, whenever one of an enzyme pair utilizes NADP and the other NAD, the former is *not* regulated and used biosynthetically, while the latter is under allosteric control and is used for energy production (17). Therefore, both the pentose phosphate shunt and the Embden–Meyerhof pathway for glycolysis supply hydrogen and carbon but the functions are different (Table 1).

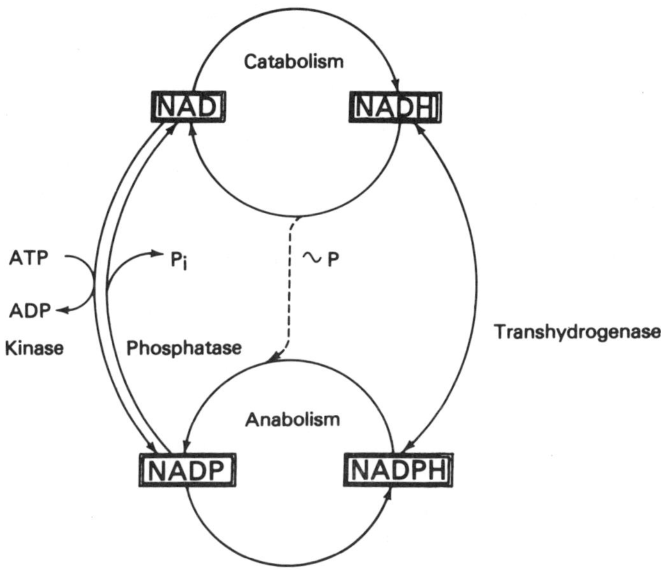

Figure 3 The mechanisms of equilibration between NAD and NADP concentrations.

Table 1 Functions of Different Pathways of Glucose Degradation

	Glycolysis + Krebs cycle	Pentose phosphate pathway
Hydrogen	NAD ATP synthesis	NADP Reductive synthesis Energy for hydroxylation reactions Peroxide generation
Carbon	Carbon skeletons for CO_2 formation Nonessential amino acids for protins, purines, pyrimidines Acetyl-coenzyme A for fat, steroids. Glycerophosphate for fats	Carbon skeletons for pentose phosphates for nucleic acids

II. THE COMPARTMENTATION OF PYRIDINE NUCLEOTIDES

Carefully prepared mitochondria do not oxidize NADH (18). Shunting of the hydrogen between cytoplasm and mitochondria via a common substrate and through the completed action of a pair of soluble and insoluble dehydrogenases was found to be the mechanism of hydrogen transport. The first such complex proposed was dihydroxyacetone/α-glycerophosphate (6,19,20). This cycle was found in insect flight muscle but was proposed as a general reaction in all tissues (15). Later the complex β-hydroxybutyrate/acetoacetate was considered to have a role in hydrogen transport (21).

However, the most general reaction is probably the malate/oxaloacetate shuttle (Fig. 4). As early as 1937 Szent-Györgyi suggested that malic and oxaloacetic acids were hydrogen carriers and the link between cytosolic carbohydrate degradation and the cytochrome system (22,23). As originally formulated, the theory predicted that succinate should be the anaerobic end product of carbohydrate degradation if enough four carbon acids were available. The theory was abandoned when Krebs proposed what is now known as the Krebs cycle or the citric acid cycle (24). Although that theory explained Szent-Györgyi's data, Krebs was referring to *carbon* metabolism while Szent-Györgyi was speaking of *hydrogen* metabolism. The modern day relevance of Szent-Györgyi's observations was pointed out independently in the mid-1960s (25,26).

The malate/oxaloacetate shuttle has all the prerequisites for the major NAD-hydrogen carrier between cytoplasm and mitochondria. First of all, the complex malate/oxaloacetate is in equilibrium with the lactate/pyruvate pool (27). The hydrogen of malate is preferentially disposed of through oxidation and is not readily used for biosynthetic purposes (28). In contrast, lactate and glycerol readily yield hydrogen for biosynthetic purposes (28,29). Malic dehydrogenase is present in both cytoplasm and mitochondria in quantities too ample to be compatible with glycolytic rates (30). The early observations have since

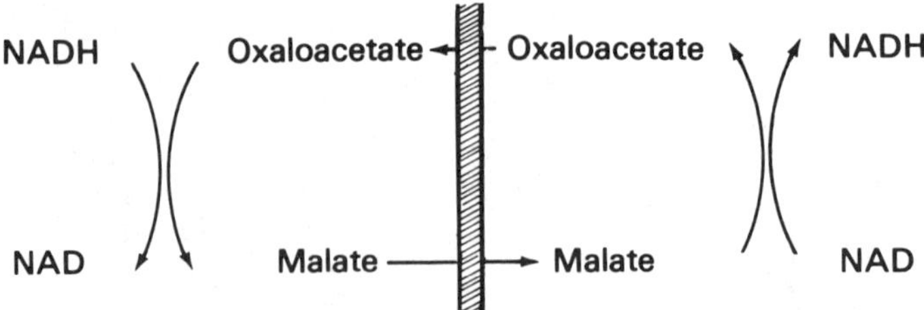

Figure 4 The malate/oxaloacetate shuttle between cytoplasm and mitochondria.

been repeatedly verified. However, the mitochondria are not as permeable to oxaloacetate as they are to malate, so that the shuttle is more complex.

Oxaloacetate levels are a key factor in efficient communication between extramitochondrial and intramitochondrial metabolism. Not only does oxaloacetate function as a hydrogen carrier, but it is also the acceptor for acetyl-coenzyme A to form citrate as the first step into the Krebs cycle. The citrate is then oxidized to isocitrate. Both citrate and isocitrate can traverse the mitochondrial membrane through a malate-activated tricarboxylic acid carrier (31). In contrast to cytosolic and mitochondrial malate dehydrogenase which are both NAD dependent, the extramitochondrial isocitrate dehydrogenase requires NADP and the intramitochondrial enzyme requires NAD. The extramitochondrially generated NADPH is used for biosynthetic purposes, especially fatty acid biosynthesis. The citrate can give rise extramitrochondrially to acetyl-coenzyme A for fatty acid biosynthesis. The pivotal role of isocitrate dehydrogenase and the near total compartmentation of NADP hydrogen with shuttle mechanisms for NADH are of paramount importance. The allosteric regulation of these reactions is reviewed by Stadtman (17), and the pertinent reactions are summarized in Figure 5.

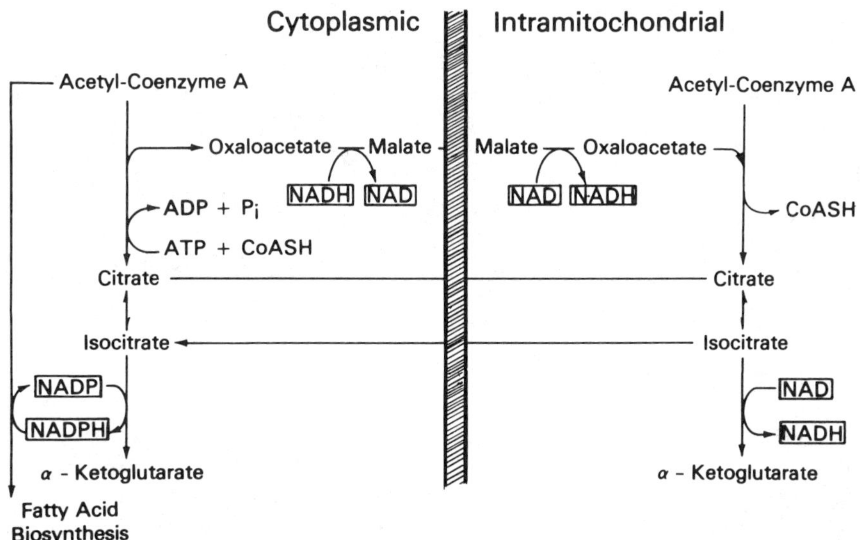

Figure 5 Metabolic equilibration between intramitrochondrial and cytoplasmic metabolism.

III. TRANSHYDROGENATION REACTIONS

There is thus a distinct division of labor between the pyridine nucleotide coenzymes. However, as already mentioned there is equilibration between the coenzymes in the enzyme-catalyzed reaction called a transhydrogenation:

$$NADH + NADP^{+} \rightleftarrows NAD^{+} + NADPH$$

This remarkable reaction was first reported by Colowick at the Annual McCollum Pratt Institute Symposium in 1951 (32). The enzyme activity was first discovered in *Pseudomonas* fluorescence studies (33,34), but was soon also identified in animal tissues (7). The enzyme is different from the usual flavin adenine dinucleotide-dependent diaphorases, which can also act as transhydrogenase between pyridine nucleotide analogues (35,36): most importantly, the transhydrogenase of *Pseudomonas* catalyzes a direct hydride ion transfer as deduced by deuterium exchange (37), while flavoproteins catalyze electron transfer (36).

The *Pseudomonas* enzyme is an easily solubilized protein while the mammalian enzyme is associated with the inner membrane of mitochondria; the latter has been purified to homogeneity from beef heart mitochondria (38,39). These two transhydrogenases are distinguishable in two major ways. First, they have a different stereospecificity. The enzyme from *Pseudomonas* is specific for the B side of both coenzymes (37,40), whereas the mitochondrial enzyme is specific for the A-side of NADH and the B side of NADPH (41). Hoek et al. proposed the distinction between AB and BB transhydrogenases (40).

The more important physiological distinction is that the enzyme from the mitochondria is energy linked. A review by Rydström traces that realization from the first discovery by Krebs that NADPH-dependent dehydrogenases are sensitive to nitrophenol (43) to our current understanding.

The overall reaction suggested by Rydström (42) is:

$$NADH + NADP^{+} + ATP \rightleftarrows NAD^{+} + NADPH + ADP + P_{I}$$

The equilibrium constant for the reaction

$$[NADPH][NAD^{+}]/[NADH][NADP]$$

has been estimated at 500 for the steady rate of energy-linked transhydrogenation (44,45). This is far removed from the equilibrium for nonenergy-dependent transhydrogenation (7), but not as far toward NADPH generation as the free energy of hydrolysis of ATP would imply.

The reaction is reversible but requires special conditions for this to occur as, for example, when NAD analogues with a more positive redox potential are used. Thus the reaction

NADPH + acetylpyridine-adenine dinucleotide →
NADP$^+$ + reduced acetylpyridine-adenine dinucleotide

can be shown to proceed with the equivalent of energy storage (46). Thionicotinamide-adenine dinucleotide can also be used.

Although in submitochondrial particles the reaction seems to be directly coupled to ATP hydrolysis, it is more accurate to write the equation as:

$$\mathrm{NADPH} + \mathrm{NAD}^+ + nH_m^+ \rightleftarrows \mathrm{NADP}^+ + \mathrm{NADH} + nH_c^+$$

where *m* and *c* represent matrix and cytosol sides of the mitochondrial membrane (42,47,48). These reactions are of profound significance in the mechanism whereby oxidative energy is harnessed into high-energy phosphate bonds (i.e., in oxidative phosphorylation). The importance of the transhydrogenases in this basic biochemicophysiological phenomenon was already demonstrated by Kaplan (49) in early experiments with mitochondria depleted of pyridine nucleotides but with those coenzymes re-added. Utilizing a substrate that employed NADP, no oxidative phosphorylation occurred, but on addition of NAD *and* NADP, oxidative phosphorylation was restored (Table 2). The presence of transhydrogenase could be demonstrated in those mitochondria, and the reaction was then thought to be a means of regulating respiration and energy-releasing reactions. However, the mechanism of oxidative phosphorylation is more complex.

IV. OXIDATIVE PHOSPHORYLATION AND THE POSSIBLE ROLE OF PYRIDINE NUCLEOTIDE TRANSHYDROGENASE

The exact mechanism by which oxidative phosphorylation occurs is one of the major unsolved problems of biochemistry. The sequence of reactions through which NADH is oxidized to yield H_2O + NAD$^+$ are well documented. There are coupling points at which the reaction generates ATP from ADP + inorganic phosphate. The oxidative reactions proceed on an organized matrix, wherein the proteins are immobilized. The participants include flavin-adenine dinucleotide, ubiquinone (coenzyme Q), and various porphyrinoproteins called cytochromes, which are designated by letters. The coupling sites for phosphorylation can be pinpointed in the sequence. Figure 6 shows the sequence and the points at which phosphorylation occurs. By using selective inhibitors, the sequence can be dissected. By using alternate substitutes the P:O ratio (μmoles ATP formed from ADP per μatoms oxygen), can be changed from 3 to 2 to 1. The process is tightly coupled, in that under optimal circumstances oxidation does not proceed without phosphorylation. Yet the search for an intermediate that would explain this interrelationship has not been successful.

Table 2 Restoration of Phosphorylation with Isocitrate Oxidation by NAD

Mitochondria	Coenzyme (1.5 μmoles)	Substrate[a] (30 μmoles)	Oxygen uptake (μatoms)	Phosphorus (μmoles)	P:O ratio
Undepleted[b]	–	glutamate	3.52	7.62	2.2
Depleted[a]	–	glutamate	0	0	0
Depleted	NAD	glutamate	5.3	15.0	2.8
Undepleted	–	isocitrate	9.95	29.9	3.0
Depleted	–	isocitrate	1.10	0	0
Depleted	NADP	isocitrate	3.2	0	0
Depleted	NADP + NAD	isocitrate	9.6	22.2	2.3

[a]Undepleted mitochondria contained 20 μg NAD + NADP/Warburg flaste; depleted mitochondria contained <1 μg/Warburg flaste.

[b]Depletion was accomplished by preincubation in phosphate buffer and sucrose for 15 minutes at 30°C.

Source: Ref. 43.

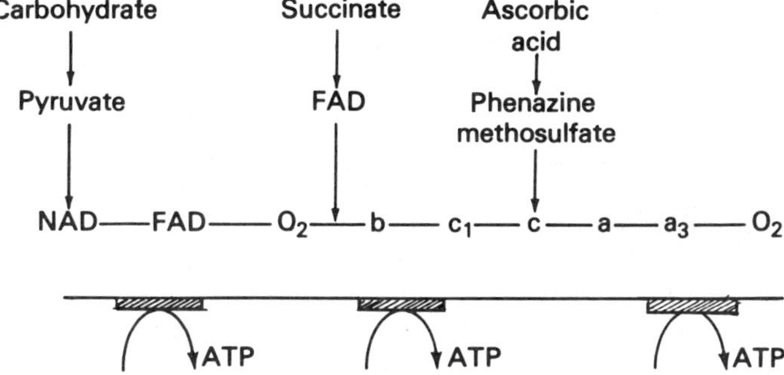

Figure 6 The electrotransport chain.

The concept of chemical coupling suggests that a substrate, A, in the chain would be converted to a high-energy intermediate, analogous to the reaction of trisephosphate dehydrogenase:

$$\text{R}-\overset{\text{O}}{\ddot{\text{C}}}\text{H} + \text{NAD} + \text{Enzyme-SH} \rightarrow \text{Enzyme-S}-\overset{\text{O}}{\ddot{\text{C}}}\text{R} + \text{NADH} + \text{H}^+$$

$$\text{Enzyme-S}-\overset{\text{O}}{\ddot{\text{C}}}\text{R} + \text{H}_3\text{PO}_4 \rightarrow \text{Enzyme-SH} + \text{R}-\overset{\text{O}}{\ddot{\text{C}}}-\text{OP O}_3\text{H}_2$$

By postulating further intermediates this concept could be applied to oxidative phosphorylation. The mitochondria are highly organized, but can be solubilized to yield a variety of proteins with catalytic functions in the oxidative sequence or in the hydrolysis of ATP to yield ADP and inorganic phosphate. A large number of reconstitution experiments have been reported which failed to elucidate the intermediate steps involved. A highly readable summary was published by Racker (50). However, no high-energy intermediate has ever been found in intact mitochondria or reconstituted particles, and while that does not rule out such intermediates, alternative hypotheses have been proposed.

In 1961 the chemiosmotic hypothesis of oxidative phosphorylation was proposed (51). It has since been extensively debated, expanded, refined, and examined. The description of Fillingame is useful (52):

> The electron transport chain is topologically organized in the inner mitochondrial membrane so that protons are translocated from the inside to the outside as electrons are passed from substrate to O_2. The inner mitochondrial membrane is relatively impermeable to H^+, OH^- and other ions. Consequently the electrogenic pumping of H^+ across the membrane generates an electrochemical potential for H^+ composed of a membrane potential and a chemical concentration gradient of H^+. This thermodynamic potential established

is used to drive the synthesis of ATP. This is carried out by a reversible H^+ translocating ATPase that synthesizes ATP as protons flow down their electrochemical gradient and enter the mitochondria.

The stoichiometry for the reaction

$$nH^+_{out} + ADP + P_i \rightleftarrows ATP + H_2O + nH^+_{in}$$

is being measured. The original hypothesis would predict a H^+/ATP ratio of 2, but recently measured values in mitochondria have exceeded that number (53). Furthermore, there must then be an equivalent translocation of protons per coupling site in electron transport. A stoichiometry of $12H^+$/O for NADH-linked substrates has been determined, and $8H^+$/O for succinate. This would allow $4H^+$ per ATP synthesized (54–56).

The specific mechanism whereby the proton translocation occurs is not clear as yet. The energy-dependent mitochondrial transhydrogenase is electrogenic and brings about a net vectorial translocation of protons (42,47,48,57,58). When submitochondrial particles were used, a coupled H^+ translocation with hydride ion transfer could be measured, though the H^+/H^- was only 0.2 (58). A theoretical ratio of 2 was proposed (58).

Nevertheless, the ability of the transhydrogenase to act as a proton pump elevates the enzyme to an important role in oxidative phosphorylation. The function of the enzyme could be tested when it was isolated in homogeneous form from beef heart. When the enzymes were incorporated into liposomes, the action of the enzyme generated a translocation of protons (59). The H^+/H^- ratios obtainable ranged from 0.77 to 0.84. The last word has not yet been written on these reactions, yet it seems clear that there are unique roles for NAD and NADP in energy generation during cellular respiration. We do not mean to imply that all proton translocation reactions definitely involve the transhydrogenase. However, it is true that the majority of electron-transferring events generated in cellular energy metabolism start with NADH as hydride ion/electron door. Studies of the cytochrome segments of oxidative phosphorylation have been recently reviewed by Wikström et al. (60). In the same volume the mechanism of ATP synthesis is discussed (61). The general reaction may be written as:

$$A_{red} + B_{ox} \rightleftarrows A_{ox} + B_{red} + \Delta\ \mu\ H^+$$

$$ADP + P_i + \Delta\ \mu\ H^+ \rightleftarrows ATP + H_2O$$

Two reactions involving the pyridine nucleotides which can generate the chemiosmotic driving force include the transhydrogenase reaction and the NADH to coenzyme Q segment of the electron transport chain. It is possible, however, that the transhydrogenase reaction is a fundamental general component of the overall mechanism of oxidative phosphorylation. The state of the knowledge is now at a point at which the biophysical observations of the Mitchell hypothesis may soon be translated into biochemical explanations (60).

REFERENCES

1. H. Holzer and S. Schneider, *Biochem. Z. 330:*240(1958).
2. F. Nazario, B. B. Ernster, and L. Ernster, *Biochim. Biophys. Acta 26:*416(1957).
3. E. Kun, J. E. Ayling, and B. V. Siegel, *Proc. Nat. Acad. Sci. 46:*622(1960).
4. G. E. Glock and P. McLean, *Exp. Cell. Res. 11:*234(1956).
5. K. B. Jacobson and N. O. Kaplan, *J. Biol. Chem. 226:*603(1957).
6. T. Bucher and M. Klingeberg, *Angew. Chem. 70:*552(1958).
7. N. O. Kaplan, S. P. Colowick, and E. F. Neufeld, *J. Biol. Chem. 205:*1(1953).
8. J. A. Olson and C. B. Anfinsen, *J. Biol. Chem. 841*(1953).
9. H. A. Krebs, *Advances in Enzyme Regulation 5:*409(1967).
10. M. Klingenberg and W. Slenczka, *Biochem. Z. 331:*486(1959).
11. H. B. Burch, M. E. Bradley, and O. H. Lowry, *J. Biol. Chem. 242:*4546(1967).
12. J. R. Williamson, A. Jakob, and R. Scholz, *Metabolism 20:*13(1971).
13. H. F. Wood, L. V. Eggleston, and H. A. Krebs, *Biochem. J. 119:* 501(1970).
14. R. H. Ylikahri and T. Leino, *Metabolism 1428:*25(1979).
15. H. A. Krebs, *Advances in Enzyme Regulation 6:*467(1968).
16. H. A. Krebs and L. V. Eggleston, *Adv. Enzyme Reg. 12:*421(1974).
17. E. R. Stadtman, *Adv. Enzymol. 28:*41(1966).
18. A. L. Lehninger, *Harvey Lectures 49:*176(1955).
19. R. W. Estabrook and B. Saktor, *J. Biol. Chem. 233:*1014(1958).
20. E. Zebe, A. Delbrück, and T. Bücher, *Biochem. Z. 331:*254(1959).
21. T. M. Devlin and B. H. Bedell, *J. Biol. Chem. 235:*2134(1960).
22. A. Szert-Györgyi, *Studies on Biological Oxidation and Some of its Catalysts.* J. A. Barth, Leipzig, 1937.
23. A. Szent-Györgyi, in *Recurrent Aspects of Biochemical Energetics* (N. O. Kaplan and E. P. Kennedy, eds.) Academic Press, New York, 1966, p. 63.
24. H. A. Krebs and W. A. Johnson, *Enzymologica 4:*148(1937).
25. J. van Eys, *J. Biol. Chem. 239:*3544(1969).
26. J. B. Chappell, *Biochem. J. 90:*225(1969).
27. M. J. Hohorst, F. H. Kreutz, and T. H. Bücher, *Biochem. Z. 332:* 18(1959).
28. H. D. Hobermann, *J. Biol. Chem. 232:*9(1958).
29. B. Bloom, *J. Biol. Chem. 234:*2158(1959).
30. A. Delbrück, H. Schimassek, K. Bartsch, and T. H. Bücher, *Biochem. Z. 331:*297(1959).

31. J. B. Chappell and B. H. Robinson, in *Metabolic Roles of Citrate* (T. W. Goodwin, ed.), Academic Press, New York, 1968, p. 123.
32. S. P. Colowick, in *Phosphorus Metabolism, Vol. I.* (W. D. McElroy and B. Glass, eds.), The Johns Hopkins Press, Baltimore, 1951, p. 436.
33. S. P. Colowick, N. O. Kaplan, E. F. Neufeld, and M. M. Ciotti, *J. Biol. Chem. 195:* 95(1952).
34. N. O. Kaplan, S. P. Colowick, and E. F. Neufeld, *J. Biol. Chem. 195:* 107(1952).
35. M. M. Weber and N. O. Kaplan, *J. Biol. Chem. 225:* 909(1957).
36. M. M. Weber, N. O. Kaplan, A. San Pietro, and F. E. Stolzenbach, *J. Biol. Chem. 227:* 27(1957).
37. A. San Pietro, N. O. Kaplan, and S. P. Colowick, *J. Biol. Chem. 212:* 941(1955).
38. W. M. Anderson and R. R. Fisher, *Arch. Biochem. Biophys. 180* (1978).
39. B. Höjeberg and J. Rydström, *Biochem. Biophys. Res. Comm. 78:* 1183(1977).
40. J. B. Hoek, J. Rydström, and B. Höjeberg, *Biochim. Biophys. Acta 333:* 237(1974).
41. C. P. Lee, N. Simard-Duquesne, L. Ernster, and H. D. Hoberman, *Biochim. Biophys. Acta 105:* 397(1965).
42. J. Rydström, *Biochim. Biophys. Act. 463:* 155(1977).
43. H. A. Krebs, *Bull. Johns Hoplins Hosp. 98:* 34(1954).
44. C. P. Lee and L. Ernster, *Biochim. Biophys. Acta 81:* 187(1964).
45. J. Rydström, A. Teixeira da Cruz, and L. Ernster, *Eur. J. Biochem. 17:* 56(1970).
46. D. C. Phelps, Y. M. Galante, and Y. Hatefi, *J. Biol. Chem. 225:* 9647(1980).
47. P. Mitchell, *Biol. Rev. Camb. Philos. Soc. 41:* 445(1966).
48. J. F. Blazyk, D. Lam, and R. R. Fisher, *Biochem. 15:* 2843(1976).
49. N. O. Kaplan, M. N. Swiartz, M. E. Frech, and M. M. Ciotti, *Proc. Natl. Acad. Sci. 42:* 481(1956).
50. E. Racker, *A New Look at Mechanisms in Bioenergetics,* Academic Press, New York, 1976.
51. P. Mitchell, *Nature 191:* 144(1961).
52. R. H. Fillingame, *Ann. Rev. Biochem. 49:* 1079(1980).
53. A. Alexander, B. Reynafarje, and A. L. Lehninger, *Proc. Nat. Acad. Sci. 75:* 5296(1978).
54. B. Reynafarje, M. D. Brand, and A. L. Lehninger, *J. Biol. Chem. 251:* 7442(1976).
55. A. Vercesi, B. Reynafarje, and A. L. Lehninger, *J. Biol. Chem. 253:* 6379(1978).
56. A. Villalobo and A. L. Lehninger, *J. Biol. Chem. 254:* 4352(1979).

57. P. Mitchell and J. Moyle, *Nature 208:* 1205(1968).
58. L. Moyle and P. Mitchell, *Biochem. J.: 132:* 571(1973).
59. S. R. Earle and R. H. Fisher, *J. Biol. Chem. 255:* 827(1980).
60. M. Wilström, K. Krab, and M. Saraste, *Ann. Rev. Biochem. 50:* 623(1981).
61. R. L. Cross, *Ann. Rev. Biochem. 50:* 681(1981).

15

CONCLUDING REMARKS

The physiological roles of nicotinamide are pervasive and pivotal. Nicotinamide coenzymes control the key steps in:

1. Cytoplasmic catabolism and generation of substrates for mitochondrial oxidation
2. Biosynthetic reactions in the cytoplasm
3. Drug metabolism in endoplasmic reticulum
4. Mitochondrial hydrogen oxidation and ATP synthesis
5. Nuclear DNA repair

Figure 1 illustrates the pervasive role of the pyridine nucleotide coenzymes. The level of coenzymes in mammals depends on the circulating nicotinamide, the nuclear turnover, and the degree of excess cytoplasmic NAD^+. Excessive intake of nicotinic acid and nicotinamide affects the levels in tissues. A deficiency of nicotinamide will affect different cellular compartments differently. There is as yet no evidence that compels the selection of one finding over another in the determination of the site of the biochemical lesion in niacin deficiency. The biosynthesis of niacin from tryptophan complicates the issue further. It is tempting to speculate that the nuclear reactions are the most sensitive to NAD deficiency and that two of the three D's of pellagra (dermatitis and diarrhea) are due to the rapid turnover of the skin and intestinal mucosa and thus the consequent demand on DNA replication and repair, but that would be as yet unfounded by fact. It is, in fact, highly unlikely that a *single* biochemical defect exists in niacin deficiency, such as is postulated for instance for thiamine deficiency.

Viewing metabolism through the physiology of niacin potentially generates a complete summary of all of cellular transformations. This review has been an attempt at a condensed summary rather than an exhaustive review.

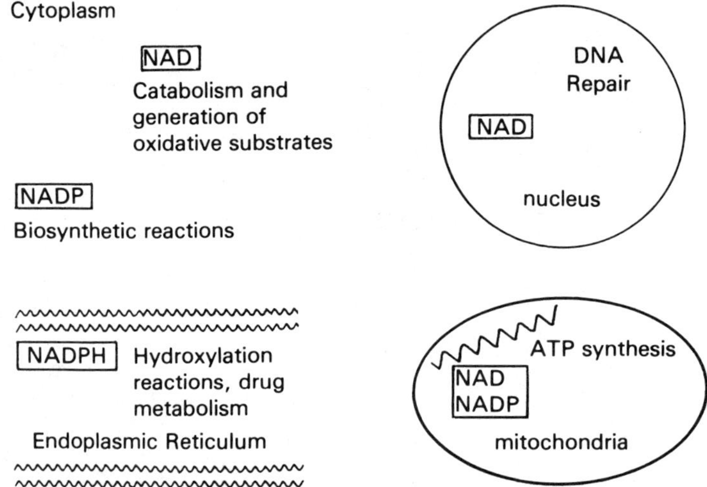

Figure 1 The major functional sets of NAD and NADP.

PART III

NICOTINIC ACID AS A DRUG

Nicotinic acid is unusual among the vitamins in that doses in excess of nutritional requirements result in a number of readily recognizable pharmacological responses, some of which have therapeutic significance. By and large, these responses can be broadly classified as vascular or metabolic, and both types may be acute or chronic, therapeutic or toxic. In this part, the pharmacokinetics of nicotinic acid (i.e., how the body handles the compound) will be reviewed with particular attention to those aspects which are perceived as being relevant to the therapeutic potential of the agent. The nature of the pharmacodynamic responses (i.e., what the compound does to body systems) will then be discussed with that same orientation, following which the specific therapeutic goals of nicotinic acid as a drug will be considered in some detail.

In view of the relationship of nicotinic acid to nicotinamide, as discussed in detail in prior sections, it is remarkable that little if any of the considerable pharmacological actions of nicotinic acid are shared with nicotinamide. Certainly, the two are quite interchangeable as nutrients, even though their in vivo interconversion is not a simple one-step transamidation (see Chap. 4). In fact, some of the pharmacological effects of nicotinic acid can be blocked by pretreatment with a large dose of nicotinamide.

The acute flush response to nicotinic acid is its most obvious but certainly not its most important vascular effect. It is frequently erroneously equated with altered blood flow, local temperature changes, and fibrinolytic activation. Responses to nicotinic acid which reflect metabolic effects can be detected promptly by following the concentration of several normal metabolic constituents of plasma, but the changes of therapeutic interest are generally best recognized after days or weeks of treatment. They are mediated predominantly through influences on hepatic mechanisms, a fact which must not be ignored in reviewing the pharmacokinetics of nicotinic acid and in designing and evaluating therapeutic formulations and programs.

Much of the work contributing to our knowledge about the pharmokinetics and pharmacology of nicotinic acid comes from studies comparing the agent with various derivatives designed in the hope of improving the therapeutic utility of nicotinic acid itself. While several derivatives have come into clinical use, such as nicotinic alcohol (Roniacol) as a vasodilator, there have not been many successful therapeutic agents derived from the extensive efforts to synthesize variants of nicotinic acid for one or another specific therapeutic purpose. Some of these will be discussed from time to time, but the bulk of Part III will be devoted to the therapeutic aspects of nicotinic acid itself.

16
PHARMACOKINETICS

While pharmacokinetics per se can be looked upon as a physicochemical science independent of the nature of the biological response induced by a drug, such a "know-nothing" approach to the pharmacokinetics of an agent like nicotinic acid is likely to result in a great deal of wasteful effort and misapplication of the pharmacokinetic data in relation to important questions concerning the pharmacodynamic and clinically significant actions of nicotinic acid. There is no compelling reason to insulate the discussion of nicotinic acid pharmacokinetics from other important facets of the nature of the drug. The following dissection of various aspects of nicotinic acid pharmacokinetics is undertaken in this spirit.

I. ANALYTICAL METHODS

Bioassays for niacin, the nutrient, were originally performed in nutritionally deficient dogs suffering from black tongue (1) and in chicks by measuring their growth rates. Eventually, more practical and quantitative bioassays of nicotinic acid were developed, using niacin-dependent microorganisms grown in a niacin-deficient medium (2). All these assay methods lack specificity, and will "read" nicotinic acid, nicotinamide, and any precursor or metabolite which retains the capacity to serve the niacin function essential to growth (see Chap. 5). Nevertheless, the bacterial growth-promoting activity in a niacin-deficient medium of blood or urine specimens taken before versus after a dose of nicotinic acid, and quantified in terms of equivalence to graded amounts of nicotinic acid added in vitro, can yield time–response and dose–response data which reflect quite well the amount and pattern of drug or conversion products reaching the systemic circulation from the formulation administered (3) (Fig. 1). However, for a fuller understanding of the pharmacokinetics of any drug, methodology of defined specificity and sensitivity for estimating the concentration of the drug in biological fluids is needed. This requirement should not be taken to imply that high specificity and sensitivity are always necessary or even desirable in approaching pharmacokinetic problems. The nature of the methodology best suited to answering a specific question is very much dependent upon the nature of the question, and is not of necessity the most sensitive or specific method conceivable. Consequently, study objective should be a key factor in selecting analytical methodology.

With most drugs, a prime factor influencing the design of an analytical method of adequate sensitivity and specificity is the nature, magnitude, and variability of the "blank" which the method yields (i.e., the value obtained when the biological specimen, usually blood, plasma, or urine, collected prior to drug administration is put through the method). Since most drugs do not exist in any amount in the body until a dose

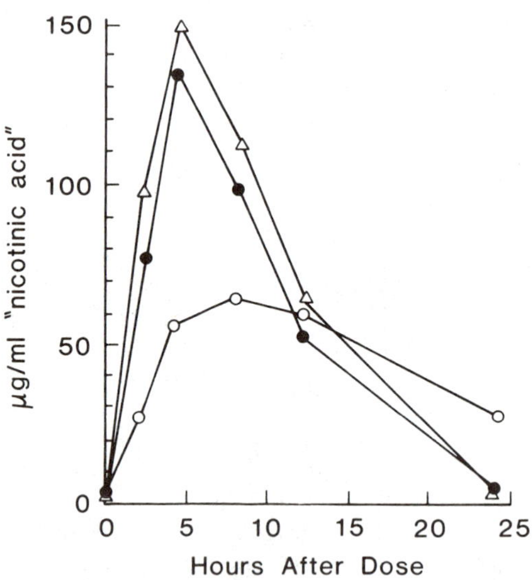

Figure 1 Plot of "nicotinic acid" blood levels in man (22) following a single 1 g (two 500 mg tablets) of several formulations using a turbinometric microbiological assay with *L. arabinosus* as a test organism (23). (●) Nicotinic acid as such; (Δ) nicotinic acid as the aluminum salt; (○) "enteric-coated" aluminum salt. Control (blank) plasma levels ranged from 2.8 to 4.7 μg per ml which is consistent with levels observed by spectrophotometric assay (20).

is administered, the "blank" reading is generally due to entities in the specimen or the reagents used which interact somewhere along the procedure to give a reading measurably more than zero in the final step. One goal of methodology development is to reduce this blank to a value below the sensitivity of the final detecting instrument which, in turn, needs to be sufficiently sensitive to estimate the range of concentrations which exist following pharmacologically significant doses. Obviously, detection sensitivity far greater than the "blank" is not likely to have much utility.

With nicotinic acid, the evaluation of the "blank" for measuring the "drug" concentration after dosage is complicated by the fact that there is a definite amount of true nicotinic acid and related metabolic substances present in essentially all biological specimens, and in concentrations which vary with the individual, food intake, time of day, etc. (4). Biochemists involved in studying quantitative changes in sugar, lipids, amino acids, etc., over a period of time are thoroughly familiar with this situation. But to pharmacokineticists, it is an unusual problem which they sometimes ignore at their own peril.

Several biochemical assays for nicotinic acid have been developed based on a variety of colorimetric techniques (4–6). While they vary in sensitivity and specificity, they are by and large considerably more sensitive and specific than the older bioassays and can be adapted to study nicotinic acid itself and niacin-related compounds individually and in parallel. Much has been learned from more or less complete studies of essentially all significant products of niacin metabolism in parallel following doses of varying magnitude. Examples of such studies include the observations in animals by Petrack et al. (11), using ^{14}C-labeled material with appropriate chromatographic separation, and the studies in humans by Mrocher et al. (7), using a computer-interfaced high-resolution liquid chromatographic UV analyzer of materials processed through coupled anion and cation exchange resin columns. However, full and parallel quantitative evaluation of all the metabolic derivatives of niacin may be impractical and unnecessary to answer some important clinical questions. What is necessary is a careful selection of the method suitable to the purpose. For example, a highly specific method which measures only nicotinic acid in systemic blood is of limited utility to determine the extent of nicotinic acid absorption, since such methodology will miss the considerable amount of absorbed drug which undergoes first-pass hepatic metabolism. Minor variations in hepatic function may determine whether 96% vs. 98% of a dose is metabolically converted before reaching the systemic circulation. This fact can result in 4% vs. 2% of unchanged drug reaching the systemic circulation, with the potential for an erroneous interpretation that only half as much "absorption" was accomplished in the second as compared to the first instance. Thus, the method which measures nicotinic acid, only, may fail to demonstrate a truth about absorption which is readily revealed by less specific methods.

Among the more commonly used analytical methods are the cyanogen bromide method of Friedman and Frazier (8) which measures both nicotinic acid and nicotinamide, and the specific method for nicotinic acid by Carlson (6). The earlier method of Stoltz (9) also excludes nicotinamide, but will read nicotinic acid, nicotinuric acid, and some oxidative metabolites. Since these account for essentially all of the materials appearing in urine after pharmacological doses of nicotinic acid, the method has been used in studies designed to evaluate the urinary excretion pattern as a reflection of the bioavailability of nicotinic acid from various oral formulations.

II. ABSORPTION

Nicotinic acid in the gastrointestinal contents from food and other sources is continuously subject to absorption and to a very considerable degree of dose-dependent, first-pass hepatic metabolism. This creates a complex "moving baseline" of systemic blood levels which may complicate studies of the bioavailability of nicotinic acid using classical methods for studying drugs in various pharmaceutical formulations, especially with relatively small doses or slow-release formulations. In untreated human subjects, nicotinic acid plasma levels generally range from about 0.1–0.3 mg % (1–3 μg/ml) by specific biochemical methods (4,5), and somewhat higher by bioassay (3). If specimens are drawn casually and without reference to food ingested, especially in this era of "fortified" cereals, etc., it is not unusual to find levels of 2 μg/ml or more, and 24-hr urine analyses commonly indicate excretion of amounts of nicotinic acid and metabolites equivalent to from 10 to 100 mg/day. With careful attention to dietary content, predrug plasma levels can be kept below 1 μg/ml.

Since food is presumably the source of nicotinic acid in untreated subjects, it may seem surprising that under controlled conditions the highest blood level is generally found in the morning *before* a meal (4). Continuous biochemical interactions at varying rates between nicotinic acid, nicotinamide, their complexes, and their metabolites may predominate over the rate and completeness of gastrointestinal absorption of dietary nicotinic acid as the determinant of the pattern of plasma nicotinic acid levels. When nicotinic acid is administered in a readily available form in a single dose of "drug" magnitude (i.e., about 1 g), the resulting peak plasma levels are of the order of 25 μg/ml of nicotinic acid itself (10) and about 150 μg/ml of "nicotinic acid" by bioassay (Fig. 1). The specific nicotinic acid peak is achieved in about a half-hour and falls with a half-life of about 45 min, while microbiological assay levels are higher after 4 hr than after 2, and fall off with a half-life of about 4 hr.

In any event, the variability of the predose blood level in a range of 1 or 2 μg/ml is of no practical consequence in evaluating the usual pharmacokinetic parameters reflecting rate and completeness of absorption, time of peak concentration, and half-life. However, the fact of "first-pass" hepatic metabolism can markedly influence specifically measured nicotinic acid levels in the systemic blood. Since this hepatic mechanism is saturable, a variable fraction of the nicotinic acid entering the portal vein will find its way into the systemic circulation in unaltered form. This fraction will depend on the amount of nicotinic acid absorbed and presented to the liver per unit time, which, in turn, is a function of both the magnitude and rate of release of the administered dose from its formulation. Consequently, with small or slowly released doses, and using specific analytical methodology, the usual pharmacokinetic calculations of "area under the curve" and slope of fall-off curve are of limited value in comparing the completeness of absorption from different formulations. Similarly, caution is necessary in considering the extent to which the time and magnitude of "peak" levels after dosage are a quantitative reflection of how the drug had been absorbed.

The time of peak levels after oral dosage varies with species, being about ½ hr in rats and 1½ hr in dogs. Values in man, using solutions or readily dissolving formulations, range from ½ to 2 hr after dosage. These values are not the sole consequence of absorption rate, but also reflect the influence of the rate of distribution out of blood into tissues, particularly the liver and fat depots, and the rate of metabolic conversion by the liver.

In spite of these complicating influences on blood levels, there is little doubt that dissolved or readily dissolving formulations of nicotinic acid are rapidly and completely absorbed. With relatively large pharamcological doses, a large fraction of the absorbed material escapes first-pass metabolism, and the plasma and urine levels of unchanged drug are consistent with rapid and complete absorption (3, 11,12). It is unlikely that smaller doses are less well absorbed, even though blood level studies by methods specific for nicotinic acid are destined to yield relatively low levels which are inconclusive and often erroneously interpreted. A more detailed review of the fate of various magnitudes of nicotinic acid dosage is presented in Sec. IV, below. The application of these lessons to the practical problem of the use of nicotinic acid as a hypocholesterolemic drug is discussed in Chap. 18.

III. DISTRIBUTION

The wide distribution of niacin in nature has been documented long ago (32) and has been reviewed in detail in Part I. Its stability in the absence of enzymatic activity is illustrated by the fact that barley over 38 centuries old, found in the tomb of Tutankhamen, had essentially no enzymatic activity and contained about half the amount of nicotinic acid found in similar grains today (14). In living mammals, however, enzymes which act on nicotinic acid abound, and the enzymatic conversion of nicotinic acid in the liver is very rapid, a fact which indirectly influences its distribution.

Standard texts record the pK_a of nicotinic acid as 4.85 and indicate that it is soluble in water and insoluble in ether, from which one might predict poor lipid penetration. However, there are two ionizable sites on the nicotinic acid molecule ($COOH \rightleftharpoons COO^-$ and $NR_3H^+ \rightleftharpoons NR_3$), and from studies of the partition coefficients in chloroform and diethyl ether (15), one might postulate a significant fraction of the zwitterion species at neutral pH, which could be a determining factor in the lipid solubility needed to achieve distribution through fatty membranes and into body fat depots.

About 15–30% of nicotinic acid in plasma is protein bound (5). This type of "percent bound" determination is frequently studied because of the hypothesis that only free (not protein-bound) drug can leave the intravascular compartment. Thus, protein binding is presumed to result in a longer half-life and persistence of a larger fraction of the "body burden" of the drug in plasma. However, correlations of "percent bound" and drug half-life in plasma are very poor (16), and there is good reason to believe that drug–protein complexes may be cleared from the plasma as such (17). In addition, the strength of a drug–protein bond as well as the kinetics of the binding and unbinding may not be reflected in the "percent bound" figure, and can considerably affect whatever influence plasma binding may have on distribution. Consequently, it is difficult to attribute any real significance to the reported degree to which nicotinic acid is bound to plasma proteins.

The rapid disappearance of an intravenous dose of nicotinic acid is largely due to its rapid entry into adipose tissue (12). Radiographs following labeled nicotinic acid administered intravenously in the mouse show most of the label in body fat and the kidney 5 min after injection. By 15 min, the liver is the predominantly labeled organ, and in 1 hr, activity is no longer detectable in fat (12). Thus, nicotinic acid appears to move freely through lipid structures. However, its ultimate distribution is also influenced by the speed and pattern of its metabolic conversion which, in turn, is dependent on the form, magnitude, and route of dosage administration. The half-life of nicotinic acid in both blood and the liver is much shorter than that of nicotinamide (11), consistent with the metabolic characteristics to be described below. At high doses, the rate of urinary clearance of the unaltered compound becomes a contributing factor to the plasma half-life.

In humans, a single oral dose of 1 g disappears with a half-life of about 45 min (10) after a plateau peak level of about 25 μg between 30 min and 1 hr after administration. Smaller doses have a shorter half-life, with peak levels being reached in less than 30 min (18). Intravenous infusion of nicotinic acid at a rate of 1 mg/min (0.014 mg/kg/min) fails to result in a detectable increase in plasma level after 30 min. When the infusion rate is increased to 5 mg/min (0.07 mg/kg/min), levels accumulate to about 3 μg/ml in half an hour, and are back to control within 30 min of discontinuing infusion (19).

In the following discussions, the relationships of the short half-life of nicotinic acid, as determined by distribution studies, to the metabolic fate, the conversion to coenzyme (NAD), and drug action will be reviewed.

IV. METABOLISM

The pathways and enzyme systems involved in niacin conversion have been reviewed in detail (Part I). It has also been pointed out that the dose-dependent character of the metabolic conversions has a marked influence on the quantitative and temporal pattern of metabolites produced. Under nondrug conditions, most of the compounds recognized as niacin metabolites are derived primarily from the breakdown of pyridine nucleotides, and vary considerably with species, as reviewed by Fumagelli in 1971 (20). However, the fate of druglike doses of nicotinic acid is quite different, as illustrated in Figure 2, derived from the animal data of Petrack et al. (11). In humans, as in other species, large doses of nicotinic acid are excreted in the urine in large measure as nicotinuric acid, a direct metabolite of nicotinic acid (not via nicotinamide) and the

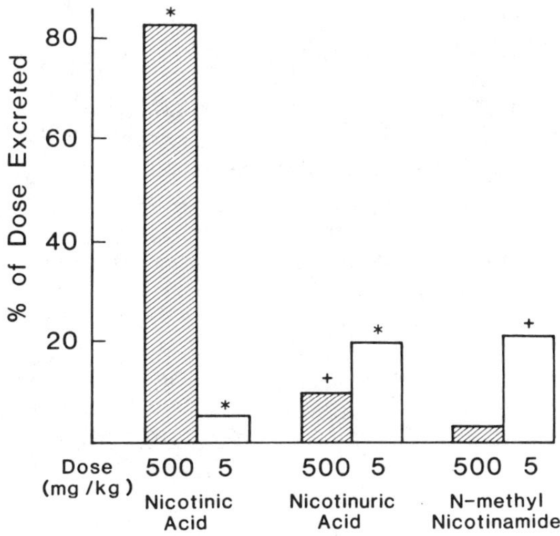

Figure 2 Urinary excretion following [^{14}C] nicotinic acid i.p. in rats. (*) = Major excretion in first 2 hr; (+) = Gradual excretion over 8+ hr. (From Ref. 30.)

fraction of the total dose which is excreted unchanged becomes larger with increasing magnitude of dose (7), suggesting saturation of the metabolic mechanism. The portion undergoing conversion to nicotinamide is further metabolized to *N*-methylnicotinamide and 2-pyridone derivatives. The urinary output of these compounds also increases after pharmacological doses in some species, including humans, and the proportion of these metabolites varies with dose (21). It is interesting that the urinary content of *N*-methyl-4-pyridone-5-carboxamide in humans does not change following a 100-mg dose of nicotinic acid, while the corresponding 2-pyridone increases several-fold (22), suggesting that the 4-pyridone derivatives come from the normal metabolic turnover of the nucleotides rather than from the freshly ingested dose of niacin. Human data suggest a sequence of conversion of pharmacological doses from nicotinic acid to nicotinamide to *N*-methyl-2-pyridone-5-carboxamide, but not the reverse (23). Consequently, in examining pharmacological effects seen with nicotinic acid but not with nicotinamide, attention should be focused on the acid itself and the biochemical events involved in the conversion processes to the amide and to nicotinuric acid, rather than the several metabolic processes and products formed via an initial transformation to nicotinamide.

The shorter half-life of nicotinic acid compared to nicotinamide may contribute to the interesting fact that pharmacological doses of nicotinamide are more efficiently converted to NAD than similar doses of nicotinic acid, even though such doses of nicotinamide are first deaminated to nicotinic acid before incorporation into NAD (24).* In addition, enhanced NAD synthesis can be accomplished by the administration of some presumably unrelated drugs in the face of severe niacin deficiency (25).

Tryptophan is clearly convertible to nicotinic acid (26,27) through a complex pathway which involves a spectrum of enzymatic conversions (Fig. 3). Rats fed B_6 (pyridoxal) -deficient diets show enhanced xanthurenic acid and reduced nicotinic acid production in response to a dose of tryptophan (27). However, if there is adequate protein supplement in the diet, there may not be a need for pyridoxal to accomplish the tryptophan-to-nicotinic acid conversion (28), and coadministration of nicotinic acid can suppress the tryptophan-enhanced production of xanthurenic acid. There are other complex interactions of nicotinic acid with other B vitamins and amino acids that influence a phosphoribosyl transferase which incorporates quinolinic acid, derived from the oxidation of 3-hydroxyantrhanilic acid into nicotinic acid-ribonucleotide (29).

---

*See Chap. 10 for a detailed discussion of alternative pathways leading to NAD synthesis.

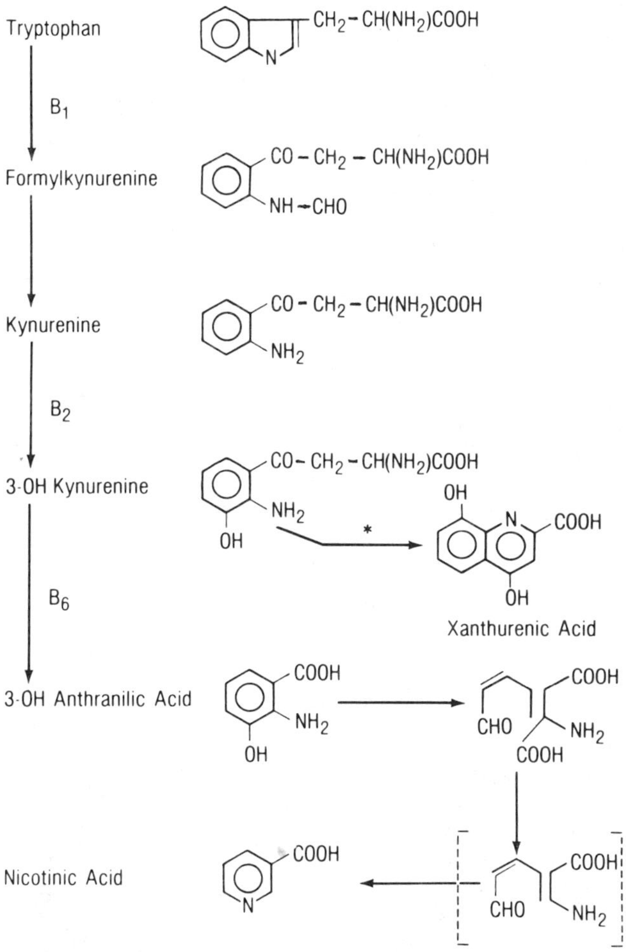

Tryptophan $CH_2-CH(NH_2)COOH$

B_1

Formylkynurenine $CO-CH_2-CH(NH_2)COOH$ $NH-CHO$

Kynurenine $CO-CH_2-CH(NH_2)COOH$ NH_2

B_2

3-OH Kynurenine $CO-CH_2-CH(NH_2)COOH$ NH_2 OH

* OH N COOH OH

Xanthurenic Acid

B_6

3-OH Anthranilic Acid COOH NH_2 OH

COOH CHO NH_2 COOH

Nicotinic Acid COOH N

COOH CHO NH_2

Figure 3 Pathway of tryptophan conversion to nicotinic acid (4), indicating sites of activity of enzymes requiring thiamine (B_1), riboflavin (B_2) and pyridoxal (B_6). The asterisk indicates pathway when tryptophan is administered to B_6-deficient animals or humans (36). Administration of nicotinic acid inhibits xanthurenic acid production in these animals. (From Ref. 28.)

These facts illustrate the complexity of establishing what is an appropriate "baseline" status for evaluating the metabolic fate and pharmacological consequences of nicotinic acid treatment. Further reference will be made to the metabolic fate of nicotinic acid as it is relevant to the discussion of some of the specific therapeutic applications of nicotinic acid.

As with any drug, the interaction of pharamcological doses of nicotinic acid with other agents might cause significant alterations in its metabolic fate. There have been very little data published concerning such drug-drug interactions which may be of potential importance in clinical treatment settings and in experiments designed to study mechanisms of action by the coadministration of selected drugs. One such study (18) aimed at evaluating the role of prostaglandins in the pharmacology of nicotinic acid included nicotinic acid blood levels to assure that there was no alteration of nicotinic acid absorption as a consequence of indomethacin cotreatment. The study was designed to determine if this prostaglandin synthetase inhibitor would alter the flush reaction to nicotinic acid. While the published data were reassuring as to the absence of an effect on absorption, calculations from a semilogarythmic plot of the tabulated postabsorption nicotinic acid levels suggest a possible change in disposition. There was a nicotinic acid half-life of 29 min when indomethacin was coadministered, as compared to 42 min without indomethacin. Neither the statistical nor the mechanistic significance of this half-life difference can be determined from these data. However, it does give pause in concluding that the pharmacological differences observed as a consequence of indomethacin coadministration are due solely to prostaglandin synthetase inhibition. Whenever two drugs are coadministered, changes in metabolism and distribution must be considered in interpreting altered pharmacological responses.

The formation of nicotinuric acid, the glycine conjugate of nicotinic acid and a prime metabolic derivative, might be competitive with the conversion and excretion of other agents which are eliminated as glycine conjugates in the urine. A particularly well-investigated example of the potential complexity of such phenomena is the data (31) showing that (1) the conversion of salicylic to salicyluric acid is zero-order rather than first-order, suggesting saturation of the capacity of the responsible metabolic system; (2) the salicylate-to-salicyluric conversion is essentially blocked by the coadministration of benzoic acid, which conjugates with glycine to form hippuric acid; and (3) the coadministration of glycine with benzoic acid enhances the rate of urinary hippuric acid excretion, suggesting that the supply of glycine is a limiting factor. In contrast, coadministration of glycine with salicylate has no influence on the rate of salicyluric acid excretion (i.e., the supply of glycine is not a rate-limiting factor).

There are as yet no definitive data indicating the effect of the coadministration of compounds such as benzoic acid, salicylic acid, or glycine on the fate of the nicotinic-to-nicotinuric conversion or the consequences of such coadministration to the pharmacological effects of nicotinic acid.

V. EXCRETION

The nature of the urinary content of nicotinic acid derivatives reflects the metabolic conversion of nicotinic acid, and consequently varies with the magnitude and nature of the dose administered. At doses of 3 g/day, about 80% of the administered dose can be accounted for in 24-hr urine collections (7). With doses of this magnitude, about one-sixth is found as unconverted nicotinic acid and one-fourth as nicotinuric acid. Neither of these compounds is found in significant amounts in the urine following similar doses of nicotinamide. By the same analytical methods, only about 55% of a 1-g dose of nicotinic acid is detected in the urine, and essentially no unconverted nicotinic acid is found. The fraction appearing as nicotinuric acid is only about one-eighth of the dose. With a 500-mg dose, the urinary recovery of acid-hydrolyzable derivatives [using the modified method of Stoltz (9)] was less than 50% in 24 hr (32), although other workers detected over 75% of a 500-mg dose in 72-hr urine collections (23). In the former studies (32), dividing a 500-mg dose into 10 doses of 50 mg administered at half-hourly intervals reduced the urinary recovery from over 40% to under 20%. These figures are consistent with other date (33) showing an average urinary recovery of about 19% following a single 100 mg dose. Thus, the urinary excretion pattern not only is dose related, but also can vary with the manner in which the oral dose is administered.

In the rat (11), nicotinic acid in a large parenteral dose (500 mg/kg) has an essentially identical short half-life (1 hr) in blood and liver. The urinary excretion pattern corresponds to the availability of unaltered nicotinic acid for excretion [i.e., 65% of the dose in the first 2 hr, 14.8% in the next 2 hr, and very little (less than 2.5%) thereafter]. In contrast, urinary nicotinuric acid levels are higher in the second 2-hr period than in the first, and almost half the urinary nicotinuric acid excretion takes place more than 4 hr after dosage. This pattern (i.e., a delay of several hours between the time a parent drug disappears from the blood and its metabolite appears in the urine) strongly suggests that the metabolite is produced in the liver, excreted into the bile, and finds its way into the urine only after being reabsorbed from the gastrointestinal tract, a mechanism clearly demonstrated for some drugs (34), and reviewed elsewhere (35).

REFERENCES

1. C. J. Koehn and C. A. Elvehyem, *J. Biol. Chem. 118:*693(1937).
2. H. Baker and H. Sobotke, *Adv. Clin. Chem. 5:*173(1962).
3. J. L. S. Holloman, C. G. Davis, and L. C. Leeeper, *J. Am. Geriatr. Soc. 10:*903(1962).

4. A. M. Diab, *Arzneimittelforsch./Drug Research 27:*2314(1977).
5. W. T. Robinson, L. Cosyns, and M. Kraml, *Clin. Biochem. 11:*46 (1978).
6. L. A. Carson, *Clin. Chim. Acta 13:*349(1966).
7. J. E. Mrochek, R. L. Jolly, D. S. Young, and W. J. Turner, *Clin. Chem. 22:*1821(1976).
8. T. E. Friedemann and E. I. Frazier, *Arch. Biochem. Biophys. 26:*361(1950).
9. E. Staltz, *J. Lab. Clin. Med. 26:*1042(1941).
10. L. A. Carlson in *Metabolic Effects of Nicotinic Acid and its Derivatives* (K. F. Gey and L. A. Carlson, eds.) Hans Huber, Bern, 1971, p. 163.
11. B. Petrack, P. Greengrad, and H. Kalinsky, *J. Biol. Chem. 241:*2367(1966).
12. L. A. Carlson and A. Hanngren, *Atherosclerosis 28:*81(1977).
13. R. W. McVicar and G. H. Berryman, *J. Nutr. 24:*235(1942).
14. E. C. Barton-Wright, R. G. Broth, and W. J. S. Pringle, *Nature 153:*228(1944).
15. K. Sandell, *Naturwissenshaften 53:*330(1966).
16. M. Weiner, P. G. Dayton, and A. G. Hendrickk, in *Use of Sub-human Primates in Drug Evaluation* (H. Vagtborg, ed.), Univ. of Texas Press, Austin, 1968.
17. M. Weiner, *Life Sciences 13:*1473(1973).
18. H. Svedmyr, A. Heggelund, and G. Aberg, *Acta Pharmacol. Toxicol. 41:*397(1977).
19. H. Svedmyr, L. Harthon, and L. Lundholm, *Clin. Pharmacol. Ther. 10:*559(1969).
20. R. Fumagelli in *Metabolic Effects of Nicotinic Acid and its Derivatives* (K. F. Gey and L. A. Carlson, eds.) Hans Huber, Bern, 1971, p. 33.
21. C. J. Walters, R. R. Brown, M. Kaihara, and J. M. Price, *J. Biol. Chem. 217:*489(1955).
22. W. M. L. Chang and B. L. Johnson, *J. Biol. Chem. 226:*799(1961).
23. W. T. M. Holman and D. J. de Lange, *Nature 165:*604(1950).
24. B. Petrack, P. Greengard, A. Cranston, and F. Sheppy, *J. Biol. Chem. 240:*1725(1962).
25. P. Greengard, E. B. Sigg, I. Fratta, and S. B. Zak, *J. Pharmacol. Exp. Ther. 154:*624(1966).
26. H. A. Krebs in *Metabolic Effects of Nicotinic Acid and its Derivatives* (K. F. Gey and L. A. Carlson, eds.) Hans Huber, Bern, 1971, p. 1115.
27. C. H. Lushbough and B. S. Schweigert, *Ann. Rev. Biochem. 27:*313(1958).
28. M. Heimberg, F. Rosen, I. G. Leder, and W. A. Perlzweig, *Arch. Biochem. Biophys. 28:*225(1950).
29. H. E. Sauberlick, *Ann. N. Y. Acad. Sci. 335:*80(1980).

30. K. Diem and G. Leuthner, eds., *Documenta Geigy,* J. R. Geigy, Basle p. 476(1970).
31. G. Levy *Drug Metab. Rev. 9:* 3(1979).
32. M. Weiner *Drug Metab. Rev. 9:* 99(1979).
33. K. K. Reddi and E. Kodicek, *Biochemistry 53:* 286(1953).
34. B. B. Brodie, M. Weiner, J. J. Burns, G. Simson, and E. Yale, *J. Pharmacol. Exp. Ther. 106:* 453(1952).
35. M. Weiner, *J. Clin. Pharmacol. 16:* 550(1976).
36. L. D. Greenberg, D. F. Bohr, H. McGrath, and J. R. Rinchart, *Arch. Biochem. Biophys. 21:* 237(1949).

17
PHARMACODYNAMICS

The dictionary defines pharmacodynamics as "the action of drugs on living organisms." This chapter on the pharmacodynamics of nicotinic acid as a drug will deal primarily with the biochemical mechanisms which are significantly influenced by pharmacologic doses, and will review some correlations of these effects with the previously reviewed pharmacokinetics. More specific disease–drug relationships will be reviewed in subsequent chapters.

The electron transfer role of niacin incorporated into the NAD group of cofactors is essential for many dehydrogenase enzymes, and has been described in some detail (Chaps. 9 and 12). In spite of, or perhaps because of this essential role, an inhibitory action by nicotinic acid in doses of drug magnitude on these same systems has to be considered as a possible mechanism of its pharmacological effects.

Certainly, massive inhibition of the synthesis of as vital a biological substance as the NAD group of cofactors throughout the organism would be disastrous. Yet, many drugs are believed to work by the inhibition of the synthesis or the destruction of one or another fundamentally important enzymatic cofactor, neurohumor, or second messenger. The key to the utility of drugs which affect such basic mechanisms is locus selectivity and dose-controllable potency.

In addition to the possibility of pharmacological action via an influence on NAD synthesis, there is a broad spectrum of mechanisms to be considered, from availability and release of precursor substrates to direct action on effector structures. These include the basic pathways of metabolism, especially those involving electron transfer–dehydrogenation enzyme systems, "second messenger" functions, such as the prostaglandins and the adenylate cyclase/cyclic AMP/phosphodiesterase system, and the neurohumoral transmitter systems, which include synthesis, storage, release, reuptake, and receptor characteristics relating to a considerable number of natural chemical mediators. Pharmacological doses of nicotinic acid have been found to influence several of these biochemical systems.

I. INHIBITION OF NAD SYNTHESIS

Large doses of nicotinic acid result in the generation of NAD which is readily detectable by periodic postdose analyses of hepatic NAD levels. The dose-time patterns of such increased formation of NAD in rat liver are particularly instructive. The *rate* of formation after 50, 150, and 500 mg/kg is actually slower with larger doses (1), so that 1 hr after dosage, the liver NAD is increased to a degree *inversely* proportional to the magnitude of the dose (Fig. 1). Since the lower hepatic nicotinic acid levels achieved during the first hour after a 50 mg/kg dose results in a greater degree of NAD synthesis, it must be concluded that higher nicotinic acid levels are actually inhibitory. At 3 hr after the 50 mg/kg dosage, the NAD level is reduced from its earlier peak, while the 150 mg/kg dose now reaches a peak distinctly higher than the earlier peak achieved with the 50 mg/kg dose, and is at a higher level than the NAD response to 50 mg/kg at that time (3 hr). At 6 hr, the response to both the 50 and 150 mg/kg doses have largely disappeared, while rats treated with 500 mg/kg have some NAD elevation, but not to the degree observed earlier with the smaller doses. In contrast, 500 mg/kg of nicotinamide resulted in a more rapid, higher, and

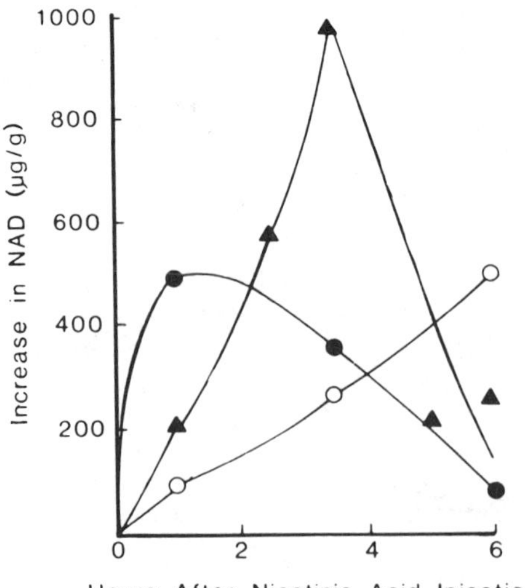

Figure 1 The increase in rat liver NAD as a function of time following the administration of nicotinic acid at 50 (●—●), 150 (▲—▲) and 500 (○—○) mg per kg. The increase in NAD was obtained by subtracting the endogenous value of 467 μg per g found in liver extracts from control rats. Each value represents the average of determination on at least three rats. [See Ref. (1)].

longer-lasting elevation of hepatic NAD levels than any of the nicotinic acid doses tested.

To the extent that the pharmacological and toxic effects of nicotinic acid versus nicotinamide may be mediated by influences on NAD synthesis, these data may reflect a critically important difference. Deamidation of nicotinamide, described by some workers (2) as the rate-limiting step in NAD synthesis, may also prevent the development of high inhibitory concentrations of nicotinic acid in the liver, and thus be protective against the consequences of such high levels.

This type of action of nicotinic acid as a drug (i.e., inhibition of its own normal incorporation into NAD) might occur at some loci and not at others, depending on drug distribution, and may have a time-course of action which can fortuitously yield a clinically useful pharmacological result.

It is not clear if and how the above described differences between nicotinic acid and nicotinamide vis-à-vis NAD relate to their many pharmacological differences. While it is only nicotinic acid which is hypolipemic, vasoactive, and fibrinolytic in modest doses, nicotinamide is about twice as toxic in classical LD_{50} studies (3). There are also some very interesting and as yet unexplained species differences. Nicotinic acid is distinctly more toxic in dogs than in rats, and shows little or no effect on cholesterol levels in either dogs or rats (4), although very large parenteral doses of nicotinic acid in rats will cause a drop in lipids, including cholesterol (5), and nicotinamide will cause a smaller but slower and less intense response, consistent with its conversion to nicotinic acid as detected in the blood in this species. A detectable fibrinolytic response to intravenous nicotinic acid readily produced in humans (6) was not observed in rats, dogs, or guinea pigs in spite of numerous attempts.

II. EFFECT ON CYCLIC AMP

The early description of cyclic AMP (cAMP) as the mediator of catechol-induced lipolysis (7) and the reports of cutaneous flushing following intravenous cAMP (8) stimulated interest in the concept that the pharmacological effects of nicotinic acid may be mediated via stimulation of the production of or otherwise potentiating cAMP. Indeed, studies of guinea-pig ear tissue (9) showed that either the in vivo administration of nicotinic acid or its addition to isolated tissue slices resulted in elevations of cAMP levels in the ear tissue. However, data from other models suggest an opposite effect.

In a rabbit gut loop model in which water and electrolyte secretion is induced by cholera toxin and is associated with elevated cAMP levels, nicotinic acid was found to prevent or reduce both the cAMP level and the water and salt secretion induced by the toxin (10).

Secretin-stimulated water and electrolyte flow in the perfused cat pancreas and carbachol-stimulated enzyme secretion in this same model are both inhibited by nicotinic acid, but without altering the enhanced adenylate cyclase activity stimulated by the activators (11). There was no discernible change in phosphodiesterase action by nicotinic acid. In the basal (unstimulated) model, nicotinic acid itself elevated adenylate cyclase activity. These authors conclude that possibly some, but certainly not all the effects of nicotinic acid observed in their models are cAMP mediated.

The relation of cAMP to the lipolytic action of nicotinic acid is also complex. In isolated fat cells, nicotinic acid suppressed the elevated cAMP and fat release initiated by glucagon, ACTH, or epinephrine (7), and was an effective inhibitor of catechol-stimulated lipolysis in adipose tissue in vitro (12). However, in the intact animal, using 500 mg/kg nicotinic acid i.p., Benito et al. (13) found the expected rise in plasma free fatty acids (FFA) and ketone bodies, but only after a rapid and marked depletion of liver glycogen. Liver cAMP was rapidly elevated, and all the effects were enhanced by the phosphodiesterase inhibitor aminophylline. The rise in plasma free fatty acid and ketones could be prevented by the administration of glucose or insulin, suggesting that the plasma FFA response was a function of the "glucose–fatty acid cycle" as influenced by the plasma insulin/glucagon ratio, rather than a direct consequence of the observed lipolysis and elevated cAMP following nicotinic acid administration. It is possible that these observations with such massive doses of nicotinic acid (about 10 times the full "therapeutic" dose in man) may not reflect the mechanisms responsible for the lipid effects seen clinically.

It is generally believed that cAMP may serve to activate a phosphorylase (19) or a local protein kinase (15). Given the involvement of nicotinic acid with NAD, etc., it is not too difficult to imagine either competitive (phosphate-consuming) or protective (slowed phosphate consumption by slowed NAD synthesis) influences on cAMP-activated phosphorylations. Thus, nicotinic acid may appear to have contrasting effects on the consequences of cAMP under varying conditions.

III. EFFECT ON PROSTAGLANDIN ACTION

As with cAMP, the intravenous infusion of several prostaglandins in humans has also been reported to result in flushes (52) described as similar to those caused by nicotinic acid. This and other observations have led to the design of experiments based on the hypothesis that nicotinic acid somehow enhances the formation, release, or persistence of endogenous prostaglandins which, in turn, cause the flushing and perhaps other pharmacological effects of nicotinic acid. Since prostaglandin E_1, for example, may stimulate adenylate cyclase and raise the level of cAMP, influences attributed to prostaglandin changes may in turn be mediated via the prostaglandin effect on cAMP. Furthermore, the activation of the prostaglandin and/or cAMP systems may be secondary to the release of neurohumors.

Andersson et al. (16) studied these relationships in their guinea-pig ear model. Modest doses of nicotinic acid raised both the skin temperature of the ear and the cAMP level of the ear tissue. Both of these effects could also be achieved with prostaglandin E_1. The prostaglandin synthetase inhibitor, indomethacin, blocked both effects when induced by nicotinic acid, but not by prostaglandin, suggesting that the nicotinic acid action was mediated by an effect on prostaglandin production.

Svedmyr et al. (17) designed a set of studies in humans to determine whether indomethacin would block the flush response to nicotinic acid. These authors carefully monitored the effect of indomethacin on nicotinic acid absorption. They concluded that the results of their skin temperature and FFT measurements were consistent with the hypothesis that prostaglandins mediate the nicotinic acid induced flush, but not lipolysis. While the prior discussion of their plasma level data (Chap. 16) leaves some doubt as to whether the indomethacin-induced reduction in ear temperature response is solely the consequence of an inhibition of prostaglandin synthesis, their conclusion is, nevertheless, reasonable.

The prostaglandin mechanisms are now known to involve the availability of the precursor, arachidonic acid, which may be converted by alternative pathways of metabolism to physiologically radically different analogues, such as thromboxane and prostacycline (18–20) (Fig. 2).

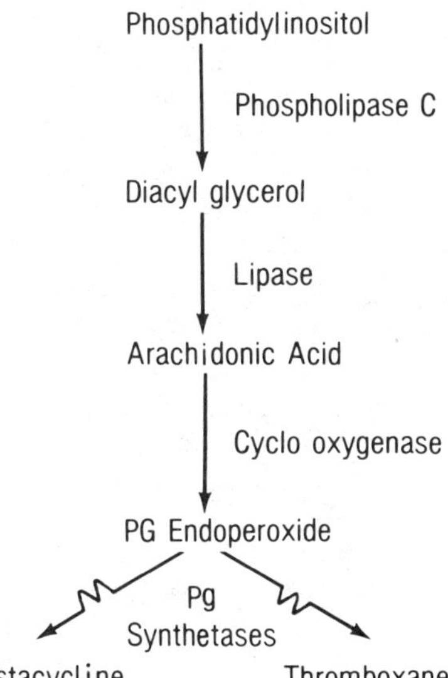

Figure 2 Simplified scheme of steps involved in the synthesis of prostaglandins from the presumed phosphatidylinositol precursor.

Glucocorticoid inhibition of prostaglandin-mediated actions has been attributed to a blocking of the release of arachidonic acid (21), while indomethacin apparently blocks the conversion of arachidonic acid into the active prostaglandins by inhibiting either prostaglandin synthetase (22) or the lipase which releases arachidonic acid from a diacylglycerol derived from the hydrolysis of phosphatidylinositol by phospholipase C (23,24) (Fig. 2).

There may be significant differences in the localization, release, and destruction of the differently active prostaglandins (25). Considering the paucity of information about the effects of nicotinic acid on these systems, it is currently premature to attempt to develop a definitive scheme of nicotinic acid action via prostaglandin mechanisms.

IV. EFFECT ON NEUROHUMORAL TRANSMITTER ACTION

A host of neurohumors have activities which bring to mind several aspects of nicotinic acid pharmacology. However, attempts to account for nicotinic acid by way of an influence on these neurohumors have been inconsistent at best. Classical adrenergic and antihistaminic blockers do not block the nicotinic acid flush (26), and the spectrum of effects of adrenergic drugs on lipolysis and intermediate metabolism resembles the effects of nicotinic acid only in some respects. Several chapters in a previous review (27) have been devoted to studies of the influence of nicotinic acid on the actions induced by adrenergic and other neurohumors. By and large, most of these neurohumor–nicotinate interactions have been best explained by considerations involving the second messengers discussed above rather than the synthesis, release, or destruction of the neurohumoral transmitters per se by nicotinic acid. The classic "nicotinic" component of cholinergic neurotransmission is, of course, based on the pharmacology of nicotine, and not nicotinic acid.

V. OTHER MECHANISMS OF ACTION

The metabolic conversion of nicotinic acid in the liver is accomplished by enzyme systems which employ cofactors and substrates that are also involved in other metabolic functions of the liver. These systems are of limited capacity, as is apparent from the previously discussed dose-dependent metabolic conversion pattern of nicotinic acid itself. It is therefore reasonable to consider that the hepatically mediated effects of nicotinic acid (i.e., its influence on hyperlipidemia) may be the result of competition between its metabolic conversion and the "normal" metabolic functions of the enzymes, cofactors, and substrates involved. These mechanisms may include not only cholesterol synthesis, but also synthesis of a variety of phospholipids, "carrier" lipoproteins, and glycogen–lipid interconversions.

Since some nicotinic acid is excreted in the form of methylated derivatives, it had been postulated that competition for the methyl pool might be one mechanism of nicotinic acid pharmacological action. However, more of a nicotinamide dose is excreted in methylated form than a similar nicotinic acid dose (28), and nicotinamide lacks these pharmacological effects. In addition, coadministration of the methyl donor, methionine, fails to interfere with the hypolipemic action of nicotinic acid (29).

The fact that nicotinuric acid, the glycyl conjugate of nicotinic acid, is one of its prime metabolites, raises the possibility of action by competition for the glycine pool. This possibility has been dismissed (4) because of a lack of effect of nicotinic acid on bile acid secretion into the stool, and the lack of nicotinic acid-like pharmacology by benzoic acid, a potent glycine conjugator. However, neither of these observations rule out competition for glycine at specific intrahepatic sites with specific reactions involved in lipid metabolism. Nicotinyl-CoA is probably an intermediate in the conversion of nicotinate to nicotinurate (30) and might compete with lipid metabolic processes through this route.

In rats, the effect of nicotinic acid on the carbohydrate metabolism of adipose tissue is insulin-like, causing a drop in blood glucose and activation of glycogen synthesis and glucose uptake in adipose tissue (31). In humans, however, nicotinic acid frequently reduces glucose tolerance promptly after administration of a large dose, apparently by mechanisms not involving altered insulin secretion (32,33) and occurring well before an effect on serum lipids becomes apparent. On the other hand, Svedmyr et al. (26) found that slow intravenous infusion of nicotinic acid (0.014 mg/kg/min), which was inadequate to cause a measurable increase in nicotinic acid plasma levels, did cause a flush and a modest *fall* in blood sugar and free fatty acid levels in half an hour. When rate of infusion was increased fivefold, there was a rise in glucose levels to above control, and a further fall in free fatty acids, all of which were reversed within a half hour of discontinuing infusion.

Essentially all of these metabolic effects after nicotinic acid, whether direct or indirect, can be presumed to take place largely in the liver. For such mechanisms, the pattern of nicotinic acid concentration reaching the liver is of critical importance. It is not just the peak level or "area under the curve" of systemic blood levels which determines hepatic exposure to the drug. Perhaps more important to determine whether a desired metabolic inhibition is achieved is the time period (hours per day) during which the liver mechanisms are exposed to a significant degree of competitive challenge by nicotinic acid reaching it directly from the gut via the portal vein. When large, promptly absorbed doses of oral nicotinic acid reach the liver at so rapid a rate that much of the drug passes into the systemic circulation unaltered, the "escaped" drug may represent material which failed to participate in the postulated therapeutic function via hepatic metabolic competition and may serve only to create unwanted systemic actions.

In contrast, the vascular and fibrinolytic actions of nicotinic acid may require that the drug reach the systemic circulation at an effective rate. A totally different pattern of drug disposition may be essential. For these therapeutic goals, both the locus and pharmacodynamics can be completely different from that involving liver-dependent lipid metabolism. The pharmacological actions of nicotinic acid in inducing fibrinolysis probably involve an activation or release of a limited store of some endogenous activator which may be rapidly depleted, leading to an "acute tolerance" to that pharmacological action (34). Some vascular effects of nicotinic acid appear to be a function of a rising systemic blood level of the drug, rather than of the drug concentration per se (26). Each of these will be reviewed in more detail in subsequent chapters. The point to be made at this juncture is that the markedly different pharmacodynamics involved may require correspondingly different interpretations and applications of bioavailability data in the interest of therapeutic objectives. These facts illustrate the folly of legalistic "requirements" for pharmacokinetic studies to prove "bioavailability" according to preconceived rules and without regard to the therapeutic intent and the variety of the possible mechanisms of action by which a drug may work.

REFERENCES

1. B. Petrack, P. Greengard, and H. Kalinsky, *J. Biol. Chem. 241:* 2367(1966).
2. B. Petrack, P. Greengard, A. Cranston, and F. Sheppy, *J. Biol. Chem. 240:* 1725(1965).
3. K. Unna, *J. Pharmacol. Exp. Ther. 65:* 95(1939).
4. O. N. Miller and J. G. Hamilton, in *Lipid Pharmacology* (R. Paolette, ed.), Academic Press, New York, 1964.
5. C. Dalton, T. C. Van Trabert, and J. X. Dwifer in *Metabolic Effects of Nicotinic Acid and its Derivatives* (K. F. Gey and L. A. Carlson, eds.), Hans Huber, Bern, 1971, p. 65.
6. M. Weiner, Redisch, and J. M. Steele, *Proc. Soc. Exp. Biol. Med. 98:* 755(1958).
7. V. W. Butcher, L. F. Baird, and E. W. Sutherland, *J. Biol. Chem. 243:* 1705(1968).
8. R. A. Levine, L. M. Dixon, and K. B. Franklin, *Clin. Pharmacol. Ther. 9:* 168(1968).
9. R. G. G. Anderson, G. Aberg, R. Brattsand, E. Ericsson, and L. Lundholm, *Acta Pharmacol. Toxicol. 41:* 1(1977).
10. N. Turiman, G. S. Gotterer, and T. R. Hendrix, *J. Clin. Invest. 61:* 1155(1978).

11. S. L. Bonting, J. J. H. H. M. De Pont, H. J. M. Kempen, R. M. Case, P. A. Smith, and T. Scratherd, *Br. J. Pharmacol. 61:*243(1977).
12. L. A. Carlson, *Acta Med. Scand. 173:*719(1963).
13. M. Benito, F. J. Moreno, J. M. Medina, and F. Mayor, *Arch. Biochem. Biophys. 188:*21(1978).
14. G. A. Robison, R. W. Butcher, and E. W. Sutherland, *Ann. Rev. Biochem. 37:*149(1968).
15. J. F. Kuo and P. Greengard, *Proc. Natl. Acad. Sci. U.S.A. 64:*1349(1969).
16. R. G. G. Andersson, G. Aberg, R. Brattsand, E. Ericsson, and L. Lundholm, *Acta Pharmacol. Toxicol. 41:*1(1977).
17. N. Svedmyr, A. Heggelund, and G. Aberg, *Acta Pharmacol. Toxicol 41:*397(1977).
18. S. Moncada, R. Grygleuski, J. Buntueg, and J. R. Vane, *Nature 263:*663(1976).
19. J. V. Vane and H. Robinson, (ed.) *Proceedings International Symposium on Prostaglandin Synthetase Inhibitors,* Raven Press, New York, 1973.
20. V. R. Gorman, G. L. Bundy, D. C. Peterson, F. F. Sun, O. V. Miller, and F. A. Fitzpatrick, *Proc. Natl. Acad. Sci. USA 74:*4007(1977).
21. N. S. Doherty, *Br. J. Pharmacol.* Submitted, 1981.
22. J. V. Vane, *Nature (New Biol.) 231:*232(1971).
23. R. L. Bell, N. L. Baenziger, and P. W. Majerus, *Prostaglandins 20:*269(1980).
24. S. Rittenhouse-Simmons, *J. Biol. Chem. 255:*2259(1980).
25. F. E. Preston, S. Whipps, C. A. Jackson, A. J. French, P. J. Wyld, and C. J. Stoddard *N. Eng. J. Med. 304:*76(1981).
26. M. Svedmyr, L. Harthon, and L. Lundholm, *Clin. Pharmacol. Ther. 10:*559(1969).
27. K. F. Gey and L. A. Carlson (eds.) *Metabolic Effects of Nicot hic Acid and its Derivatives,* Hans Huber, Bern, 1971.
28. O. N. Miller, J. G. Hamilton, and G. A. Goldsmith, *Am. J. Clin. Nutr. 8:*480(1960).
29. O. N. Miller in *Metabolic Effects of Nicotinic Acid and its Derivatives* (K. F. Gey and L. A. Carlson, eds.), Hans Huber, Bern, 1971, p. 83.
30. V. Fumagelli in *Metabolic Effects of Nicotinic Acid and its Derivatives* (K. F. Gey and L. A. Carlson, eds.), Hans Huber, Bern, 1971, p. 83.
31. E. Shafrir, M. Oreviand, and A. Gutman in *Metabolic Effects of Nicotinic Acid and its Derivatives* (K. F. Gey and L. A. Carlson, eds.), Hans Huber, Bern, 1971, p. 763.
32. Z. N. Gaut, H. M. Solomon, and O. N. Miller in *Metabolic Effects of Nicotinic Acid and its Derivatives* (K. F. Gey and L. A. Carlson, eds.), Hans Huber, Ber, 1971, p. 923.
34. M. Weiner, *Am. J. Med. Sci. 246:*72(1963).

18

TREATMENT OF THE HYPERLIPIDEMIAS

A major interest in nicotinic acid as a drug relates to its effect on patterns of hyperlipidemia associated with atherosclerosis. While there is little doubt that real associations exist between lipid metabolism and thrombovascular disease, they are complex and still poorly understood. The nature of these correlations as determined during the past few decades has contributed to the demise of former beliefs that atherosclerosis is a natural and inevitable facet of the aging process rather than a potentially controllable disease.

I. BLOOD LIPIDS AND ATHEROSCLEROSIS

Ever since the original demonstration that the atherosclerotic plaque contains the lipid cholesterol (1), there has been keen interest in the levels of circulating cholesterol and other lipids as a reflection of the atherosclerotic process. A great deal of experimental effort has been expended in attempts to find or create animal models with hypercholesterolemia and atherosclerosis in order to better understand the relationship between these conditions and to study methods of therapeutic interference. The vascular changes induced in chickens and rabbits have been much employed and much criticized as to their relevance to human disease. More recently, studies in swine (2), and particularly the "mini pig" (3), have confirmed the possibility of achieving hyperlipidemia and lipid vascular lesions more or less in parallel, which more closely resembles what is perceived to be the clinical problem in humans.

Since the ultimate evaluations must be done in humans with the disease, and since the clinical quantitation of atherosclerosis is difficult at best, clinicians are obliged to seek some indirect manner of attacking the problem. The approach generally employed may be expressed in simplistic logic as follows:

1. Atherosclerosis is associated with hyperlipidemia.
2. The effect of a drug on atherosclerosis per se is difficult to measure.
3. The effect of a drug on hyperlipidemia is easy to measure.
4. The effect of a drug on blood lipids is determined and it is considered effective if it controls hyperlipidemia.

Many refinements of the above logic have been made by characterizing subclasses of atherosclerosis and patterns of hyperlipidemia (4). In each instance, however, the scenario remains a tenuous substitute for empirical proof of control of the clinical manifestations of the disease, and the physician knows it. However, the physician also knows that to postpone decisions pending generation of uncontestable morbidity-mortality data is not an option. Thus the physician is compelled to decide whether, at a given time with a given patient and the current state of the art, the patient's chances are or are not improved by normalizing hyperlipidemia. Obviously, the risks and problems occasioned by the measures to be taken to accomplish this end must enter into the equation.

There are many approaches to the therapy of the group of chronic diseases covered by the descriptive term "atherosclerosis." In general, a treatment modality is likely to evolve on the basis of some theory about the nature of the disease coupled with a deductive rationale for the proposed treatment. In the last analysis, trial and error (i.e., the empirical observation of the clinical consequences of treatment) is the inescapable method of judging the value of any treatment. It is the method that yields the most critical information and requires the least imagination and intelligence. Unfortunately, there are severe limitations of the extent to which controlled empirical clinical studies of efficacy can be performed with a disease as chronic and varied as atherosclerosis.

Early studies of the clinical consequences of effectively lowering blood cholesterol levels were quite encouraging (5–7). One of the more massive controlled studies was the 15,000 subject evaluation of clofibrate in hypercholesterolemics under WHO sponsorship (8) which demonstrated reduced cholesterol levels and a reduced incidence of ischemic heart disease, but not reduced mortality. The mere size of the report (50 pages) reflects the complexities of interpretation, some of which will be discussed below.

Another large study (9) which started with five hypolipemic agents, concluded with two (clofibrate and nicotinic acid), both of which favorably influenced the incidence of myocardial infarction, but not the overall mortality of the treated group. Other less massive studies have yielded similar results (10). On the whole, dietary studies were equally unimpressive as regards survival (11).

These data do not represent proof that correcting hyperlipidemia has no clinical value. They merely indicate the difficulties of studies in which quite heterogeneous populations must be pooled and observed for prolonged periods during which numerous covariables are subject to change. It is therefore understandable that in coming to conclusions regarding therapy for individual patients, the physician is often compelled to depend upon less direct but more readily quantifiable criteria of therapeutic utility. The U.S. Food and Drug Administration allows the use of agents to correct hyperlipidemia but currently requires the following notice in the labeling of most such drugs (12):

> It has not been established whether drug-induced lowering of serum cholesterol or triglyceride levels has a detrimental, a beneficial, or no effect on the morbidity or mortality due to atherosclerosis, including coronary heart disease. Investigations now in progress may yield an answer to this question.

The wisdom of this caution is accentuated by a recent update report of the WHO study (13) which further clouds the question of the clinical utility of clofibrate in spite of its distinct effect on lipid blood levels. As of 4.3 years after the 5.3-year study period, there was a greater subsequent death rate (by 25%) in the clofibrate-treated as compared to the control group, with no significant change in pattern of category of disease causing death.

In view of this situation, the discussion of the therapeutic use of nicotinic acid for its hyperlipidemic effect will be preceded by a brief review of current concepts concerning the various patterns or syndromes of atherosclerosis and hyperlipidemia and the possible mechanisms behind currently employed methods for their control.

II. PATTERNS OF HYPERLIPIDEMIA

Hyperlipidemia is generally assessed by three techniques: chemical analysis, electrophoresis, and ultracentrifugation (14). In general, the high-density lipoproteins (HDL) move electrophoretically as a component called *alpha*, the low-density lipoproteins (LDL) as *beta*, and the very low-density lipoproteins (VLDL) as *prebeta*. Chylomicrons represent a fourth class of lipoproteins usually seen after meals and in conditions associated with low lipoprotein-lipase activity. Cholesterol, triglycerides, and free fatty acid levels are generally determined by chemical methods.

The so-called Frederickson classification of hyperlipidemias (4,15) is based on patterns of lipids, clinical characteristics, genetic trends, life styles, dietary habits, obesity, etc., which seem to occur together, as summarized in abbreviated form in Table 1. While the classification remains basically an empirical statement of observed patterns, it has proved helpful in considering the various etiological possibilities and responses to attempts at treatment of hyperlipidemia-associated disease. There is a high incidence of various types of hyperlipidemia associated

Table 1 Clinical Correlates of Classic Types of Hyperlipidemia

Frederickson classification and prevalence	Probable causal associations	Clinical characteristics	Serum characteristics
I Rare	Inherited	Pancreatitis, xanthoma Coronary disease rarely	Low lipoprotein lipase High triglycerides, and chylomicrons
II High	Both familial and dietary factors (especially high lipid (dairy) diet)	Coronary disease	High cholesterol, low HDL High β-lipoproteins High triglycerides in IIb, not IIa
III Rare	Both familial and dietary factors (excessive caloric intake)	Peripheral vascular and coronary disease Palmar xanthoma	Abnormal broad-β and pre-β-lipo-protein High cholesterol
IV High	Excessive caloric intake (overweight) and diabetes)	Coronary artery and peripheral vascular disease	Increased pre-β-lipoproteins (VLDL) Modest cholesterol elevations High triglycerides
V Moderate	Excessive alcohol intake	Coronary disease rarely	High triglycerides, pre-β and chylomicrons Modest cholesterol elevations

Source: Adapted from Ref. 131.

with several more or less clearly defined clinical entities, such as diabetes, myxedema, alcoholism, and nephrotic and glycogen storage syndromes. The mechanisms and correlations of these so-called secondary hyperlipidemias with vascular disease are no better understood than in the "primary" or idiopathic hyperlipidemias.

Clinical interest in blood cholesterol has gradually shifted from total cholesterol levels to free versus esterified cholesterol, and now to cholesterol distribution among the serum lipoproteins and lipid fractions of varying density. This interest stems from increasing evidence that such patterns of distribution are at least of diagnostic and prognostic, if not etiologic, importance. Elevated cholesterol (16,17), triglycerides (18), and cholesterol in LDL and VLDL (19–21) have long been associated with coronary atherosclerosis, while more recent data (22–25) suggest that HDL-cholesterol levels are *negatively* correlated with coronary artery disease, in contrast to LDL and VLDL levels. In familial HDL deficiency (Tangier disease), LDL cholesterol is low or normal, and yet there is increased heart disease, although it tends to develop later in life and is not nearly as prevalent as in familial hypercholesterolemia in which LDL cholesterol is high (26). Cell culture studies (27) suggest that cell uptake and feedback control of cholesterol synthesis is largely related to LDL, while removal of cholesterol from cells and tissue (efflux or cholesterol excretion) is HDL related. Some interesting hypotheses concerning the receptors of LDL (28) and interrelationships of specific apoproteins with chylomicron size, lipid content and delivery are beyond the scope of this review.

These observations are causing interest in the influence of numerous environmental factors, including the incidental effect of presumably unrelated drug treatment for other conditions, on the lipid pattern. For example, hypertensive subjects treated with the beta-blocker propranolol show what is currently regarded as an undesirable effect on the blood lipid pattern (i.e., a reduction in HDL cholesterol and in the HDL to LDL and VLDL ratio, and an increase in total triglycerides) (29). Similar observations have also been reported recently as a consequence of thiazide diuretics (30). Such effects might influence the overall morbidity/mortality consequences of treatment and thus compromise the impact of blood pressure control on cardiovascular morbidity and mortality.

III. DIETARY CONTROL OF LIPID PATTERNS

Since little can be done to alter genetic factors contributing to hyperlipidemia, and since there are obvious changes in blood lipids following each meal, therapists have taken a keen interest in what might be accomplished by diet as a means of correcting abnormal lipid patterns and the clinical diseases with which they are associated. While there is no doubt that the amount and kind of fat ingested can influence concentrations of cholesterol and other lipids, including lipoproteins (31), the pattern of influence is quite variable and frequently of minor degree. In general, substituting vegetable for animal fats lowers cholesterol levels (32); however, there are marked differences between animal fats, and the generalization that polyunsaturates lower cholesterol while mono-unsaturates have no effect, and that saturated fatty acids raise cholesterol is an oversimplistic statement of the facts. Dietary carbohydrate also influences the response of serum lipids to dietary fats. The consequences of dietary cholesterol content is also quite variable, having some (33) or no effect (34), depending on other factors in controlled studies. In swine experiments, high cholesterol intake did not lead to increased absorption and hypercholesterolemia unless high milk fat was added, and even then the average body content of cholesterol did not go up because endogenous cholesterol synthesis went down. Massive surveys in Framingham and in Israel have failed to show a significant correlation between dietary and serum cholesterol (31), and evidence showing a definite correlation of change in diet with cardiovascular mortality and morbidity, other than that attributable to control of body weight, is yet to be developed.

The dietary approach to the hyperlipidemias has gone through periods of preoccupation first with dietary cholesterol and then with saturated fatty acids. Evaluation of success in terms of serum lipid normalization has also shifted in emphasis from cholesterol levels per se to the amounts and ratios of various lipids, physical fractions, and lipoprotein complexes. Unfortunately, by any of the plasma lipid criteria, many patients in Frederickson categories II and IV fail to normalize their hyperlipidemia pattern by diet alone (35). As a rule, the contributions of dietary efforts in many of these patients is perhaps judged as well by the adequacy of weight control as it is by change in some facet of the serum lipid pattern. The dietary approach purely on the basis of cholesterol is further confounded by the fact that hepatic cholesterogenesis can be stimulated by fatty acids (36), and the body's capacity to synthesize cholesterol is about an order of magnitude greater than the amount of cholesterol usually ingested. Thus, there is a strong likelihood that the predominant disturbance in most patients is metabolic and may require metabolic interference beyond that achievable by dietary control alone (i.e., drug therapy).

IV. DRUG THERAPY FOR HYPERLIPIDEMIA

A wide variety of drugs influence blood lipid patterns by mechanisms which are clearly understood in some instances and poorly understood in others. In broad terms, these mechanisms include interference with the absorption of intestinal cholesterol and its precursors of both dietary and biliary origin, interference with hepatic cholesterol synthesis, and enhanced cholesterol catabolism.

Several agents previously used to lower cholesterol blood levels are no longer employed. These include estrogen therapy, which has little influence on the lipoproteins and was dropped from a major study because of an increase in nonfatal cardiovascular events (34), and triparanol, which gave side effects presumably associated with the accumulation in the blood of late-stage precursors of cholesterol synthesis. Several antibiotics, particularly neomycin, have been observed to influence cholesterol levels (37), presumably by changing the intestinal flora, and possibly by binding bile acids.

Dextrothyroxine (choloxin) and other forms of thyroxin increase catabolism and excretion of cholesterol, resulting in lower cholesterol blood levels in spite of a compensatory increase in the rate of cholesterol synthesis. These agents have largely fallen into disrepute as a result of manifestations of hyperthyroidism and the decision of the Coronary Drug Project Research Group to discontinue its inclusion in controlled studies because of excessive deaths over placebo (38).

Agents which interfere with the intestinal absorption of cholesterol or its precursors include sitosterols (Cytellin) and the anion exchange resins, cholestyramine (Questran) and colestipol (Colestid). These agents block bile salt reabsorption, resulting in increased oxidative conversion of cholesterol. This effect outweighs the enhanced cholesterol synthesis stimulated by reduced feedback inhibition. In contrast to drugs with a broader influence on lipid metabolism, these agents generally do not lower triglyceride levels (39). Their use is often limited by intolerance of the required bulk doses.

Several effective agents are presumed to lower cholesterol by interfering with its synthesis early in the biosynthetic chain. These include probucol (Lorelco), clofibrate (Atromid-S), and nicotinic acid, which will be discussed in more detail below. These tend to correct total cholesterol elevations as well as abnormal lipoprotein and triglyceride levels, but to varying degrees, depending on the initial pattern. Clofibrate, for example, has only a modest LDL-lowering effect in type II hyperlipidemia as compared to its efficacy in type IV patients, while cholestyramine lowers cholesterol effectively only in type II patients.

V. NICOTINIC ACID FOR HYPERLIPIDEMIA

The initial interest in the action of nicotinic acid in atherosclerosis was related to its effect on total serum cholesterol levels which was presumed to reflect the "body burden" of this substance as well as the likelihood of deposits in the walls of arteries. Empirically, there is no doubt that adequately large doses of nicotinic acid can reduce serum cholesterol levels. Even single doses can produce detectable effects (40), though more impressive data are generated by weeks of treatment with divided daily doses of the order of 2–4 g, and continuing treatment remains effective for years. It is no longer tenable to claim that reduced dietary intake secondary to the gastrointestinal side effects of large doses of nicotinic acid is responsible for the observed fall in cholesterol blood level (41). A majority of subjects in many studies (9, 42,43) develop significant falls in serum cholesterol with doses of nicotinic acid which cause only insignificant side reactions.

A. Effect on Lipid Pattern

The effects of nicotinic acid on serum lipids include a lowering of free fatty acid and total lipid levels, as well as a reduction of hypercholesterolemia. In normal subjects, 1 g of nicotinic acid t.i.d. for a month lowered serum cholesterol by 15% and triglycerides by 27%, and the major drop was in lipids transported as LDL and VLDL, while HDL cholesterol actually increased by 23% (44). There was also a lowering of the HDL_2/HDL_3 ratio of HDL subfractions. These investigators concluded that "Our present understanding of the association between HDL and atherosclerosis indicates that such changes may have prophylactic value in the prevention of coronary artery disease" (44). How the subfractions of HDL and their corresponding apoproteins are altered by nicotinic acid treatment in patients continues under investigation. Some reports (44–46) suggest a net transfer of apolipoprotein A-I from HDL_3 to HDL_2, and a reduction in apolipoprotein A-II synthesis.

It has been reported (47) that nicotinic acid lowered total cholesterol and triglycerides in both type II and type IV patients, with significant falls in VLDL cholesterol. However, nicotinic acid treatment of type IIa, but not IIb, caused a rise in HDL_2/HDL_3 ratio and a marked fall in plasma triglycerides. In type IV patients, HDL_2 and HDL_3 both rose, as did plasma apoprotein A-I, apparently because of a reduced rate of catabolism. The changes noted are not all parallel in time (48). High versus low carbohydrate diet may significantly influence the serum lipid response to nicotinic acid (49).

Thus, the effect of nicotinic acid on the pattern of blood lipid subfractions with significant disease relationships is generally consistent with the goal of preventing arterial disease by normalizing blood lipids. These more or less fortuitous realities of more recent studies may be responsible for the fact that "niacin in pharmacological doses has been one of the hardiest of hypolipidemic drugs, having thus far withstood the test of time better than any other agent" (42).

Clinical impressions suggest that nicotinic acid is at least as effective as clofibrate in type IV hyperlipidemia, and that the dose necessary in type IV patients is less than that required in type II. It is clearly important to employ doses which are adequate but no greater than what is necessary to accomplish the hypolipidemic objective, since side effects and potentially important toxic reactions are dose related. The sucess of a physician who understands the characteristics of the agent, and is free to individualize dosage regimens, is bound to be better than that achieved by investigators who are obliged to avoid "deviations from protocol" for the sake of accomplishing study objectives.

Combined nicotinic acid and clofibrate therapy is reported to yield an additive effect on hyperlipidemia (50). The clinical data concerning this combined treatment in coronary patients are limited (10). On the other hand, a distinctly enhanced lowering of cholesterol was observed when the bile acid-binding resin cholestipol was augmented by nicotinic acid (51). This combination reversed the hypertriglyceridemia seen with cholestipol alone, and also significantly lowered LDL cholesterol, increased HDL cholesterol, and reduced tendinous xanthoma (51, 52).

B. Relation of Lipid Response to Pharmacokinetics

Whether the prime therapeutic utility of nicotinic acid lies in its overall interference with cholesterol synthesis or in a more subtle alteration in lipid distribution, perhaps by interfering with apolipoprotein synthesis, the effect can be presumed to occur in the liver. In view of the pharmacokinetic fate of nicotinic acid and the pharmacological difference from nicotinamide previously discussed, it is apparent that the ideal method of nicotinic acid administration may not yet have been achieved. Mieling (53) found that small doses of nicotinic acid in slow-release formulation with a vasodilator increased HDL cholesterol and lowered total cholesterol in primary type IIa hyperlipoproteinemia. N. Svedmyr et al. (54), referring to the observations of Wolfram and Zollner in that same volume (55), point out that 1.2 g/D of a slow-release preparation of nicotinic alcohol β-pyridylcarbinol, (Roniacol) has cholesterol-lowering action equivalent to 4–5 g nicotinic acid, and suggest that "this may indicate that duration of an increased content of nicotinic acid may be of as great importance as the height of the increase." They then go on to study this possibility by measuring the plasma levels and observing the responses after i.v. infusion of various doses, pointing out that "...if nicotinic acid could be administered so as to give a sustained but moderate increase in the plasma level, the pharmacologic actions might be as pronounced as after larger but fluctuating plasma concentrations, while side effects, such as flushing, could be expected to be reduced." This same concept is also presented by Svedmyr and Harthon (56) in their comparison of nicotinic acid with the pentaerythritol tetranicotinic acid ester, niceritrol (Perycit, Cardiolipol).

Studies in the mini-pig have also found nicotinic alcohol and niceritrol to be more potent as antilipemics when administered with food in twice daily feedings, even though the plasma nicotinic acid levels studied 1 hr after the feeding were highest with nicotinic acid, as might be expected.

It is tempting to conclude from such studies that, while nicotinic acid plasma level patterns can guide development of an optimal formulation and dosage schedule for nicotinic acid itself, there are derivatives which are more potent, and are active moieties as such, rather than merely serving as "prodrugs" of nicotinic acid. However, one should not overlook the possibility that oral nicotinic acid may exert its effect during its passage through the liver by way of the portal circulation, so that systemic blood levels of nicotinic acid may not be necessary at all to achieve its hepatic action. Thus, while little or no nicotinic acid from a slowly absorbed formulation or "prodrug" may reach the systemic circulation in an unaltered form, pharmacological effects mediated in liver cells (e.g., on cholesterol or apolipoprotein synthesis) may nevertheless be achieved by such doses. Exposure of the liver to prolonged and low but pharmacologically significant amounts of the drug may be accomplished with little or no systemic levels of unchanged nicotinic acid as a consequence of a dose form and pattern which results in essentially complete first-pass metabolism. Systemic levels of nicotinic acid may be irrelevant to pharmacological effects mediated by the liver and observed after oral doses. In fact, the same slow-release formulation of nicotinic acid (Nicalex) which yielded distinct hypolipidemic effects in an extensive V.A. study (60) also failed to show significant amounts of unaltered nicotinic acid in the urine (57). Whereas patients in this study achieved an average cholesterol fall of about 16% (from 245 mg/dl to 205 mg/dl) after a year of treatment, the Coronary Drug Project Research Group, using similar doses of plain, powdered nicotinic acid, reported an average cholesterol level reduction of only 8% (58). While such comparisons may not be valid, they are suggestive that, to control hyperlipidemia, more prolonged low level exposure via controlled-release formulation is preferable to the infrequent, high peak, short exposures resulting from ordinary formulations of nicotinic acid 2–4 times a day. Diurnal variations in hepatic lipoprotein secretion (59) may also deserve consideration in the design of dosage forms and patterns of treatment.

C. Side Effects

Given the variability of doses and formulations of nicotinic acid used to control hyperlipidemia, it is difficult to arrive at figures reflecting the incidence of drug intolerance and various side effects. In the V.A. study (60), there was some degree of dyspepsia or diarrhea reported by 20% of the placebo patients, and 10% of them developed liver function abnormalities. Charman et al. (61) reported discontinuance due to intolerance (100 mg t.i.d. gradually built up to 1 g t.i.d.) in 38 of 160 drug-treated patients (24%). Eleven of 106 patients who had taken the drug more than one year subsequently dropped out because of a later development of intolerance. Consequently, the conclusions of Christenson et al. (62), who reported poor tolerance of "sustained action" preparations in patients who previously tolerated plain nicotinic acid, must be viewed critically. The incidence of intolerance he observed with his "new" preparations was considerably higher than that reported in other studies of controlled-release nicotinic acid formulations (60,63,64).

In the V.A. study (60), only 28% of the patients remained on the full initial target dose of 4 g per day, but this 28% included 8 of the 9 patients with liver function abnormalities who also had elevated bilirubin levels. All jaundiced patients recovered on withdrawal of the drug. In a recent study which combined the use of nicotinic acid with colestipol (51), a group of 22 patients treated with large doses of nicotinic acid (average of 6.8 g per day) showed no intolerable side effects. The flush reaction was reduced in some patients by taking 0.3 g aspirin half an hour before each dose.

Jones (42), citing the experience of the Coronary Drug Project (9), states: "Of particular interest, there was no unusual incidence of jaundice or elevation of liver enzymes. The earlier suspicions of hepatic dysfunction can now be laid to rest." The project leading to this conclusion employed nicotinic acid in doses of 3 g per day.

Besides hepatic effects, other side effects of nicotinic acid include the possible activation of peptic ulcer and other generally minor and less well defined gastrointestinal disorders. An acute flushing of the blush area is very common and is discussed in detail in Chap. 19. Much less common and less acute skin complaints are dryness, pruritis, and keratosis nigricans. Some degree of increased uric acid level and decreased glucose tolerance is seen occasionally. Hypotension and transient headaches may also occur. These are rarely of long-standing significance and generally are dose related.

VI. OTHER CONSEQUENCES OF DRUG TREATMENT OF HYPERLIPIDEMIA

All hypolipidemic drugs have side effects. These may involve mechanisms of drug action totally independent of the hypolipidemic response, or they might be a consequence of the very mechanisms by which the drug corrects the abnormal lipid pattern. The increased mortality from noncardiovascular causes in subjects treated with clofibrate (8), which seems to involve an increased incidence of biliary calculi and malignancy, is a case in point.

Several structurally unrelated hypolipidemic agents have been reported to induce liver cell proliferation and carcinoma (65). Increased intracellular populations of peroxisomes, observed as a feature of the reaction to these drugs, are believed to be involved in the β-oxidation of long-chain fatty acids via a fatty acyl-CoA oxidizing system and carnitine acetyltransferase. Chronic treatment with large doses of clofibrate and other hypolipidemics produce increased hepatic cell peroxisomes, hepatomegaly, and hepatic carcinoma in the rat, while nonhypolipidemic analogues did not stimulate peroxisome proliferation or its enzymes. In man, clofibrate has not been found to increase hepatic cell peroxisomes, even though hyperlipidemia is clearly altered.

Teleologists and Darwinians have wondered whether the lipid abnormalities which develop so commonly with age might not have some unrecognized useful purpose, and perhaps their "correction" might be counterproductive to that purpose. In a recent note in *The Sciences* entitled "Cholesterol, diet, drugs and death," Lasagna (66) reviewed the data suggesting that the successful reduction of elevated serum cholesterol, triglycerides, and nonfatal ischemic heart incidents, by whatever means, seems to be associated with an increase in fatal carcinoma. While studies of hypolipemic drugs in animals (65) and humans (8) may be interpreted to indicate "drug carcinogenesis," one of the first studies to raise the issue (67) was an 8-year, controlled diet trial in which no drug was involved. However, no plausible biochemical hypothesis has been presented from which an inverse relationship between athersclerosis and carcinogenic potential might be inferred. In fact, a recent review by Cruse et al. (68) suggests that a higher than normal incidence of rectal cancer occurs in atherosclerotic populations, possibly reflecting a local cocarcinogenic effect of bowel cholesterol. This observation seems to be in conflict with a recent report (69) indicating that the incidence of cancer, and particularly rectal cancer, is higher in adult males with the lowest serum cholesterol levels. Since serum levels and bowel exposure to cholesterol are not necessarily correlated and may in fact be inversely influenced by some treatment modalities, there is no real conflict between the hypothesis presented by Cruse and the observations summarized by Lasagna. Consequently, it seems reasonable to search for a biochemical hypothesis to explain why an overall increase in cancer deaths might accompany a variety of therapeutic measures which are successful in reducing hyperlipidemia. In conducting such a search, it is not necessary to be confined to hypotheses which deal only with hypercholesterolemia, or hypertriglyceridemia, or even the currently popular HDL/LDL ratio, etc. Other less frequently discussed but nevertheless real correlates with atherosclerosis and hyperlipidemia, such as elevated uric acid or homocysteine levels, also deserve attention even though they are rarely seriously evaluated in current studies of antilipemic drug effects.

Severe atherosclerotic disease in young children has been associated specifically with metabolic defects involving high levels of homocysteine (70,71). Recently, there has been a renewed interest and popularization (72) of the idea that even presumably minor accumulations of homocysteine may contribute to the common forms of adult atherosclerosis (73). Accumulations of homocysteine are postulated to result from a high protein diet rich in methionine, the metabolic precursor of homocysteine, plus a relative deficiency of vitamin B_6 (pyridoxine), an essential cofactor for many metabolic processes, including the enzymatic "neutralization" of homocysteine, the presumed atherogenic toxin. According to this concept, the favorable effect of hypolipidemic diets and other effective antiatherogenic programs may be largely the result of a reduced dietary supply of methionine and/or an enhanced supply or more efficient use or sparing of pyridoxine, deficiency of which has been linked experimentally to atherosclerosis (74) and to impaired conversion of tryptophan to nicotinic acid (75).*

Homocysteine-induced experimental atherosclerosis is being studied from the point of view that high levels of sulfhydryl compounds may be disruptive to the endothelium, triggering platelet adhesion and mesothelial smooth muscle migration with eventual atherosclerosis (77, 78). Drugs such as sulfinpyrazone and dipyridamole which reduce platelet aggregation are effective in this model (79), even though they may not influence the triggering mechanism, which is believed to be via methionine metabolism (78).

From the vantage point of investigators of carcinogenic mechanisms, thiols derived from methionine have been described as "traps" for the active intermediates of carcinogenic hydrocarbons formed by the liver (80,81). A deficiency in the availability of such thiols might enhance the likelihood of hydrocarbon-induced malignancy, while a rich store of thiols would be protective. Thus, a treatment which reduces the supply of thiol precursors and speeds their degradation and elimination might tend to limit exposure to presumably atherogenic homocysteine but would also limit the supply of free-radical trapping (i.e., anticarcinogenic) thiols (Fig. 1).

To help evaluate this speculation, it would be useful to know how normal and atherosclerotic subjects compare as regards blood levels of pyridoxine and homocysteine, or the cysteine–homocysteine mixed disulfide (73) before and after effective niacin or other hypolipidemic therapy. Unfortunately, such data are not available.

Another problem involving the consequences of lipid metabolism shifts is that related to the likelihood of gallstone formation. Much has been learned about the spectrum of steroids and bile salts which make for so-called "lithogenic" bile. There is strong evidence that gallstone problems are increased with clofibrate therapy (9) and some evidence that clofibrate changes the nature of the bile content so as to make it more lithogenic (82). The data concerning nicotinic acid in this regard are considerably less incriminating.

---

*The need for pyridoxine in the tryptophan to nicotinic acid conversion is reduced by a high protein diet (75), and the metabolic fate of tryptophan is altered by the coadministration of nicotinic acid. The fate of tryptophan may also be altered by clofibrate which competes strongly with many substances, including pyridoxine, thyroxin, and tryptophan for anionic binding sites on serum albumin (76).

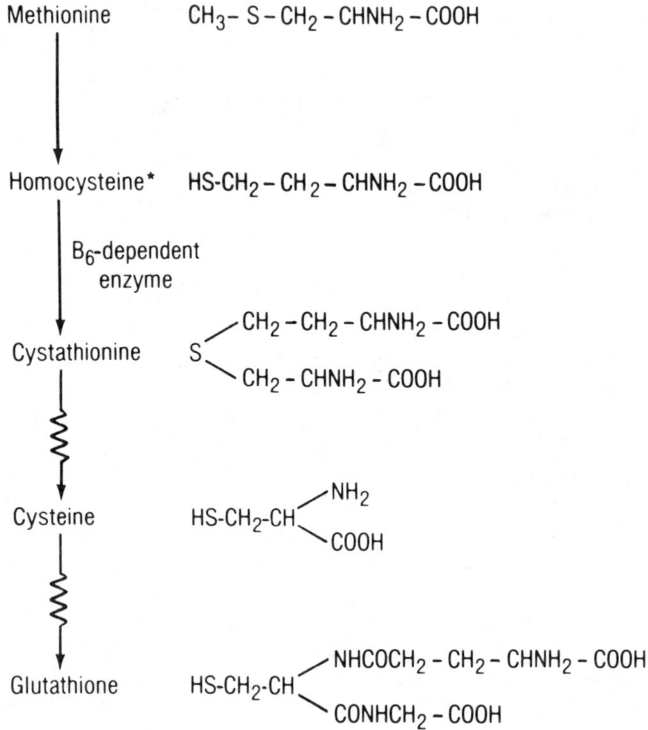

Figure 1 Homocysteine is the presumed "atherogenic toxin" that accumulates when capacity for enzymatic conversion to cystathionine is inadequate.

The normalization of hyperlipidemia may influence blood viscosity which may in turn significantly influence circulatory patterns. In type III patients, characterized by peripheral vascular disease and hyperlipoproteinemia, diet and clofibrate treatment, which effectively reduced serum lipids and improved the hyperemic reactive response to ischemia, also reduced whole blood viscosity (83). The overall contribution of this variable to morbidity-mortality data is difficult to access.

The role of hyperlipidemia in relation to the thrombotic component of thrombovascular disease is unclear. There is a considerable literature of long standing about the concept of hypercoagulability and endothelial surface thrombosis as a cause as well as a consequence of atherosclerotic vascular lesions. Certainly, phospholipids play a critical role in the clotting process, and there is increasing interest in surface mechanisms, particularly on platelet and endothelial cell membranes in relation to clotting (84). Hyperlipemic diets are also thrombogenic (85), and the observed thrombovascular consequences seem to correlate better with alterations induced in platelets than in serum lipids.

A more direct therapeutic approach to thrombovascular disease is based on the alteration of coagulability by anticoagulants (86). To this day, there is controversy about the rationale for various types of anticoagulants and the empirical evidence of their efficacy in thrombovascular disease. Interest seems to be shifting increasingly to agents which alter platelet (87,88) and other membrane surface mechanisms (84) involved in the initiation of thrombosis, rather than more drastic changes in the activity of circulating clotting enzymes.

It is interesting that heparin, the first major therapeutic anticoagulant and an overall inhibitor of coagulation, is also a potent activator of plasma lipoprotein lipase (LPL) activity, presumably by releasing the enzyme from cell membrane surfaces. The enzyme "clears" grossly hyperlipidemic, milky plasma. Heparin has been used for that purpose in doses which are too small to cause detectable inhibition of coagulation or alteration of cholesterol levels. However, chronic low-dose heparin therapy is not commonly used, possibly because its prime antilipemic effect is to clear chylomicrons, and these are a characteristic of the less common types I and V patients, and not of types II and IV. By careful study, it has been shown that the heparin effect on coagulation is at least temporarily reduced during the process of LPL activation (89). The potency of various heparinoids as LPL activators does not parallel their anticoagulant potency (90), even though some heparinoids are believed to act by causing the release of endogenous heparin (91).

While nicotinic acid therapy may somewhat improve the intravenous (92) and oral (55) fat tolerance test (an index of LPL activity), such therapy does not change the LPL response to a test dose of heparin, a response which tends to be subnormal in patients with hyperlipidemia-associated diseases. In animals, where LPL can readily be studied in tissues as well as plasma, other LPL enzyme moieties seem to be involved, and there are distinct responses, particularly in fat tissue, to the administration of nicotinic acid, consistent with its insulin-like effect as regards carbohydrate metabolism. Plasma triglyceride levels in rats are lower and have a shorter half-life following nicotinic acid treatment.

These considerations illustrate the complexity of developing judgments as to the optimal approach to treating clinical disturbances associated with hyperlipidemia. On balance, nicotinic acid remains high on the list of agents deserving further evaluation to determine how and when it should be used to control hyperlipidemia and its related clinical conditions.

REFERENCES

1. H. T. Blanenthal (ed.) *Cowdry's Arteriosclerosis*, C. C. Thomas, Springfield, Illinois, 1967.
2. A. Marsch, D. N. Kim, D. T. Lec, J. M. Reiner, and W. A. Thomas, *J. Lipid Res. 13*:600.
3. L. Lundholm, L. Jacobson, R. Brattsand, and O. Magnusson, *Artherosclerosis 29*:239(1978).
4. D. S. Frederickson, R. I. Levy, and R. S. Lees, *N. Engl. J. Med. 276*:34(1967).
5. Newcastle study, *Br. Med. J. 4*:757(1971).
6. Scottish society, *Br. Med. J. 4*:775(1971).
7. L. R. Krasno and G. J. Kidera, *JAMA 219*:845(1972).
8. M. F. Oliver, J. A. Heady, J. N. Morris, and J. Cooper, *Br. Heart J. 40*:1069(1978).
9. Coronary Drug Project Research Group, *JAMA 231*:360(1975).
10. L. A. Carlson, M. Daniolson, I. Ekberg, B. Klintemar, and G. Rosenhamer, *Atherosclerosis 28*:81(1977).
11. E. H. Ahrens, *Ann. Int. Med. 85*:87(1976).
12. *Physicians' Desk Reference* Med. Economics, Oradell, N. J., 1980.
13. M. F. Oliver, J. A. Heady, J. N. Morris, and J. Cooper, *Lancet 2*:379(1980).
14. H. A. Eder, *Am. J. Med. 23*:269(1957).
15. J. Beaumont, L. A. Carlson, G. R. Cooper, Z. Fejfer, D. S. Frederickson, and T. Strasser, *Bull. WHO 43*:891(1970).
16. M. M. Gertler, S. M. Garn, and J. Lerman, *Circulation 2*:205 (1950).
17. W. B. Kannel, W. P. Castelli, and P. M. McNamara *J. Occup. Med. 9*:611(1967).
18. M. J. Albrink and E. B. Mann, *Arch. Intern. Med. 103*:4(1959).
19. E. H. Strisower, G. Adamson, and B. Strisower, *Am. J. Med. 45*: 488(1968).
20. D. P. Barr, E. M. Russ, and H. A. Eder, *Am. J. Med. 11*:480 (1951).
21. J. W. Gofman, H. B. Jones, F. T. Lindgren, T. P. Lyon, H. A. Elliott, and B. Strisower, *Circulation 2*:161(1950).
22. N. E. Miller, O. J. Forde, D. S. Thelle, and O. D. Mjos, *Lancet 1*:965(1977).
23. T. Gordon, W. P. Castelli, M. D. Hjortland, W. B. Kannel, and T. R. Dawber, *Am. J. Med. 62*:707(1977).
24. G. G. Rhoads, C. L. Gulbrandsen, and A. Kagan. *N. Engl. J. Med. 294*:293(1976).
25. W. P. Castelli, J. T. Doyle, T. Gordon, C. G. Haines, J. C. Hjortland, S. B. Hulley, A. Kagan, and W. J. Zukel, *Circulation 55*:767(1977).
26. E. J. Schaefer, L. A. Zech, D. E. Schwarthz, and H. B. Biewer, *Ann. Intern. Med. 93*:261(1980).

27. J. D. Wu and J. M. Bailey, *Arch. Biochem. Biophys. 202*:467 (1980).
28. J. L. Goldstein and M. S. Brown, *Metabolism 26*:1257(1977).
29. P. Leren, P. O. Foss, A. Helgeland, I. Holme, I. Hjermann, and P. G. Lund-Larsen, *Lancet 0*:4(1980).
30. R. H. Grimm, A. S. Leon, V. D. Hunninghake, K. Lenz, P. Hannan, and H. Blackburn, *Ann. Int. Med. 94*:7(1981).
31. E. V. Newman, DHEW Publication No. (NIH) 72-219, *2*:57(1971).
32. E. H. Akrens, Jr., D. H. Blankenhorn, and T. T. Tsaltas, *Proc. Soc. Exp. Biol. Med. 86*:872(1954).
33. I. D. Frantz, Jr. and R. B. Moore, *Am. J. Med. 46*:684(1969).
34. E. H. Ahrens, Jr., J. Hirsch, W. Insull, Jr., T. T. Tsaltas, R. Blomstand, and M. L. Peterson, *Lancet 1*:943(1957).
35. J. L. Witztum, M. A. Dillingham, W. Giese, J. Bateman, C. Diekman, E. K. Blaufuss, S. Weidman and G. Schonfeld, *N. Eng. J. Med. 303*:907(1980).
36. E. H. Goh and M. Heimberg, *Gastroenterology 65*:542(1973).
37. P. Samuel, C. M. Holtzman, and J. Goldstein, *Circulation 35*:938 (1967).
38. Anon., The coronary drug project *JAMA 220*:996(1972).
39. R. P. Howard, O. J. Brusco, and R. H. Furman, *J. Lab. Clin. Med. 68*:12(1966).
40. R. Altschul, A. Hoffer, and J. D. Stephen, *Arch. Biochem. Biophys. 54*:558(1955).
41. O. N. Miller, and J. G. Hamilton, in *Lipid Pharmacology* (R. Paoletti, ed.), Academic Press, New York, 1964.
42. R. H. Jones, *Adv. Exp. Med. Biol. 82*:656(1977).
43. J. N. Nestords, T. A. Ban, and H. E. Lehman, *Int. Pharmacophychiatry 12*:215(1977).
44. J. Shepperd, C. J. Packard, J. R. Patsch, A. M. Grotto, Jr., and O. D. Taunton, *J. Clin. Invest. 63*:858(1979).
45. J. R. Patsch, D. Yeshurun, R. L. Jackson, and A. M. Gotto, *Am. J. Med. 63*:1001(1977).
46. C. B. Blum, R. L. Levy, S. Eisenber, M. Hall, R. H. Goebel, and M. Berman, *J. Clin. Invest. 60*:795(1977).
47. C. J. Packard, J. M. Steward, J. Third, H. G. Morgan, T. D. V. Lawrie, and J. Shephard, *Biochim. Biophys. Acta 618*:52(1980).
48. L. A. Carlson, A. G. Olsson, and D. Ballantyne, *Atherosclerosis 26*:603(1977).
49. B. J. Kupchodkar, H. S. Sodhi, L. Herlier, and D. T. Mason, *Clin. Pharmacol. Ther. 24*:354(1978).
50. B. Vessby, H. Lithell, I. B. Gustafsson, and J. Boberg, *Atherosclerosis 33*:457(1979).
51. J. P. Kane, J. M. Malloy, P. Tun, N. R. Phillips, D. D. Freedeman, M. L. Williams, J. S. Rowe, and R. J. Harel, *N. Eng. J. Med. 304*:251(1981).

52. D. R. Illingwerth, B. E. Phillipson, J. H. Raff, and W. E. Connor, Connor, *Lancet 1*:296(1981).
53. L. Meiring, *S. Afr. Med. J. 57*:345(1980).
54. N. Svedmyr, in *Metabolic Effects of Nicotinic Acid and its Derivatives* (K. F. Gey and L. A. Carlson, eds.) Hans Huber, Bern, 1971, p. 1085.
55. E. A. Nikkila, in *Metabolic Effects of Nicotinic Acid and its Derivatives* (K. F. Gey and L. A. Carlson, eds.) Hans Huber, Bern, 1971, p. 488.
56. N. Svedmyr and L. Harthon, *Acta Pharmacol. Toxicol. 28*:66 (1970).
57. M. Weiner, *Drug Metab. Rev. 9*:99(1979).
58. Coronary Drug Project Research Group *Atherosclerosis 30*:239 (1978).
59. E. H. Goh and M. Heimberg, *Fed. Proc. 36*:1142(1977).
60. H. K. Schock, in *Drugs Affecting Lipid Metabolism* (W. L. Holmes, L. H. Carlson, and R. Paoletti, eds.), Plenum Press, New York, 1969, p. 405.
61. R. C. Charman, L. B. Matthews, and C. Braeuler, *Angiology 23*: 29(1972).
62. N. A. Christenson, R. W. P. Achor, K. G. Berger, and H. L. Mason, *JAMA 177*:546(1961).
63. W. B. Parsons, Jr., *Am. J. Clin. Nutr. 8*:471(1960).
64. W. B. Parsons, Jr., *Arch. Int. Med. 107*:653(1961).
65. J. K. Reddy, M. S. Rao, D. L. Azarnoff, and S. Sell, *Cancer Res. 39*:159(1979).
66. L. Lasagna, *The Sciences* May-June: 29(1979).
67. M. L. Pearce and S. Dayton, *Lancet 1*:464(1971).
68. P. Cruse, M. Lewin, and C. G. Clark, *Lancet 1*:752(1979).
69. R. R. Williams, P. D. Sorlie, M. Feinlieb, P. M. McNamara, W. B. Kannel, and T. R. Dawber, *JAMA 245*:247(1981).
70. K. S. McCully, *Am. J. Path. 56*:111(1969).
71. K. S. McCully and B. D. Ragsdale, *Am. J. Path. 61*:1(1970).
72. E. R. Ruber and S. R. Raymond, *Atlantic Monthly* 000:59(1979).
73. D. E. L. Wilcken and B. Wilcken, *J. Clin. Invest. 57*:1079(1976).
74. J. F. Rinehart and L. D. Greenberg, *Am. J. Path. 25*:481(1949).
75. M. Heimberg, F. Rosen, I. G. Leder, and W. A. Perlzweig, *Arch. Biochem. Biophys. 28*:225(1950).
76. K. D. Edwards, *Biochem. Pharmacol. 19*:2719(1970).
77. R. T. Wall and L. A. Harker, *Ann. Rev. Med. 31*:361(1980).
78. K. N. von Kaulla, *Circulation 17*:187(1958).
79. L. A. Harker, R. J. Wall, J. M. Harlan, and R. Ross, *Clin. Res. 26*:554(1978).
80. B. B. Brodie, W. Reid, A. K. Cho, G. Sipes, G. Drishna, and J. R. Gillette, *Proc. Natl. Acad. Sci., 68*:160(1971).

81. D. J. Jollow, J. R. Mitchell, and J. R. Gillette, *Pharmacology, 11*: 151(1974).
82. S. Grundy, *Effect of Drugs on Metabolic Parameters of Tick in Man*, 81st Annual Meeting, ASCPT, San Francisco, 1980.
83. R. Zelis, D. T. Mason, E. Braunwals, and R. I. Levy, *J. Clin. Invest. 49*:1007(1970).
84. R. F. A. Zwaal, *Biochim. Biophys. Acta 515*:163(1978).
85. S. Renaud, R. L. Kinlough, and J. F. Mustard, *J. Lab. Invest. 22*:339(1970).
86. S. Shapiro and M. Weiner, *Coagulation, Thrombosis, and Dicoumarol* Brooklyn Med. Press, New York, 1949.
87. E. Genton, M. Gent, J. Hirsh, and L. A. Harker, *N. Eng. J. Med. 293*:1174(1975).
88. H. J. Weiss, *Am. Heart J. 92*:86(1976).
89. R. L. Hirsch, A. Keller, and R. Ircland, *J. Exper. Med. 112*: 699(1960).
90. C. H. Duncan, M. M. Best, and C. J. McGraff, *Lancet 0*:1014 (1959).
91. R. Pulver, *Arzneimittelforsch. 12*:582(1962).
92. J. Boberg, L. A. Carlson, S. Froberg, A. Olsson, L. Oro and S. Rossner, in *Metabolic Effects of Nicotinic Acid and its Derivatives* (K. F. Gey and L. A. Carlson, eds.), Hans Huber, Bern, 1971, p. 465.

19

NICOTINIC ACID EFFECTS ON VASCULAR TONE

During World War II, drug supplies were scarce in many areas of the world and local populations frequently sought medications from American military medical facilities. One of us (M.W.) was surprised to find that there was a high demand for injectable nicotinic acid around Manila in the Philippines, and to learn later from a colleague that there was a similar demand for the drug in Greece. The result of inquiries locally suggested that the potent flush induced by the drug was perceived as evidence that the medicine was "strong," even though the rationale for its use was tenuous. Other flush-inducing preparations, particularly calcium gluconate, were also in similar demand.

In current medical practice, the flush reaction to nicotinic acid is generally considered an undesirable side effect, although several nicotinic acid derivatives which were designed as "prodrugs" to deliver nicotinic acid have been employed as vasodilators to enhance blood flow in various types of vascular disturbances.

The clinical description of a drug as a vasodilator is of little value unless one can define which vascular beds are dilated and with what consequences. In general, redness, local temperature elevations, and increased blood flow-rate measurements are taken as evidence of vasodilatation. But these changes in response to a drug may not occur in parallel in different vascular beds, and may even be in opposite directions. In fact, any significant increase in blood flow to one tissue must be accompanied by either (a) a corresponding decrease in flow to other tissues, (b) an increase in cardiac output, and/or (c) a fall in blood pressure, in which case the drug is labeled a "hypotensive agent" rather than a "vasodilator" (1).

The development of a flush is a manifestation of increased blood in the superficial cutaneous capillary—venular bed. It may or may not be accompanied by altered sweating, with a corresponding variability in heat loss and skin temperature. Postvenular constriction with capillary dilatation can lead to an increased blood volume in the superficial cutaneous bed without a significant increase in the rate of blood flow through the skin. Under these conditions, the "blush area" may take on a deep color but with no increase in skin temperature. In contrast, a blood flow through the finger, for example, can be largely a function of arteriolar—venular anastamoses, which may result in considerably enhanced blood flow with only a minor change in the volume of the unoccluded finger and blood supply to the tissue. It requires considerable understanding of the nature of a drug's vaso-action to rationally evaluate its potential clinical utility.

The reactions observed to nicotinic acid must be considered against this background of variables. The methods and pitfalls of measuring drug effect on blood flow will be reviewed briefly before examining the data concerning the vascular effects of nicotinic acid.

I. MEASUREMENT OF VASOMOTOR RESPONSES

A. Skin Temperature

By and large, skin temperature is influenced by ambient conditions (temperature, humidity, and air currents), both directly via the effect on surface temperature, and indirectly by influencing vasomotor tone. If ambient conditions are well controlled, changes in skin temperature become a reliable reflection of changes in skin blood flow. To study a potential vasodilator, it is important that the basal state be one with an adequate degree of constriction. Thus, at a room temperature of 26° or 28°C it may not be possible to detect a dilator response readily demonstrable in a 20°C environment (2). One must also be aware that some vascular beds, such as the forehead, come into equilibrium with ambient conditions quite rapidly (within minutes), while others, such as the toes, may take over an hour, and still others, such as the fingers, often show erratically varying skin temperature unless conditions are such as to cause near maximal dilation.

B. Plethysmographic Measures

In using this fundamentally simple noninvasive measure of rate of change in volume resulting from venous (but not arterial) occlusion, one must bear in mind (1) the importance of A-V shunting in the fingers, and (2) the fact that foot plethysmography (up to the ankles) reflects primarily skin flow, while leg plethysmography (up to the knee) gives mostly muscle flow. These two are often differently influenced by drugs.

C. Conductance Measures

Electromagnetic flow meters, used so effectively in animal studies, can rarely be applied to clinical drug evaluations. Less invasive techniques, such as the Hensel needle measure of heat conductivity as a function of blood flow, have been used clinically to register flow changes in a bulk structure such as the leg and arm muscles (3,4). Cutaneous electrical conductance measures are noninvasive but generally are of limited use in drug studies. They are used mostly in psychological tests (5) and seem to better reflect neurological rather than drug vasodilator activity.

D. Radiolabel Methods

Several techniques have been developed for measuring flow rates through deep structures such as coronary (6) and cerebral (7) vessels, as well as limbs (3), by measuring the radioactivity over selected areas after administering an appropriate radioactive agent. Some of these are highly sophisticated and can "map" flow patterns in the brain, for example, to the point of detecting an increase in blood flow in the visual cortex or speech area when the subject undertakes visual or verbal tasks (8). The nature and intensity of such short-term vascular variations help explain the difficulties of evaluating attempts to induce and measure relatively long-term blood flow changes with drugs.

II. VASCULAR RESPONSE TO NICOTINIC ACID

Before reviewing the flush reaction to nicotinic acid, it may be useful to examine the objective characteristics of "hot flashes" seen in postmenopausal women. While the triggering mechanism is not fully understood, the subjective "flash" lasts only about 1 or 2 min, and is accompanied by a flushing of the face and *drop* in the temperature of the forehead (9). A measurable increase in skin conductance (sweat-related?) and finger skin temperature follows. The conductance effect peaks in about 4 min, and peak finger skin temperature change is noted in about 10 min (10) after the flush. As one might predict, the rise of skin temperature, especially with sweating, is accompanied by a slight drop in core body temperature after each flush (9).

When cutaneous finger temperature was recorded continuously in menopausal women, the temperature fell steadily over a period of hours between flashes, each of which was followed by an abrupt finger temperature rise (11). These rises were correlated with, but not preceded by, abrupt elevations in blood levels of luteinizing hormone (LH) (12). The pattern is reminiscent of the observations of Hellman et al. (13), which suggest that some endocrine organs secrete at a constant rate or not at all, so that blood levels are either rising or falling at a characteristic rate at any given moment. Thus, homeostasis is by periods of "on" or "off," rather than by variation in the degree of activity. In this instance, it is postulated that the proximity of the hypothalamic LH releasing center and the thermoregulatory center may explain the coordination of these phenomena, with central catecholamines mediating the effect (14). The initiating peripheral stimulus could be the degree of fall of skin temperature. In any event, the "baseline status" is quite difficult to define, since, under these conditions, skin temperature is either rising or falling at any given moment.

Two aspects of the flush reaction of nicotinic acid deserve attention: (1) the nature of the vascular response, and (2) its biochemical mechanism.

A. Characteristics of the Vascular Reaction

The flush reaction to nicotinic acid (15) can be seen in humans seconds after as little as one milligram intravenously, and within minutes of a somewhat larger dose orally. At times, a distinctly visible flush is not accompanied by a detectable change in skin temperature. While 100 mg nicotinic acid induced a large (2.4°C) increase in blush area skin temperature, it did not alter toe temperature (16). By foot plethysmography, there is a distinct increase in skin blood flow following nicotinic acid, but the increase begins well after the visible blush area flush has peaked, and may not reach its maximum until the flush has largely subsided 15 to 20 min after i.v. dosage (17). It is reasonable to assume that any clinically useful vasodilator effect which nicotinic acid might have is better reflected by the increased flow rate seen by foot plethysmography than by the flush.

It is difficult to determine a "threshold blood level" of nicotinic acid necessary to result in a flush reaction in humans. In fact, the flush seems to be a consequence of a *rising* blood level rather than the level itself (18,19). In Svedmyr's studies in humans (18), an intravenous infusion of nicotinic acid at a rate of 0.014 mg/kg/min for 30 min was not adequate to result in a measurable rise in blood nicotinic acid levels, but did initiate a flush within a few minutes after starting the influsion, and with essentially no change in blood flow to the forearm. When, after half an hour of this low rate of infusion, the dose per minute was increased fivefold, there was a distinct and progressive increase in forearm blood flow over the next half hour infusion period, parallel to an increase in nicotinic acid blood level, and with no increase in hand blood flow by the method employed.

Considering these complexities of drug level, vascular beds involved, and time differences seen in clinical studies, one can anticipate difficulty in interpreting the meaning to humans of animal skin temperature changes in response to nicotinic acid. Guinea-pig ear temperature elevations after nicotinic acid have been observed by Andersson et al. (19), with a reasonable relationship of the intensity and duration of the response to the dose administered. However, intraperitoneal doses of 0.3–1.0 mg/kg yielded only equivocal guinea-pig ear temperature responses, while such doses are distinctly flush-producing in humans. The response to 10 mg/kg i.p. in anesthetized animals was definite but less than the response to a similar oral dose in unanesthetized animals. Whether this difference relates to route of administration, the "stress" of being awake, or interaction with the anesthetic (urethane) is unclear. The biochemical correlations made with these guinea-pig temperature measurements, to be reviewed below, may or may not be relevant to the vascular effects of nicotinic acid in humans.

B. Biochemical Mechanisms

The biochemical mechanism of the vascular response to nicotinic acid is not well understood. It is generally presumed that vascular tone in the skin is under adrenergic and histaminergic control, since the cutaneous injection of sympathomimetics causes blanching (constrictive) and histamine causes a wheal and flare (dilatory). However, the nicotinic acid flush reaction does not appear to relate directly to these mechanisms, since it is not altered by beta blockade or antihistaminics (18,20). There is no evidence that nicotinic acid is vasodilator by α-adrenergic blockade (like phenoxybenzamine, which is primarily dilator to skin), or by β-adrenergic stimulation (like nylidrin, which is dilator to skeletal muscle arteries). Nor can nicotinic acid be classified with agents which act centrally to reduce sympathetic tone, such as reserpine or methyldopa. Its vascular action is still described as "nonspecific" (21).

These observations do not rule out a mechanism of action later in the chain of events initiated by adrenergic or other neurohumors. Among the local "second messengers" which are interposed between the stimulus of the neurohumor and the ultimate response, and may be affected by nicotinic acid, are the prostaglandins and phosphorylated adenosines, particularly cAMP and its relation to calcium ions. The effect of indomethacin on the nicotinic acid flush (see Chap. 17) strongly suggests prostaglandin involvement. Some other types of flushes, such as that induced by alcohol in diabetics taking chlorpropamide (22), are also poorly prevented by antihistaminic agents, but are suppressed by aspirin, another drug presumed to act via interference with the prostaglandin "second messenger" system.

The fate of the other "second messenger," cAMP, has been studied in some depth to explain nicotinic acid action (see Chap. 17, Sec. II). In some systems, the cAMP action seems to be mediated via shifts in calcium ions (23). Interestingly, intravenous calcium gluconate, like nicotinic acid, is notorious for causing an acute flush reaction. The guinea-pig ear responds to nicotinic acid with both vasodilatation and an increase in cAMP (19). While it has been proposed as the mechanism of several actions of nicotinic acid, in none is the cause—effect cAMP relationship clear.

Adenosine itself is a strong local vasodilator and may be the basis of a mechanism of rapid and highly localized vascular tone control. Intracellular phosphorylated adenosine derivatives cannot cross the uninjured cell membrane, while adenosine can (24). Whenever local cellular energy consumption goes up, the adenosine—AMP—ADP—ATP balance tends to shift toward the release of more free adenosine which can leave the cell and exert a local vasodilator influence to bring more blood to that area. It is then destroyed by the adenosine deaminase in blood before it can exert a more widespread vascular effect. Thus, the local oxygen and nutrient supply is responsive to the local energy utilization. Considerable attention has been paid to this postulated mechanism in relation to the coronary dilator action of dipyridamole (24), but it is seldom considered in relation to nicotinic acid or other vasoactive drugs.

Since the flush response to nicotinic acid is seen primarily with a rising nicotinic acid blood level and may not persist during a maintained plateau level, it has been suggested that the drug may act by releasing some exhaustible endogenous vasodilating agent in the skin (19). Such a mechanism should result in an "acute tolerance" to nicotinic acid, as is indeed seen in the fibrinolytic response to this drug (Chap. 20). The pattern of tolerance to the flush reaction is not nearly as complete or long lasting as with the fibrinolytic response, and is not quite what one would expect from the exhaustion of an endogenous supply of the active moiety.

III. CLINICAL UTILITY OF THE VASCULAR RESPONSE TO NICOTINIC ACID

One major limitation to the usefulness of vasoactive drugs in occlusive vascular disease is the fact that they attack only the spastic component of the occlusion. Atherosclerotic or thrombotic obstruction per se can hardly be expected to be overcome by vasodilation. While a vasodilator may enhance anastomotic circulation to a compromised tissue, it may also "steal" blood from the compromised tissue by diverting more blood through uninvolved vessels to uninvolved tissues. This concern is set aside by the argument that the flow rate through a particular vessel is a function of its caliber and fluid pressure. As long as pressure is maintained, flow through other vessels is irrelevant. But local pressure is likely to be influenced by the status of neighboring vessels which share a common blood source. If the coronary supply to the myocardium can serve as an example, there is real concern about highly localized differences in flow impairment and the differences in degree of response to dilators which are averaged out in many methods of overall blood-flow estimations (25). A vessel made rigid by disease may be less responsive to a dilator than a normal vessel whose internal caliber is still a function of the vascular muscle tone.

Results of objective studies of coronary vasodilators in the treatment of angina probably reflect this situation. Intravenous dipyridamole clearly increases overall coronary blood flow but fails to alleviate the symptoms or ECG pattern characteristic of angina (26). On the other hand, sublingual nitroglycerin relieves these signs and symptoms of "coronary spasm" without a measurable increase in blood flow, probably by reducing the cardiac work load. While chronic treatment with the dilator dipyridamole does seem to enhance collateral coronary circulation (27), the overall evidence suggests that drug-induced vasodilatation is not the answer to acute coronary insufficiency.

The general evidence in regard to common peripheral artery disease warrants a similar conclusion. There is no better local vasodilator than tissue anoxia, and if a given bed is lacking adequate circulation, the local vessels are very likely to be maximally dilated beyond additional influence by vasodilating drugs. Vessels to an exercising muscle will quickly become maximally dilated, a situation demonstrated to occur when patients with peripheral vascular disease develop intermittent claudication.

There remains for treatment by vasodilators those conditions in which there is exaggerated sensitivity or stimulation of vascular tone. Such conditions range from constricted microcirculation to major vessel spasm. There are instances where acute, severe arterial spasm causes seriously reduced blood flow. At times, local conditions, such as inflammatory venous thrombosis, can throw a neighboring artery into serious spasm. These situations are generally better relieved by sympathetic nerve block than by systemic vasodilators. At the other extreme, there are a variety of ill-defined minor vessel spastic circulatory disturbances, such as frostbite and Raynaud's phenomenon. Carefully monitored studies of nicotinic alcohol in patients with "trench foot" suggest that it is useful in this chronic syndrome (28). It has also been used in patients with a variety of conditions presumed to be due to impaired circulation, such as decubitus and varicose ulcers, chilblains, Meniere's disease, and vertigo. Its true effectiveness, and the mechanism of such effectiveness, remains very much in doubt.

REFERENCES

1. W. Redisch, L. Wertheimer, G. Delise, and J. M. Steele, *Circulation 9*:63(1954).
2. W. Redisch, E. Sheckman, and J. M. Steele, *Circulation 6*:862 (1952).

3. M. J. Allwood, I. Birchall, and J. S. Staffwith, *J. Physiol. 143*: 332(1958).
4. H. Rottenstein, G. Peirce, E. Russ, D. J. Gelder, and J. Montgomery, *Circulation 20*:760(1959).
5. M. S. Myslobodsky and N. Horesh, *Biol. Psychol. 6*:111(1978).
6. F. J. Klocke, *Prog. Cardiovasc. Dis. 19*:117(1976).
7. R. J. Cowan, C. D. Mzynard, I. Meschan, R. Janeway, and K. Sjigeno, *Nuclear Med. 107*:111(1973).
8. N. M. Lassen, and D. H. Ingvar, and E. Skinhoj, *Sci. Amer. 239*:62(1978).
9. G. W. Molnar, *J. Applied Physiol. 38*:449(1975).
10. D. R. Meldrum, I. M. Shamonki, A. M. Frumar, I. V. Tataryn, R. J. Chang, and H. L. Judd, *Am. J. Obstet. Gynecol. 135*:713 (1979).
11. I. V. Tataryn, D. R. Meldrum, K. H. Lu, A. M. Frumar, and H. L. Judd, *J. Clin. Endocrinol. Metab. 49*:152(1980).
12. D. R. Meldrum, I. V. Tataryn, A. M. Frumar, Y. Erlik, K. H. Lu, and H. L. Judd, *J. Clin. Endocrinol. Metab. 50*:685(1980).
13. L. Hellman, F. Nakada, J. Curti, E. D. Weitzman, J. Kream, H. Roffwarg, S. Ellman, D. K. Fukushima, and T. G. Gallaghes, *J. Clin. Endocrinol. 30*:411(1970).
14. B. Cox and P. Lomax, *Ann. Rev. Pharmacol. Toxicol. 17*:341 (1977).
15. R. Altschul, *Niacin in Vascular Disorders and Hyperlipidemia,* C. C. Thomas, Springfield, Illinois, 1964.
16. L. Wertheimer, W. Redisch, K. Hischarn, and J. M. Steele, *Circulation 11*:110(1955).
17. M. Weiner, K. Krinis, W. Redisch, and J. M. Steele, *Circulation 19*:845(1959).
18. N. Svedmyr, L. Harthon, and L. Lundholm, *Clin. Pharmacol. Ther. 10*:559(1969).
19. R. G. G. Andersson, G. Aberg, R. Brattsand, E. Ericsson, and L. Lundholm, *Acta Pharmacol. Toxicol. 41*:1(1977).
20. G. Aberg, *J. Int. Res. Common. 10*:13(1973).
21. *AMA Drug Evaluations 3rd ed.*, Publishing Sciences Group, Littleton, Mass, 1977.
22. R. C. Strakosch, D. B. Jefferys, and H. Keen, *Lancet 1*:394 (1980).
23. E. Morkin and P. J. La Raia, *N. Engl. J. Med. 290*:445(1974).
24. R. M. Berne, *Am. J. Physiol. 204*:317(1963).
25. F. J. Klocke and A. K. Ellis, *Ann. Rev. Med. 31*:489(1980).
26. D. Kinsella, W. Troup, and M. McGregor, *Am. Heart J. 63*:146 (1962).
27. A. M. Vineberg, *Can. Med. Assoc. J. 87*:336(1962).
28. W. Redisch and O. Brandman, *Angiology 1*:312(1950).

20

INDUCTION OF FIBRINOLYSIS BY NICOTINIC ACID

The natural mechanisms of clot formation (coagulation), clot dissolution (fibrinolysis), and their relation to hemostasis and thrombovascular disease have been reviewed repeatedly and in detail (1–8). In general, inadequate fibrinolytic capacity is believed to aggravate thrombotic problems, while enhanced fibrinolytic activity may help control thrombosis and may also induce bleeding. Descriptions of a "cascade" of reactions involving activators, proactivators, inhibitors, etc., of fibrin formation and dissolution grow increasingly complex (5,7), and these interact with other enzyme-based physiological processes, such as those involving inflammation and immune mechanisms.

Many proteolytic enzymes will digest fibrin, the cross-linked insoluble protein which is the clot. Some of these will also digest its soluble precursor, fibrinogen, at times converting it into the monomer which may polymerize to form a fibrin clot or be further digested directly into soluble fragments. The tryptic enzymes will dissolve clots but not without seriously altering other proteins if administered intravenously. The usual fibrinolytic activity which evolves from plasma components is attributed to the proteolytic agent called plasmin or fibrinolysin, derived from an inactive circulating precursor (plasminogen or profibrinolysin). Some diseases are associated with plasma activity which is both fibrinolytic and fibrinogenolytic. Hemorrhagic disease can result when fibrinogen levels become markedly reduced (10,11).

The consequences of congenital abnormalities associated with altered fibrinolytic states have been described and are increasingly well understood (12). Some acquired diseases, and particularly liver disease and shock (perhaps through altered hepatic function secondary to circulatory deficiency), are frequently associated with enhanced fibrinolytic activity (13). Assuming that hepatic metabolism of fibrinolysin is an important factor in the control of the level of this enzyme in plasma, the observed high fibrinolytic activity in various types of shock and hepatic disease (14), including hepatectomy (15), is not unexpected. On the other hand, normal hepatic function may be essential to the production of circulating inhibitors, and a deficiency or absence from the plasma of these liver-produced inhibitors may be the prime cause of increased fibrinolysis in hepatic disease patients (16). All fibrinolytic inhibition is not the result of a single substance. The glycoprotein in plasma called α_2 plasmin inhibitor (17) is active against plasmin by interfering with plasmin adsorption onto the fibrin clot. It also cross-links with fibrin so as to reduce the susceptibility of fibrin to plasmin action, but has little or no influence on the fibrinogenolytic action of plasmin. In contrast, an α_2-macroglobulin inhibits plasmin in the plasma milieu, thus blocking its fibrinogenolytic action, but has little effect on the ability of plasmin to bind to and digest fibrin (12). The clotting inhibitor, antithrombin III, is also antifibrinolytic, particularly in the presence of heparin (18). Thus, there are several mechanisms of varying specificity and significance by which fibrinolytic activity can be inhibited.

While the identified inhibitors of fibrinolysis appear to be produced in the liver, fibrinolytic activators seem to be primarily of extrahepatic origin (19). A variety of physiological alterations, such as acute ischemia (20), or exposure to cold and other stresses (21), have been shown to cause modest increases in fibrinolytic activity. The mechanisms of these effects appear to involve the release of activator substances but are otherwise ill-defined.

A few relatively simple compounds which were observed to induce some weak fibrinolytic activity were noted to be agents which influence vasomotor tone (22). Some studies have found more fibrinolytic activity in venous than arterial blood (23,24), suggesting that the tissues, rather than blood itself, are the sources of the activator activity which appears in plasma after these stimuli. A neurovascular component to the stimulant–fibrinolytic response is suggested by some experimental data (7).

Consistent with the clinical importance of thrombotic disease, induced enhancement in fibrinolytic activity has long been a therapeutic goal. In vivo activation of the clot-dissolving capacity of blood has been accomplished by a variety of enzymes and extracts prepared from urine, plasma, and tissues (1,3), and several have been considered as potential therapeutic agents. This therapeutic approach received a major stimulus when the bacterial fibrinolytic activator, streptokinase, was prepared in a highly purified form. However, the clinical use of even these pure materials was handicapped by immune phenomena which made dosage regulation very difficult. A direct attack on the problem was attempted by activating human plasminogen into plasmin in vitro and administering purified preparations of such directly active human plasmin. This approach also had its problems, since the doses required often had an undesirable effect on circulating fibrinogen, and the penetration of preformed plasmin into existing clots proved to be quite limited compared to the penetration of activator preparations. Since such activators, on penetrating the clot, could convert the plasminogen contained within the clot into the active enzyme (7,15), they could more efficiently dissolve preformed clots. The activator urokinase, a polypeptide extracted from human urine, has largely replaced the prior interest in bacterial streptokinase and human active plasmin. All these agents have had some degree of success in clinical trials (25–28), but their widespread use remains handicapped by the problems cited, particularly the difficulties associated with their biological origin.

In the face of this history, it is apparent that it would be highly desirable to have a relatively simple, well characterized, low molecular weight compound that significantly enhanced fibrinolytic activity and could be administered as a drug. Simple compounds including ε-aminocaproic acid (EACA) and an analogue, tranexamic acid (12), have been found which are fibrinolytic inhibitors and are useful in the control of the clinical manifestations of excessive fibrinolysis. Could an activator of equally simple structure and utility be found to enhance fibrinolysis and treat thromboembolic disease?

Weak fibrinolytic responses to drugs have been observed frequently, but are often detectable only by assays which are also sensitive to minor degrees of stress. The clinical utility of such minor and variably reproducible changes in fibrinolytic activity is open to question (7). In evaluating the potential utility of a presumably fibrinolytic drug, the nature of the fibrinolytic assay and the degree and pattern of the fibrinolytic response must be considered.

Methods of fibrinolytic assay are numerous and varied in complexity and specificity (29). Some are as simple as observing the speed of dissolution of clotted, buffer-diluted whole blood (30), or recalcified citrated plasma (31), but assay results may vary with minor differences in design and manipulation (32). Others use a purified protein or synthetic peptide as substrate. The most common methods employ a standard fibrin preparation clotted in a plate or tube to which native plasma, treated plasma, plasma fractions, or tissue extracts are added and the degree or speed of fibrin digestion determined (33). These methods allow studies to distinguish activator, inhibitor, or directly enzyme-dependent activity. Assays can determine separately the amount of plasminogen available, or the rate or completeness of its activation (1), and the presence of inhibitors (19,34). The stimulation or release of activator, the control of inhibitor, or direct effects on the action and fate of fibrinolysis are all potential mechanisms of attacking thromboembolic disease (35). The "bottom line" is enhanced thrombolytic activity.

I. NICOTINIC ACID-INDUCED FIBRINOLYSIS

One of the less sensitive and therefore more severe tests of fibrinolytic activity is the simple observation of whether a whole blood or recalcified plasma clot dissolves within a few hours after it is formed. A sophisticated variant of the simple test tube observation of this phenomenon is the coagulagraph or "thrombelastograph" (36,37), an instrument which continuously records the firmness of a clot, from which recording several aspects of the speed of clot formation as well as of dissolution—should it occur—can be readily evaluated.

During the use of this instrument in a clinical investigation of the effect of vasoactive drugs on the clotting characteristics of circulating blood, a marked fibrinolytic activation by intravenous nicotinic acid was noted (38). The observed effect was far more potent than that following other vasoactive drugs (39). An essentially identical observation of this response to parenteral nicotinic acid was reported independently in 1958 by Meneghini and Piccinini in Italy (40), using different methods. The observation was eagerly followed up in the hope that it might lead to a therapeutic approach to thromboembolic disease by thrombolysis without the need to use tissue extracts with the frequent problems of purification, reproducibility, antigenicity, etc., to achieve in vivo fibrinolytic activity.

Two characteristics of nicotinic acid-induced lysis proved to be major impediments to its potential clinical utility: (1) The duration of the fibrinolytic response, in spite of its potency, is very short (rarely as long as 1 hr). (2) An "acute tolerance" develops after the initial response to a single dose, so that a second dose is ineffective until days to weeks after the initial effect dose (41,42).

While there is no doubt about the absorption and vascular response to oral nicotinic acid (see Chaps. 16 and 17), oral doses do not induce fibrinolysis except in some patients with liver disease (see below). In fact, a prior oral dose will block the fibrinolytic response to a subsequent parenteral dose by the methods of assay employed. These and other significant observations about the characteristics of nicotinic acid-induced fibrinolysis may be summarized as follows:

1. Blood taken from most (but not all) human subjects 5–20 min after an intravenous or intraarterial dose of 10–100 mg nicotinic acid, will clot and then lyse spontaneously within minutes. Blood drawn 1 hr after dosage no longer has this fibrinolytic activity. Intramuscular doses are more weakly and less consistently effective. Oral doses are ineffective, even in large quantities (6 g/day) for months.

2. The in vitro addition of nicotinic acid to blood or plasma has no influence on fibrinolytic mechanisms.

3. If blood drawn shortly after intravenous nicotinic acid, is allowed to clot, and the clot is transferred to control (pretreatment) serum, the clot will still lyse.

4. Serum from postnicotinic acid specimens will lyse normal clots but at a much slower rate than the postnicotinic acid clot itself.

5. Citrated blood specimens with lytic properties after parenteral nicotinic acid can be stored for days and retain their lytic activity when eventually clotted by recalcification.

6. The lysis of postnicotinic acid specimens is not inhibited by mixing with nonlytic specimens obtained after a second dose of nicotinic acid.

7. Slow intravenous infusion of nicotinic acid fails to cause the lysis readily demonstrable with the same dose given intravenously as a bolus.

8. A prior dose of nicotinic acid, whether oral, parenteral, infusion, or bolus, will inhibit the usual lytic response to a subsequent parenteral bolus of nicotinic acid for days. Blood level patterns of nicotinic acid after a second intravenous dose of nicotinic acid are not different from those observed after the initial dose (43).

9. Nicotinamide does not induce lytic activity, but its prior administration blocks the usual response to a subsequent parenteral dose of nicotinic acid.

10. Studies in several other species (rabbit, dog, rat, guinea pig) failed to reproduce the phenomenon seen in humans.

11. Unlike other subjects, some patients with hepatic disturbance may have a lytic response to oral administration of nicotinic acid and also to two successive parenteral doses in the same day.

II. RELATION TO HEPATIC DISEASE

In view of the long-recognized involvement of hepatic function with fibrinolytic phenomena, it is not surprising that some hepatic disease subjects were found to differ from normal in their fibrinolytic response to nicotinic acid. These differences were studied carefully in search of clues as to the mechanism of the nicotinic acid phenomenon (43).

It has been noted that there were occasional exceptions to the rule that subjects having a lytic response to an intravenous dose of nicotinic acid failed to respond to a second similar dose a few hours later. These exceptions proved to be patients with liver disturbances. Further study of patients with proved hepatic cirrhosis confirmed this observation and also showed that over half the cirrhotic subjects tested responded to oral nicotinic acid with lysis, in contrast to normal subjects. The timing of the response was generally slower in onset and longer-lasting than the response to intravenous nicotinic acid in normal subjects.

By chance, the first cirrhosis patient observed to respond to oral nicotinic acid had recently undergone portacaval shunt surgery. As a result of this procedure, drugs absorbed into the portal vein may undergo a unique pattern of distribution, bypassing the liver and reaching peripheral tissues without prior exposure to hepatic mechanisms. It was found, however, that 5 of 14 cirrhotics who underwent shunt surgery failed to develop lysis after oral dosage, while 6 of 13 cirrhotics without shunt surgery did show the lytic response. Thus, the surgery per se was not the factor responsible for the responsiveness to oral nicotinic acid. Since some functional bypass may develop via esophageal varices without surgery, and surgically produced bypasses may fail to function (44), an attempt was made to correlate splenic pulp pressure, an index of relief of portal obstruction, with oral lysis response. No correlation was found. Thus, large vessel hepatic bypass apparently is not the determinant of lytic response to oral nicotinic acid in cirrhotic subjects.

A spectrum of liver function studies in cirrhotic "responders" versus "nonresponders" was compared. One parameter, serum bilirubin, was uniquely well correlated with the phenomenon. Subjects with severe cirrhosis and high serum bilirubin (over 1.7 mg/100 ml) regularly responded to an oral dose of nicotinic acid with lysis, in contrast to nonicteric subjects with otherwise equally severe disturbances of hepatic function associated with cirrhosis. Adding nicotinic acid to icteric plasma in vitro, or adding bilirubin to blood samples taken after nicotinic acid which were nonlytic, did not make them lytic.

Ratnoff (16) reported a high incidence of spontaneous fibrinolytic activity by his techniques in cirrhosis. He found this to be due to a more rapid than normal loss of inhibitory activity on incubation. The phenomenon did not occur in acute hepatitis or obstructive jaundice, so that bilirubin per se is not the responsible factor. By the techniques used in the nicotinic acid studies, lysis is reported to occur only if it is clearly evident within 24 hr at 37°C. By these criteria, only 2 of 27 cirrhotic subjects showed spontaneous lysis. It is not known whether there is a common basis for the lytic phenomena in chronic hepatic disease described by Ratnoff, and the increased responsiveness of cirrhotics to nicotinic acid.

III. MECHANISMS AND CLINICAL CONSEQUENCES

From the accumulated data, it is easier to state some of the things nicotinic acid does not do than to clearly define what it does to induce fibrinolysis. Clearly it is not itself an activator of plasminogen or a neutralizer of a circulating inhibitor, since it is essentially devoid of in vitro action in any system tested. Consequently, its in vivo effect must result from an action on some tissue. The absence of an animal model which imitates the phenomenon seen in humans has handicapped efforts to study the mechanisms involved.

Since nicotinic acid is obviously a vasoactive drug, and several such drugs are associated with in vivo fibrinolytic activation, the vascular effect would appear to be a prime candidate for its mechanism of action. In fact, the discovery of its fibrinolytic effect was a consequence of a project which sought to determine whether increased blood flow through one or another vascular bed measurably altered the clotting or fibrinolytic properties of venous blood. The fact that nicotinamide, which has essentially no vasoactive effect, also has no fibrinolytic action is consistent with the interdependence of vascular and fibrinolytic responses. However, attempts to correlate the intensity of visible or measurable vascular response with fibrinolytic induction were not fruitful, and not all potent vasoactive agents induce fibrinolysis. The possibility remains that an endogenous fibrinolytic activator, possibly of endothelial origin, is rapidly released into the blood stream as a consequence of nicotinic acid action.

The evidence that the circulating activators originate in the walls of blood vessels is quite strong, and measures of the "endothelial reserve" of activator are estimated as an index of thrombotic potential by measuring the fibrinolytic response to an occlusive challenge or a dose of epinephrine (7). Even the regional injection of nicotinic acid has been observed to result in a correspondingly regional release of fibrinolytic activator (45).

Assuming this mechanism of nicotinic acid action, the "acute tolerance" phenomenon might reflect an exhaustion of the activator stores which require days to regenerate. One might also postulate that slowly infused or oral doses of nicotinic acid fail to cause fibrinolysis because they fail to cause an adequately precipitous release of sufficient activator to overcome antifibrinolytic mechanisms. Nevertheless, they can exhaust the stores by slow, prolonged release so as to prevent a detectable response to subsequent intravenous bolus doses of nicotinic acid.

This presumed mechanism leaves unexplained the greater and repeated lytic response in cirrhotic subjects. One could presume that an element of hepatic bypass and deficiency in fibrinolysin-inhibiting proteins of hepatic origin may allow oral doses to be effective in cirrhotic subjects. But the capacity of such subjects to respond to successive doses 2–3 hr apart is difficult to understand, and throws into doubt the hypothesis of exhaustion of activator stores as the sole mechanism responsible for acute tolerance.

Conceivably, the fibrinolytic activation by nicotinic acid triggers a protective generation of fibrinolysis inhibitor, which persists for days in normal subjects, but is deficient in hepatic subjects. While evidence for the generation of such an inhibitor was not found in postnicotinic acid plasma, the postulated inhibitor might remain localized within the hepatic structure. The correlation of enhanced responsiveness to nicotinic acid of cirrhotics with elevated bilirubin levels suggests that the function responsible for the conjugation and clearing of bilirubin may also be involved in producing fibrinolytic inhibitor or in the clearance of fibrinolysin or circulating activator, so that plasma bilirubin and fibrinolytic responsiveness run parallel. Recent studies (18) have shown that patients with advanced liver disease tend to have reduced α_2-plasmin inhibitor (but not reduced α_2-macroglobulin) and defective clearance of activator. Elevated bilirubin levels related to obstructive biliary disease have not been found to be associated with hypersensitivity to nicotinic acid-induced fibrinolysis, reminiscent of the observation of Ratnoff that spontaneous fibrinolysis, frequently seen by his techniques in cirrhotics and attributed to a more rapid than normal loss of inhibitory activity (i.e., a "defective" inhibitor), does not occur in acute hepatitis or obstructive jaundice.

In the use of anticoagulants to control thromboembolism, it is now abundantly clear that the overall clotting time of blood is not necessarily the only, or even the best, index of antithrombotic action (14). The effect of hypoprothrombinemic agents on simple blood clotting or plasma recalcification time is difficult to detect, and the action of inhibitors of platelet aggregation on the usual measures of clotting is almost nil. Special tests, matched to the mechanism of the drug effect, are required. By analogy, the failure of oral nicotinic acid to yield a strong effect on the speed of clot lysis, or of repeated parenteral doses to yield a repeatable effect, does not rule out some persistent alteration of the fibrinolytic balance that might have therapeutic utility. Unfortunately, our understanding of the fibrinolytic mechanism, its pathophysiological alterations in disease, and the consequences of continuous nicotinic acid exposure is too meager to make a case for or against its clinical utility. Conceivably, a high endothelial reserve of activator, capable of locally preventing a thrombus from taking hold at a site of vascular injury, may be more important than the fibrinolytic potential of circulating blood (46). On the other hand, some studies suggest that strongly enhanced fibrinolytic activity in circulating blood for as short a time as 2 hr can have favorable consequences on experimental thrombosis (47).

Empirical clinical observations of the utility of nicotinic acid to treat thromboembolism are scant indeed. Given the current limited understanding, it would be difficult to make an impressive case for undertaking the mammoth type controlled clinical studies which would be required to evaluate empirically the consequences of nicotinic acid treatment of thromboembolic diseases.

In a sense, the studies of coronary reinfarction undertaken with cholesterol in mind might be considered such an empirical test. To the extent that such studies have shown a favorable response to nicotinic acid treatment, one might ask what evidence there is to credit lower cholesterol rather than an unrecognized change in fibrinolytic balance for the observed reduction in reinfarction (see Chap. 18).

Some attempts have been made to link the lipid and fibrinolytic effects of nicotinic acid. LDL has been found to be inhibitory to the fibrinolytic system, and Bang (41) has suggested that the effects on such facets of cholesterol metabolism might explain the fibrinolytic response. However, nicotinic acid-induced fibrinolysis is of rapid onset and short-lived (minutes), while the lipid effects of nicotinic acid are generally slower in development and much more persistent. Nevertheless, the normal pattern of fibrinolytic activators and inhibitors which might be altered by nicotinic acid may include lipoproteins.

There is increasing recognition of the differences in the nature of arterial versus venous thrombosis and embolic phenomena. Studies showing efficacy of fibrinolytic therapy, like heparin therapy, are most impressive in relation to embolism of venous origin (26), in contrast to platelet aggregation inhibitors which are employed largely to control small artery thrombosis (48). Cirrhotics as a group, in spite of their presumed high fibrinolytic status, commonly have venous leg thrombosis, often attributed to the venous stasis associated with ascites. However, it has been observed that pulmonary embolism occurs with remarkably low incidence in such patients (49) despite the frequent lower limb thrombophlebitis.

As of this moment, the phenomenon of nicotinic acid-induced fibrinolysis remains a medical curiosity rather than a therapeutic weapon of known utility.

REFERENCES

1. S. Sherry, A. P. Fletcher, and N. Alkjaersig, *Physiol. Rev. 39*: 343(1959).
2. R. Biggs, in *Human Blood Coagulation, Hemostasis, and Thrombosis*, Blackwell Scientific Publications, Oxford, 1976.
3. F. J. Castellino and B. N. Violand, *Prog. Cardiovasc. Dis. 21*: 241(1979).
4. R. F. A. Zwaal, *Biochim. Biophys. Acta 515*:163(1978).
5. J. W. Suttig and C. M. Jackson, *Physiol. Rev. 57*:1(1977).
6. H. C. Kwaan, *Prog. Cardiovasc. Dis. 21*:239(1979).
7. J. F. Davidson and I. D. Walker, *Prog. Cardiovasc. Dis. 21*:375 (1979).
8. *Progress in Chemical Fibrinolysis and Thrombolysis*, (J. F. Davidson, M. M. Samama, and P. C. Desnoyers, eds.), Ravin Press, New York, 1975.
9. F. J. Castellino and B. N. Violand, *Prog. Cardiovasc. Dis. 21*: 241(1979).
10. H. G. Lasch and D. L. Heene, *Thromb. Diath. Haemorrh. 26*: 351(1967).
11. M. Marder, M. Weiner, P. Shulman, and S. Shapiro, *N. Y. State J. Med. 49*:1197(1949).
12. N. Aoki, Y. Sakata, M. Matsuda, and K. Tateno, *Blood 55*:483 (1980).
13. P. Pises, R. Bick, and B. Siegel, *Am. J. Gastroenterol. 60*:280 (1973).

14. M. Weiner, *Adv. Pharmacol.* *1*:277(1962).
15. K. N. von Kaulla, *Drug Research* *15*:246(1965).
16. O. D. Ratnoff, *J. Clin. Invest.* *27*:552(1948).
17. M. Maroi and N. Aoki, *J. Biol. Chem.* *251*:5956(1976).
18. N. Aoki, *Prog. Cardiovasc. Dis.* *21*:267(1979).
19. T. Astrup, *Thromb. Diath. Haemorrh. Suppl.* *4*:47(1961).
20. H. C. Kwaan, R. Lo, and A. J. S. McFadzean, *Clin. Sci.* *17*:361 (1958).
21. R. Macfarlane and R. Biggs, *Lancet* *ii*:862(1946).
22. K. N. von Kaulla, *Circulation* *17*:187(1958).
23. H. J. Tagnon, *J. Lab. Clin. Med.* *27*:1119(1942).
24. H. J. Tagnon, S. M. Levinson, C. S. Davidson, and F. H. L. Taylor, *Am. J. Med. Sci.* *211*:88(1946).
25. V. J. Marder, *Prog. Cardiovasc. Dis.* *21*:327(1979).
26. E. Genton, *Prog. Cardiovasc. Dis.* *21*:333(1979).
27. F. Duckert, *Prog. Cardiovasc. Dis.* *21*:342(1979).
28. M. Martin, *Prog. Cardiovasc. Dis.* *21*:351(1979).
29. G. H. Barlow, *Prog. Cardiovasc. Dis.* *21*:315(1979).
30. G. R. Fearnley, G. Blammforth, and E. Fearnley, *Clin. Sci.* *16*: 645(1957).
31. M. Weiner, *Clin. Chem.* *9*:182(1963).
32. M. Weiner, *Nature* *184*:1937(1959).
33. K. K. Reddi and E. Kodicek, *Biochemistry* *53*:286(1953).
34. I. Astrup and A. Stage, *Proc. 4th Int. Congress Biochem.*, 1958.
35. T. E. Friedemann and E. I. Frazier, *Arch. Biochem. Biophys.* *26*:361(1950).
36. H. Hartert, *Z. Gesamte Inn. Med.* *117*:189(1951).
37. K. N. von Kaulla and M. Weiner, *Blood* *10*:362(1955).
38. M. Weiner, W. Redisch, and J. M. Steele, *Proc. Soc. Exp. Biol. Med.* *98*:755(1958).
39. M. Weiner, K. de Krinis, W. Redisch, and J. M. Steele, *Circulation* *19*:845(1959).
40. P. Meneghini and F. Piccinini, *Arch. E. Maragliano Patol. Clin.* *14*:69(1958).
41. M. Weiner, W. Redisch, and K. de Krinis, *Proc. 7th Cong. Eur. Soc. Haematol. Lond. part II*:890(1960).
42. D. R. Rosing, D. R. Redwood, P. Brakman, T. Astrup, and S. E. Epstein, *Thrombosis Res.* *13*:419(1978).
43. M. Weiner, *Am. J. Med. Sci.* *246*:72(1963).
44. L. M. Rousselot, A. H. Moreno, and W. F. Parke, *Ann. Surg.* *150*:384(1959).
45. M. Tesi in *Progress in Chemical Fibrinolysis and Thrombolysis* (J. F. Davidson, M. M. Samama, and P. C. Desnoyers, eds.) Raven Press, New York, 1975, p. 255.
46. J. G. Vermylen and D. A. F. Chamone, *Prog. Cardiovasc. Dis.* *21*:256(1979).

47. P. W. Boyles, *J. Lab. Clin. Med. 54*:551(1959).
48. E. Genton, M. Gent, J. Hirsh, and L. A. Harker, *N. Engl. J. Med. 293*:1174(1975).
49. J. Spitzer, N. Rosenthal, M. Weiner, and S. Shapiro, *Ann. Int. Med. 31*:884(1949).

21

MISCELLANEOUS PHARMACOLOGICAL AND CLINICAL ACTIONS

Over the many years that niacin has been recognized as a vitamin, it has been the subject of numerous concepts and observations leading to suggestions or claims of clinical utility as a drug even beyond the wide spectrum already discussed. Some of these perceived actions are old, and have been essentially abandoned, while others are recent and untried. Some are suggested by the consequences of nicotinic acid deficiency, and some represent joint action with other agents. This chapter will review the more prominent of these miscellaneous areas of interest in nicotinic acid as a drug.

I. NICOTINIC ACID IN MENTAL DISEASE

Even before the discovery of the role of vitamins, it had been observed that mental disturbances associated with diseases such as beriberi and pellagra were alleviated when an improved diet became available (1). With the discovery and study of the effects of individual vitamins, various patterns of symptoms related to nutritional deficiency started to separate out. Thus Jolliffe et al. (2) found that encephalopathic symptoms seen with nutritional deficiency, and particularly pellagra, could be cured by niacin but not by thiamine. He and others reported favorable effects even in the absence of frank pellagra. Large doses of niacin were said to be useful in toxic psychoses related to states of exhaustion (3) and later to schizophrenia (4). It was suggested that psychoses were the result of toxins which disturbed enzyme systems involved in brain metabolism, and that these disturbances might be overcome if essential B vitamins were presented to the brain in high enough concentration to osmotically bypass defective energy-requiring transport mechanisms (5). The responsible toxins were presumed to be metabolites of endogenous agents. In particular, toxic adrenalin derivatives (6) with mescaline-type toxicity, and products of tryptophan metabolism were suspect. Pharmacological studies of various tryptophan metabolites showed some of them to be neuropharmacologically active, and nicotinic acid antagonized some of these effects (7).

The search for psychogenic endogenous products included hallucinogenic methylated derivatives of biogenic amines (8). It was postulated that the production of these harmful derivatives in the brain could be reduced by the "methyl acceptor" action of high doses of nicotinic acid (9). Large doses of B vitamins, and especially of nicotinic acid, often were given intravenously in many kinds of confusional states. This practice is much less common now because no disturbances in nicotinic acid levels (1), nicotinic acid metabolism (10) or tryptophan metabolism (11) in mental disease were found and, more importantly, controlled clinical studies failed to verify the presumed improvement in the mental state of patients who were not suffering from deficiency. A careful overall evaluation and review of the relevant individual controlled studies resulted in the currently generally accepted conclusion that "substantive evidence indicates that nicotinic acid is not an antipsychotic agent" (12). This conclusion holds for both nicotinic acid and nicotinamide. In this as in other areas of study, the flushing effect of nicotinic acid makes the double blinding of controlled studies most difficult. Nevertheless, it seems fair to conclude that the administration of nicotinic acid, nicotinamide, and even NAD (13) all failed to live up to their original billing as antischizophrenic agents.

II. EFFECT ON INTESTINAL SECRETIONS

Since diarrhea is one of the "three D's" which characterize pellagra, and since fluid replacement has been considered an important and possibly life-saving component of the treatment of severe pellagra (14), it has been of interest to consider whether large doses of nicotinic acid might have a utility in treating diarrhea associated with other diseases. To make a judgment in this regard, it may be useful to consider the pathobiochemistry of pellagra which is presumed to be responsible for its diarrhea, and then review whether similar mechanisms are involved in other chemical diarrheas.

Recent studies concerning the mechanism of diarrhea induced by cholera toxin indicate an association with elevated cyclic AMP (cAMP) levels. Nicotinic acid blocks both this toxin-induced cAMP elevation and the marked dose-related enhancement in secretion which has been observed in the rabbit isolated intestinal loop model (15). However, the doses of nicotinic acid involved in these studies are not likely to encourage clinical trials for such use, and essentially no definitive clinical observations have been reported concerning the treatment of diarrhea, other than that of pellagra, with nicotinic acid.

III. INTERACTION WITH CARCINOGENS

The interactions reported between niacin and a potent natural carcinogenic toxin, streptozotocin, represent an excellent example of the value of understanding mechanisms as an aid to interpreting the significance of experimental observations.*

Streptozotocin is described (16) as a naturally occurring nitrosourea compound that exerts antibiotic, oncogenic, oncolytic, and diabetogenic activity, the latter mediated by selective destruction of pancreatic beta cells. It consists of a known carcinogen, 1-methyl-1-nitrosourea (MNU), attached to the carbon 2-position of glucose, and is taken up by pancreatic islet cells where it markedly inhibits pyridine nucleotide synthesis and destroys the beta cells. A single dose can depress the rat growth curve (17) and create severe diabetes (18), but a concomitant large dose of nicotinamide will prevent these responses, including the acute reduction of islet NAD (19) and beta cell necrosis. However, the DNA of the surviving beta cells remain subject to methylation by MNU which may be the mechanism responsible for the later appearance of islet cell adenomas (17). Thus, an empirical evaluation of the incidence of pancreatic adenomas may show more such adenomas after nicotinamide plus streptozotocin than after streptozotocin alone among the long-term surviving animals (20). Consequently, nicotinamide is at times referred to as a "cocarcinogen" when, in fact, its real effect is probably to ameliorate the action of the carcinogen streptozotocin so as to allow the survival of cells which would otherwise be killed by that action.

The same investigators who reported the "utility" of nicotinamide coadministered with streptozotocin as a way of generating a pancreatic tumor model also reported that nicotinamide *prevented* the renal tumors induced by streptozotocin (17). In this more recent study, surviving animals (81–92% of those treated) showed smaller and much fewer renal tumors in the nicotinamide cotreated groups than in those treated with the carcinogen alone (18% vs 77% of all treated animals, or 19% vs. 95% of surviving animals).

In a study using a range of streptozotocin doses, Kazumi et al. (21) reported a dose-related reduced survival among streptozotocin-treated rats, but an incidence of pancreatic tumors among surviving animals which seemed higher with the lower doses. Of 27 animals treated with 50 or 65 mg/kg, only 6 survived, and of these, 3 had tumors (50%). Of 23 animals treated with smaller doses, 16 survived, and of these, 14 (88%) had pancreatic tumors. Two other groups of animals given

---

*See Section V, Chapter 7 for a detailed discussion of possible biochemical mechanisms of streptozotocin–nicotinamide actions.

large doses (65 mg/kg) were cotreated with nicotinamide or its isomer, picolinamide. The results in these animals resembled those treated with lower doses of the carcinogen alone [i.e., they showed better survival (about 40%) and an apparently higher tumor incidence among the survivors (all but one of 22)]. Since the isomer picolinamide is presumably not convertable to NAD, its apparent capacity to influence survival of streptozotocin-treated rats is not easily explained.

The Kazumi study also reported that renal tumors were seen only when nicotinamide was coadministered, a result which they describe as "diametrically opposite" to the results of Rakieten. However, coadministration was studied only at their highest (65 mg/kg) streptozotocin dose, and at that dose only 2 of 17 animals survived the carcinogen alone, while there were 14 survivors among the 22 animals cotreated with nicotinamide. The renal tumor incidence (3 of 14 survivors) in this latter group is very similar to that reported by Rakieten et al. (17) with 50 mg/kg of the carcinogen plus nicotinamide. In view of the poor survival of Kazumi's large dose, noncotreated animals, a valid comparison cannot be made.

In a report by Volk et al. (16) which did not include control groups, 41 rats treated with 50 mg/kg streptozotocin plus nicotinamide showed gross islet tumors in 16 rats plus microscopic tumors in an additional 4 animals (total of 49%). This study makes no mention of renal tumors.

The interpretations placed on these reports have considerable potential consequences in terms of public fears and legal requirements if categorical terms are used. The label "cocarcinogen" applied to an agent as important and widely used as niacin is bound to cause consternation in some circles. In the last analysis, any energy-generating nutrient is a "cocarcinogen," since some tumors grow less vigorously in a starved animal than in one which is well fed. It would be preferable to describe niacin as an agent capable of suppressing some, but not all effects of the toxic carcinogen, streptozotocin, rather than entering it into the ill-defined and often meaningless category of "cocarcinogen."

IV. ANTINIACIN (PELLAGRAGENIC) DRUGS

In seeking to understand nicotinic acid as a drug, it may be useful to review the nature of some drugs which have been found to be "antiniacin" or pellagragenic. Considering the complexities of the relationship between niacin, tryptophan, and other nutrients in regard to the synthesis of pyridine nucleotide, one can anticipate that working out the mechanisms of pellagragenic drugs will not be a very simple task (see Chap. 7).

Classically, inhibitors of the action of natural precursors to the essential body chemistry are found among analogues of the natural

substance, and niacin is no exception. The simple analogue, 6-aminonicotinamide (6-AN), has been used for over 20 years as a readily controllable tool to produce the effects of niacin deficiency, such as teratology (22,23), and fetal resorption in rodents and black tongue in dogs (24).

The widely used antituberculosis drug, isoniazid (isonicotinic acid hydrazide, INH) can cause clinical pellagra in man apparently by inhibiting the synthesis of nicotinamide nucleotides from tryptophan. On the other hand, two other hydrazine derivatives, benserazide and carbidopa, used clinically to inhibit dopa decarboxylation, cause a marked biochemical antinicain effect (25), demonstrable as a marked drop in urinary *N*-methylnicotinamide and the products of oxidative metabolism of tryptophan, yet they have not been reported to cause clinical pellagra.

In vitro enzyme inhibition studies (26) fail to explain why these two agents do not cause the overt clinical disease noted with isoniazid. It is doubtful that differences in the dietary habits of the types of patients treated with these drugs (tuberculosis versus Parkinson's disease) are responsible. The decarboxylase inhibitors essentially do not cross the blood–brain barrier, in contrast to isoniazid which is readily found in the cerebrospinal fluid (27). Might the brain thus be protected from the metabolic consequences of the decarboxylase inhibitors and so avoid the overt clinical disease seen with isoniazid?

Isoniazid markedly increases the urinary excretion of pyridoxine (28) and can precipitate pyridoxine-deficiency anemia in predisposed individuals (29). On the other hand, pyridoxine in large doses has no detectable influence on the fate of isoniazid (30), although its administration can prevent isoniazid-induced peripheral neuritis with little, if any, interference with its antituberculosis action. The ability of hydrazines to react with pyridoxal phosphate is well established, and kynurenase, essential to tryptophan–niacin conversion, is a pyridoxal-dependent enzyme.

V. NIACIN-LIKE DRUG ACTIONS

It is beyond the scope of this volume to review all the agents which have been studied on the basis of their nicotinic acid-like structure in the hope of achieving an intensification of a desired pharmacological effect with reduced undesired effects. The spectrum of such preparations range from simple tablet formulations aimed at controlling dissolution rate of its nicotinic acid through chemically rather distant analogues which have their own unique but somewhat nicotinic acid-like pharmacology. This range of clinically studied nicotinic acid derivatives may be classified as follows:

A. Special Salts and Formulations of Nicotinic Acid

Since nicotinic acid frequently causes gastrointestinal symptoms attributed to local irritation, attempts have been made to avoid this symptom by administering the drug with antacids, including aluminum hydroxide which forms aluminum nicotinate salts (31). Such preparations can be tableted with compression in such manner as to yield a relatively slow and prolonged release of nicotinic acid over a period of hours (32). The pharmacological consequences of such formulation may be considerable, and have been discussed in some detail earlier (Chaps. 16 and 18).

B. Nicotinic Acid "Prodrugs"

These are agents conceived as metabolic precursors of nicotinic acid which have the potential advantage of being differently distributed and converted to nicotinic acid at desired places and in more effective concentrations at these places. Such materials may be nonabsorbable, large molecular weight nicotinic acid-binding polymers which slowly liberate their multiple units of nicotinic acid following oral administration. Other, smaller complexes of nicotinic acid may be absorbed as such and may contribute an element of their own distribution.

Derivatives which are oxidized to nicotinic acid in vivo, such as nicotinic alcohol (β-pyridylcarbanol, Roniacol), probably exert activity over and above that derived from their oxidation to nicotinic acid. To the extent that they also have their own intrinsic activity, they would deserve to be classified with the analogues discussed in Sec. C, below. Even some of the much studied esters of nicotinic acid, such as pentaerythritol tetranicotinate, sorbitol nicotinate, and inositol hexanicotinate, may not act solely by and after liberation of nicotinic acid from its ester. The problem resembles that of aspirin (acetylsalicylic acid) versus the salicylates. While aspirin was first conceived and made as a "prodrug" of salicylates to reduce gastric irritation, there is now convincing evidence that the ester per se is pharmacologically active.

C. Nicotinic Acid Analogues*

The degree of "analogy" which warrants inclusion in this group is somewhat arbitrary. Certainly picolinic acid, which is 2-pyridine carboxylic acid, in contrast to the 3-derivative (nicotinic acid itself) deserves this classification. However, it cannot be incorporated into NAD and has not achieved prominence as a drug, even though it has

---

*Many analogues have been discussed in detail in Chap. 11, and will not be covered here.

some actions in common with niacin. On the other hand, the antidepressant imipramine and other tricyclic drugs, which can hardly be considered analogues of nicotinic acid, resemble niacin in inducing a prompt stimulation of NAD synthesis after a single dose and can prevent some of the consequences of niacin deficiency.

Perhaps the analogue of nicotinic acid most prominently in clinical use is Roniacol, the alcohol derivative described above as oxidizable to nicotinic acid. While its clinical use is primarily as a vasodilator (33,34), effects on hyperlipidemia have been described (35).

Nikethamide, which is simply the *N*-diethyl derivative of nicotinamide, might be expected to be an active analogue and even a "prodrug" of niacin. However, its predominant pharmacological action, for which it is widely used, is that of respiratory stimulation. Except for the faint sensation of warmth which may resemble the flush reaction seen with nicotinic acid (but not nicotinamide), there is little about its known pharmacology to suggest a resemblance to nicotinic acid.

By virtue of its name and historic status as a source of nicotinic acid, the classical pharmacological agent nicotine, which mimics one type of cholinergic neurohumoral transmission, is often confused with nicotinic acid, with which it has little in common pharmacologically. These examples point up the danger of assuming that the action of a compound which happens to be a nicotinic acid analogue is necessarily involved with one of the many biochemical functions of nicotinic acid.

REFERENCES

1. P. W. Kershaw, *Br. J. Psychiatry. 113*:387(1967).
2. N. Julliffe, K. M. Bowman, L. A. Rosenbann, and H. D. Fein, *JAMA 114*:307(1940).
3. V. P. Sydenstriker and H. M. Cleckley, *Am. J. Psychiatry. 98*: 83(1941).
4. A. Hoffer, H. Osmond, M. J. Callbed, and I. Kihan, *J. Clin. Exp. Psychol. 18*:131(1957).
5. J. Gould, *Recent Prog. Psychiatry, 3*:213(1959).
6. H. Osmond and J. R. Smythies, *J. Ment. Sci. 98*:309(1952).
7. I. P. Lapin, *Psychopharmacologia 26*:237(1972).
8. J. N. Nestords, T. A. Ban, and H. E. Lehman, *Int. Pharmacopsychiatry 12*:215(1977).
9. A. Huffer, *Niacin Therapy in Psychiatry,* 1962.
10. J. E. Mrocher, R. L. Jolly, D. S. Young, and W. J. Turner, *Clin. Chem. 22*:1821(1976).
11. F. C. Brown, J. B. White, and J. K. Kennedy, *Am. J. Psychiatry 117*:63(1960).
12. H. I. Kaplan, A. M. Freedman, and B. J. Sadock, *Comprehensive Textbook of Psychiatry* 3rd Ed. Williams & Wilkins, Baltimore, 1980.

13. N. S. Kline, G. L. Barcley, J. O. Cole, A. H. Esser, H. Lehmann, and J. R. Wittenborn, *Br. J. Psychiatry 113*:731(1967).
14. N. F. Pierce, R. B. Sack, R. G. Mitra, K. L. Baughman, J. G. Banwell, D. S. Fedson, and A. Mandel, *Ann. Intern. Med. 70*: 1173(1969).
15. N. Turiman, G. S. Gotterer, and T. R. Hendrix, *J. Clin. Invest. 61*:1155(1978).
16. B. W. Volk, K. F. Wellman, and F. Brancato, *Diabetologia 10*:37 (1973).
17. N. Rakieten, B. S. Gordon, A. Beaty, D. A. Cooney, and P. S. Schein, *Proc. Soc. Exp. Biol. Med. 151*:356(1976).
18. P. S. Schein, A. D. Cooney, and M. L. Vernon, *Cancer Res. 27*: 2324(1967).
19. T. Anderson, P. S. Schein, M. G. McMenamin, and D. A. Cooney, *J. Clin. Invest. 54*:672(1974).
20. N. Rakieten, B. S. Gordon, A. Beaty, D. A. Cooney, R. D. Davis, and P. S. Schein, *Proc. Sci. Exp. Biol. Med. 137*:283(1971).
21. T. Kazumi, Yoshino, S. Fujii, and S. Baba, *Cancer Res. 38*:2144 (1978).
22. L. Pinsky and F. L. Fraser, *Biol. Nauki. 1*:106(1959).
23. F. G. Biddle, *Teratology 16*:301(1977).
24. P. Greengard, E. B. Sigg, I. Fratta, and S. B. Zak, *J. Pharmacol. Exp. Ther. 154*:624(1966).
25. D. A. Bender, C. J. Earl, and A. J. Lees, *Clin. Sci. 56*:89(1979).
26. D. A. Bender, *Biochem. Pharmacol. 29*:707(1980).
27. *U.S. Pharmacopeia Dispensing Information*, 1981 ed., p. 364.
28. J. P. Biehl and R. W. Vilter, *Proc. Soc. Exp. Biol. Med. 85*:389 (1954).
29. R. Vilter, J. P. Biehl, J. F. Miller, and B. Friedman, *Fed. Proc. 131*:776(1954).
30. D. K. Riemensnider and R. S. Mitchell, *Am. Rev. Respir. Dis. 82*:412(1960).
31. J. L. S. Holloman, C. G. Davis, and L. C. Leeper, *J. Am. Geriatr. Soc. 10*:903(1962).
32. L. Meiring, *S. Afr. Med. J. 57*:345(1980).
33. W. Redisch and O. Brandman, *Angiology 1*:312(1950).
34. *AMA Drug Evaluations 3rd ed.* Publishing Sciences Group, Littleton, Mass. 1977.
35. L. Landholm, L. Jacobson, R. Brattsand, and O. Magnusson, *Atherosclerosis 29*:217(1978).

SUBJECT INDEX